Low-Rank Coals for Power Generation, Fuel and Chemical Production

Related titles

Advances in the Science of Victorian Brown Coal
(ISBN 978-0-08-044269-3)

Clean Coal Engineering Technology
(ISBN 9781856177108)

The Coal Handbook: Towards Cleaner Production: Vol 1: Coal Production
(ISBN 978-0-85709-422-3)

The Coal Handbook: Towards Cleaner Production: Vol 2: Coal Utilisation
(ISBN 978-1-78242-116-0)

Oxy-fuel Combustion for Power Generation and Carbon Dioxide Capture
(ISBN 978-1-84569-671-9)

Woodhead Publishing Series in Energy

Low-Rank Coals for Power Generation, Fuel and Chemical Production

Edited By

Zhongyang Luo

Michalis Agraniotis

ELSEVIER

WP
WOODHEAD
PUBLISHING
An imprint of Elsevier

Woodhead Publishing is an imprint of Elsevier
The Officers' Mess Business Centre, Royston Road, Duxford, CB22 4QH, United Kingdom
50 Hampshire Street, 5th Floor, Cambridge, MA 02139, United States
The Boulevard, Langford Lane, Kidlington, OX5 1GB, United Kingdom

British Library Cataloguing-in-Publication Data
A catalogue record for this book is available from the British Library

Library of Congress Cataloging-in-Publication Data
A catalog record for this book is available from the Library of Congress

ISBN: 978-0-08-100895-9 (print)
ISBN: 978-0-08-100929-1 (online)

For information on all Woodhead Publishing publications
visit our website at https://www.elsevier.com/books-and-journals

Publisher: Joe Hayton
Acquisition Editor: Maria Convey
Editorial Project Manager: Natasha Welford
Production Project Manager: Priya Kumaraguruparan
Cover Designer: Greg Harris

Typeset by MPS Limited, Chennai, India

Contents

List of Contributors

Michalis Agraniotis Mitsubishi Hitachi Power Systems Europe, Duisburg, Germany

Sukhvinder P.S. Badwal CSIRO Energy, Clayton, VIC, Australia

Christian Bergins Mitsubishi Hitachi Power Systems Europe, Duisburg, Germany

Sankar Bhattacharya Monash University, Clayton, VIC, Australia

Hokyung Choi Korea Institute of Energy Research, Daejeon, South Korea

Donghyuk Chun Korea Institute of Energy Research, Daejeon, South Korea

Panagiotis Drosatos Centre for Research & Technology Hellas, Ptolemais, Greece; School of Mechanical Engineering, NTUA, Athens, Greece

Guenter Harp Harp Process Chemistry Consulting, Duesseldorf, Germany

Klaus Jähnig DBI-EWI GmbH, Freiberg, Germany

Emmanouil Kakaras Mitsubishi Hitachi Power Systems Europe, Duisburg, Germany

Sangdo Kim Korea Institute of Energy Research, Daejeon, South Korea

Nikolaos Koukouzas Centre for Research and Technology-Hellas, Athens, Greece

Janusz Lasek Institute for Chemical Processing of Coal, Zamkowa, Zabrze, Poland

Sihyun Lee Korea Institute of Energy Research, Daejeon, South Korea

Zhongyang Luo Zhejiang University, Hangzhou, PR China

Jozo Mamić ARCUS Technologie GmbH & Co. KG, Cottbus, Germany

Mohammed Morsy ARCUS Technologie GmbH & Co. KG, Cottbus, Germany

Nikolaos Nikolopoulos Centre for Research & Technology Hellas, Ptolemais, Greece

Halina Pawlak-Kruczek Wrocław University of Technology, Wrocław, Poland

Herbert Richter International Gasunion (IGU), IL, United States

Daniel G. Roberts CSIRO Energy, Brisbane, QLD, Australia

Harold Schobert Schobert International LLC, Jordan, MN, United States

Malgorzata Stein-Cichoszewska Mitsubishi Hitachi Power Systems Europe, Duisburg, Germany

Wang Tao Zhejiang University, Hangzhou, PR China

Christian Trautmann ARCUS Technologie GmbH & Co. KG, Cottbus, Germany

Ioannis Violidakis Centre for Research & Technology Hellas, Ptolemais, Greece

André Werner ARCUS Technologie GmbH & Co. KG, Cottbus, Germany

Louis J. Wibberley CSIRO Energy, Newcastle, NSW, Australia

Jiho Yoo Korea Institute of Energy Research, Daejeon, South Korea

Jarosław Zuwała Institute for Chemical Processing of Coal, Zamkowa, Zabrze, Poland

Part I

Low rank coal properties and preparation

Introduction to low-rank coals: Types, resources, and current utilization

1

Harold Schobert
Schobert International LLC, Jordan, MN, United States

1.1 The concept of low-rank coal

Many of the major coal-producing countries have developed standards and procedures for classifying coals by rank. The national system used most widely outside of its country of origin is that established by the American Society for Testing and Materials (ASTM). In addition to the numerous national systems, the International Organization for Standardization (ISO) also has a coal classification system. Additional information on the rationale for classifying coals by rank and on the various systems of rank classification is available in various monographs [1–3], as well as on the web sites of international or national standards-setting organizations.

In the ASTM system, low-rank coals are lignites and subbituminous coals. This classification is based on the calorific value, expressed in Btu per pound on a mineral-matter-free basis, but with the bed moisture included, following the procedures of ASTM D 388-84 [1,2,4]. This is shown in Table 1.1. Brown coals, such as those in Australia, would be classified as lignite B in the ASTM system.

In the ISO system, coals having a mean random vitrinite reflectance less than 0.4 are classified as brown coal and lignite [5]. Coals having mean random vitrinite reflectance ≥ 0.4 but less than 0.5 are classified as subbituminous. The ISO system for brown coals and lignites assigns a code number based on two parameters: the total moisture content in run-of-mine coal, used to assign a class number, and on tar yield from low-temperature carbonization, reported on dry, ash-free basis, and used to assign a group number [1,3]. These assignments are shown in Table 1.2.

As an example, a coal having 20%–30% moisture and a dry, ash-free tar yield of 10%–15% would be classified with a code number 1110.

Numerous other systems for coal classification exist, often based on national standards-setting organizations such as in Germany or Britain. Convenient descriptions of these systems are available in various monographs [3,6,7]. In some systems, lignite is also known as brown coal, and represents low-rank coal, while subbituminous coals are often included with the hard coals (i.e., anthracite and bituminous). A further complication is the occasional reference to subbituminous coal as black lignite, with all low-rank coals then being called lignite [8]. International Energy Agency reports on coal show the lack of consistency from country to country;

Low-Rank Coals for Power Generation, Fuel and Chemical Production
DOI: http://dx.doi.org/10.1016/B978-0-08-100895-9.00001-2

Table 1.1 Classification of low-rank coals in the ASTM system

Class	Group	Heating value (Btu/lb), moist but mineral-matter-free basis
Subbituminous		
	Subbituminous A	10,500–11,500
	Subbituminous B	9500–10,500
	Subbituminous C	8300–9500
Lignite		
	Lignite A	6300–8300
	Lignite B	< 6300

Modified from H.H. Schobert, Chemistry of Fossil Fuels and Biofuels, Cambridge University Press: Cambridge, 2013 (Chapters 17, 21).

Table 1.2 Classification parameters in the ISO classification system

Moisture in freshly mined coal (%)	Class number	Tar yield, dry, ash-free basis (%)	Group number
≤ 20	10	≤ 10	00
20–30	11	10–15	10
30–40	12	15–20	20
40–50	13	20–25	30
50–60	14	≥ 25	40
60–70	15		

Modified from J.G. Speight, The Chemistry and Technology of Coal, Marcel Dekker, New York, 1994 (Chapters 1, 17).

subbituminous coal is sometimes included in the categories of *other bituminous coals*, or of *lignite*, or *brown coal*, or *hard coal* [9].

Most of the world's low-rank coals were deposited in various epochs of the Tertiary period, so are less than 65 million years old. A few are indicated to be of Pleistocene age, from the beginning of the Quaternary period.

1.2 The types of low-rank coals

This section provides brief descriptions of the low-rank coals. Two points should be kept in mind. First, values of any physical or chemical property of any rank of coal vary over wide ranges. Coals, as a group, vary widely among themselves, to such an extent that it has sometimes been suggested that the world *coal* should never be used in the singular. Second, while it is useful and convenient to establish general descriptions of coals (as in the subsections below), their great variability means that anomalies or exceptions likely will be encountered somewhere in the literature. The terms *brown coal* and *lignite* are not used consistently in the existing literature; sometimes they used interchangeably, sometimes used to imply that brown coal is a separate classification of rank lower than that of lignite. *Lignite* is the term preferred by the International Commission for Coal and Organic Petrology [10]. As noted above, further variations of nomenclature exist. Throughout this chapter, the terminology used in the original sources is retained.

Low-rank coals have higher moisture contents than bituminous coals and anthracite. Low-rank coals also are more friable and softer. Some lignites display a clearly evident woody texture. Many low-rank coals have low sulfur content, though there are some spectacular exceptions, with some lignites having around 10% sulfur. Low-rank coals do not appear to be glossy or shiny. As rank progresses from lignite B (or brown coal) to subbituminous A, several general trends are evident. Moisture content is greatly reduced. On a moisture- and ash-free basis, fixed carbon, total carbon, and calorific value all increase. The woody texture characteristic of some lignites is not evident in the subbituminous coals. Rather, subbituminous coals begin to show bright and dull bands, which are often observed in bituminous coals, and are more characteristic of the bituminous rank. A low sulfur content, combined with higher calorific values relative to lignites, makes subbituminous coals attractive fuels for electricity generation. The coals of the Powder River Basin in the United States are a noteworthy example.

The progressive increase in carbon content with increasing rank is accompanied by a monotonic decrease in oxygen, as well as by changes in the distribution of oxygen among its various functional groups. Most notably, the amount of oxygen incorporated in carboxylic acid groups decreases significantly as rank increases, these groups being absent from bituminous coals. Carboxylic acid groups serve as ion-exchange sites in low-rank coals, accumulating ions of the alkali and alkaline earth elements (i.e., sodium, potassium, magnesium, and calcium). During combustion, these ions are liberated to participate in ash-forming reactions; low-rank coal ashes have proportionally higher amounts of these elements than ashes of higher rank coals. Ions of alkali and alkaline earth elements act as fluxes in ash chemistry, so the ash fusion behavior and fluid properties of slags are different than those of higher rank coals. Usually ash fusion temperatures and slag viscosities at a given temperature for low-rank coal ashes are lower than for ashes of high-rank coals.

1.3 Distribution, reserves, and resources of low-rank coals

Coals of some kind are found in almost every country of the world. Low-rank coals occur from Portugal [11] across to Thailand [12], and from Canada [13] south to Chile [14]. Even Antarctica contains low-rank coal [15]. Proved reserves of low-rank coals worldwide are 488,332,000 tonnes [16], representing 55% of world coal reserves. Proved reserves are those that have been identified by detailed geological

Table 1.3 Comparison of worldwide reserves of fossil fuels, in units of million tonnes of oil equivalent

Resource	Reserves (million tonnes of oil equivalent)
Coal, anthracite and bituminous	268,799
Coal, subbituminous and lignite	162,777
Petroleum	239,400
Natural gas	168,958

Data from BP, BP Statistical Review of World Energy, 65th edition. <http://www.bp.com/content/dam/bp/pdf/energy-economics/statistical-review-2016>, 2016.

Table 1.4 Proved reserves of low-rank coals by world region [16]

Region	Reserves (million tonnes)
Europe and Eurasia	217,981
North America	132,253
Asia-Pacific	130,525
South and Central America	7359
Middle East and Africa	214
Total world	*488,332*

exploration and that can be extracted economically using present-day technology, or modifications of current technology that will be available in the near future [17]. The potential contribution of low-rank coals in the world energy scene is illustrated by the comparison of reserves shown in Table 1.3. (Three tonnes of lignite are approximately equivalent to one tonne of oil equivalent [18].)

The largest reserves occur in Europe and Eurasia, mainly in a broad belt running from Germany through many of the countries of central and eastern Europe. This region accounts for nearly half of the proven reserves of low-rank coals in the world. Most of the rest of the world's low-rank coals occur in roughly equal amounts in the Asia-Pacific region, in which China, Australia, and India dominate; and in North America, primarily in the United States. Proved reserves by world region are summarized in Table 1.4.

In terms of low-rank coal production, the top ten nations are shown in Table 1.5 [19].

1.3.1 Europe and Eurasia

The proved reserves of lignite for countries in this region are shown in Table 1.6 [20]. Only Ukraine has significant reserves of coal reported as subbituminous rank, at 16,577 million tonnes, with a small amount, 190 million tonnes, in Bulgaria [20].

The major deposits of subbituminous coals in Russia are found in the Kansk-Achinsk (eastern Siberia) and Kuznetsk basins. The Kansk-Achinsk coals are

Table 1.5 Top ten nations for production of low-rank coals, based on 2013 data [19]

Nation	Production (million tonnes)	Nation	Production (million tonnes)
Germany	183	Australia	63
Russia	73	Greece	54
United States	70	India	45
Poland	66	Czech Republic	40
Turkey	63	Serbia	40

Table 1.6 Proved reserves of lignite in countries in the European and Eurasian region [20]

Nation	Proved reserves (million tonnes)	Percent of world total
Germany	40,500	20
Serbia	13,400	6.7
Kazakhstan	12,100	6.0
Russia	10,450	5.2
Ukraine	8380	4.2
Greece	3020	1.5
Bulgaria	2174	1.1
Turkey	1945	1.0
Poland	1287	0.64
Czech Republic	871	0.43
Republic of Macedonia	332	0.16
Romania	280	0.14

typically of very low sulfur content, often below 1% [1]. Generally, Kuznetsk coals are also low-sulfur coals. A very large deposit of lignite occurs in the Moscow basin, but currently there seems to be little, if any, mining activity there.

The Lower Rhine Basin in Germany is the largest brown coal deposit in Europe [21]. Most of this coal formed during the Miocene [21]. It is noteworthy for a very low sulfur content, about 0.35% on a moisture- and ash-free basis [22], as well as a very high calcium content, about 40%–60% CaO in the ash (corresponding to about 2% calcium on a whole, dry coal basis) [21]. This lignite occurs in the western part of Germany, west of Cologne, stretching toward the confluence of its borders with Belgium and the Netherlands. Additional brown coal deposits, somewhat older geologically (Eocene to Oligocene age), occur further to the east in the Central German and the Lusatian basins [21]. A recent analysis of the future of the industry in Germany suggests that the use of lignite in power generation should continue to be economically viable at least until 2030 [23]. However, it is important to note the potential impact of the German energy transition policy, *Energiewende*, on future use of coal [24]. This policy emphasizes continued sustainable development with increasing reliance on renewable energy sources. Greenhouse gas emissions are to be reduced to 80%–95% of 1990 levels by 2050. As a result, the contribution of lignite and brown coal to German electricity production is likely to diminish, and indeed be severely reduced, in the future.

Lignite deposits in Ukraine occur in the central part of the country, west of the better-known high-rank coals in the Donets Basin. Little, if any, lignite mining is going on at present, even though Ukraine is obliged to import coal to meet its overall energy needs.

Lignite is located in the Maikuben, Turgay, and Nizhne-Iliyskiy basins of Kazakhstan, and represents about 25% of the country's total coal reserves [25]. Production in 2012 was about 5.5 million tonnes [26]. Moisture content of the as-mined lignite is up to 45%.

Lignite constitutes most of the coal reserves in Turkey [27]. Lignites of various geological ages occur extensively throughout Turkey [28]. Lignites also occur in the Muğla Basin, on the Aegean coast, in southwestern Anatolia [29]. Some of the Eocene-age lignites of northwestern Turkey have extraordinary sulfur contents, up to about 9.6% [28]; several of them contain about 8% sulfur on a dry, ash-free basis [30]. In the recent past some Turkish lignite production was used in domestic heating (about 16%), and a similar percentage for industrial heating [28]. Nearly the same percentage is used currently in industry, but the remainder is used almost entirely in electricity production [31], particularly the high-ash and high-sulfur lignites.

Lignite occurs throughout much of Greece, in seven major coal basins [32], in the Ptolemais basin of northern Greece, and is actively mined for electric power production [33]. A mine has also been opened in the Florina basin in Greek Macedonia [34]. About 63% of Greek electricity production is from lignite-fired plants [35]. Most production is of lignite of Pliocene age, noteworthy for the lignite seams alternating with clay partings. As a group, Greek lignites are highly variable in quality, ranging from 10% to above 40% in ash value, from 0.1 to over 4% sulfur content, and with calorific values of 4–14 MJ/kg [32,36]. Coals of lignite to subbituminous B rank are found on Crete [37]. Somewhat similar to the situation in Germany, the Greek National Emissions Reduction Plan, and Greek compliance with the European Union's Large Combustion Plants Directive, it is likely that the present dominant position of lignite in electricity supply will decrease [38]. Also, there is significant potential for future contributions from solar and wind energy for electricity generation in Greece.

Lignites of very low rank occur in the Republic of Macedonia [39]. Ash values can be as high as 21.5%. Some of this lignite is of fairly high sulfur content, 3.3%–5.6%, although for most of the seam the sulfur contents are more typical of low-rank coals, i.e., 0.7%–1.3%.

Lignites represent a major energy resource in Bulgaria [40]. Some are notable for high ash values, up to about 43% [40]. High moisture contents are also observed, up to about 63% for Maritza lignite [41]. Lignite in the Maritza basin in eastern Bulgaria [42], and another further west, in the Sofia region [43], represent the principal fuel source for the Bulgarian power industry [41]. Some of the lignites are difficult to handle, especially when wet, because they incorporate significant amounts of sticky clays such as montmorillonite. Sulfur contents can be up to 13.5% (dry basis), for the Maritza West lignite [44]. Subbituminous coals occur in the Pernik and Bobov Dol deposits in western and southwestern Bulgaria [43,45]. The Bobov Dol coal has about 6%−12% moisture and highly variable ash value, ranging from 15% to 65% [45].

Essentially all of the coal produced in Romania is lignite [46]. The Oltenia coal basin is in the southwestern part of the country, in the region between the Olt and the Danube rivers [47].

Brown coal deposits in northern and northeastern Serbia occur in an area of several hundred square kilometers, and are important for electric power generation [48].

The most important Polish brown coal deposits are an extension of the brown coals in the eastern part of Germany [21]. Lignite occurs in southwestern and central Poland [21,49]. Considerable variation in ash value is found, around 25% in the highest cases, but only about 4% at the low end. These lignites are generally low in sulfur content, sometimes only about 0.2%, and seldom exceeding 1.5% [21].

In the Czech Republic, production of lignite peaked in 1985 [50]. The brown coal mining area in the North Bohemian Basin is the most important brown coal basin in the Czech Republic [51,52]. Considered from the perspective of landscape ecology and earth science, the North Bohemian Basin is thought to be one of the most damaged landscapes in the world [53]. The coal occurs in seams of about 30 m thickness on average [52], with exceptional values of up to 70 m [51].

1.3.2 North America

Table 1.7 provides information on the proved reserves of low-rank coals in North America [20]. Unlike Europe and Eurasia, North America has substantial reserves of subbituminous coal.

Subbituminous coal provides 37% of U.S. coal reserves, with lignite adding another 9% [1]. The most important subbituminous coal area is the Powder River Basin of the Northern Great Plains Province, in parts of Wyoming and Montana. These very desirable coals represent the largest recoverable reserves in the United States, and typically have ash values of less than 10% and only about 1% sulfur.

Table 1.7 Proved reserves, million tonnes, of low-rank coals in North America [20]

Nation	Subbituminous coal		Lignite	
	Proved reserves	Percent of world total	Proved reserves	Percent of world total
United States	98,618	34	30,176	15
Canada	872	0.30	2236	1.1

Additional subbituminous coal occurs in the Rocky Mountain Province, also in Wyoming. The United States is the world's largest producer of subbituminous coal, all used for electric power generation. About 40% of U.S. coal production currently comes out of the Powder River Basin [54]. On an as-mined basis, these coals range from 7600 Btu/lb to 9400 Btu/lb calorific value, and 0.2% to 0.8% sulfur [55].

Virtually all low-rank coals mined in the United States are used in electric power generation. Two large lignite deposits occur in the United States, one in the Fort Union Region of the Northern Great Plains Province, in parts of North Dakota and Montana, and the other in portions of six southern states in the Gulf Province, dominated by Texas [21,56,57]. The Fort Union is considered to be the largest lignite deposit in the world [58,59]. In Texas, lignite occurs in three units: the Jackson and Wilcox Groups and the Yegua Formation [56,57]. Gulf Coast lignites, which are the lowest rank of coal found in the United States, generally have ash values higher than Fort Union lignites [21]. Ash values for Jackson lignites average about 26% [56]. Sulfur contents are typically <1.4% [56]. All U.S. lignites have high proportions of alkali and alkaline earth elements. Fort Union lignites can sometimes exceed 9% Na_2O in the ash [21]. Small amounts of lignites occur in other regions of the United States [57].

The major lignite deposit of Canada occurs in Saskatchewan in the Ravenscrag Formation [13,60]. This lignite represents about one-third of the recoverable reserves of coal in Canada [57]. It is used entirely for electricity generation. Additional lignite deposits are found in Ontario, British Columbia, Manitoba, and in the Canadian Arctic; these lignites are not used extensively. Subbituminous coals in British Columbia represent some of the thickest low-rank coal seams in the world [61], but present mining activities seem confined to high-rank coals [62]. Active mining of subbituminous coal occurs in Alberta, especially for supplying nearby electric power stations.

1.3.3 Asia-Pacific

Table 1.8 summarizes data for proved reserves of low-rank coals in the major nations of the Asia-Pacific region [20]. Subbituminous coal reserves in India, and

Table 1.8 Proved reserves, in millions of tonnes, of low-rank coals in the Asia-Pacific region [20]

Nation	Subbituminous coal		Lignite	
	Proved reserves	Percent of world total	Proved reserves	Percent of world total
China	33,700	12	18,600	9.2
Australia	2100	0.73	37,200	18
India	NR	–	4500	2.2
Indonesia	28,107	10	NR	–

NR, Not Reported.

lignite reserves in Indonesia, were not reported separately in this source. However, this does not imply that the values are necessarily zero, since, for example, Indonesia enjoys a thriving trade in lignite.

Though coal deposits can be found in much of China, most of the coal is in the north. In China, coal is classified by the standard GB/T5751-1986. The Chinese classifications can be compared to the ASTM system [63]. About 29% of China's coal reserves are subbituminous, and 16% are lignite [64]. Lignites occur throughout China, though primarily in the northeast (Jilin and Yunan provinces [65]) and in eastern Inner Mongolia [66]. Production in 2012 was about 200 million tonnes [67]. Additional mining occurs in several other provinces [65]. China is the second-largest producer of subbituminous coal.

Brown coals underlie approximately two-thirds of the land area of the state of Victoria [68], nearly 160,000 km^2, in the Gippsland basin. Seams are remarkably thick, up to about 140 m. As mined, the coal often has very high moisture content, up to 70%. Ash values are low, typically in the range of 1%–3.5%; sulfur is also low, often less than 1% [69]. A small amount of mining still occurs at Leigh Creek in South Australia [70], where there is a deposit of Lower Triassic lignite. The state of Western Australia also has lignite deposits [21], which are not being mined at present.

The state of Tamilnadu has about 80% of India's lignite, of which 3700 million tonnes are proved reserves [71]. Although the total lignite resource is immense, about 43,000 megatonnes, approximately 80% lies below 150 m, making it unattractive for mining with current technology. The largest lignite mines occur near Neyveli, which is in the Cuddalore District (formerly known as South Arcot) of Tamilnadu [21]. This coal is relatively low ash, typically $\leq 12\%$ sometimes approaching 2%, but has moisture contents up to 55%. Lignite also occurs in five other states in India. The Akrimota lignite, from the Kutch District of Gujarrat, is noteworthy for a sulfur content of 8.8% on a dry, ash-free basis [30].

Subbituminous coal accounts for about 54% of Indonesia's proved reserves, with lignite adding another 20% [72]. By far the greatest reserves are on Kalimantan, over 80% of proven reserves [72]. The low-rank coals of Indonesia are attractive from the perspective of being low-sulfur and low-ash coals. Typical values, on an as-received basis, are $<1\%$ sulfur and $<10\%$ ash [73]. About 80% of Indonesia's coal production is exported [72]. Total annual production of lignite is about 160 million tonnes [74], of which 58 million tonnes are exported to China [75]. This export trade in lignite provides a counter-example to the general rule that low calorific value and high moisture content militate against shipping lignites for any significant distance.

1.4 Current utilization of low-rank coals

1.4.1 Electric power generation

Combustion of coal for electric power generation is by far the dominant use of coal world-wide, and certainly true of the low-rank coals. Pulverized-coal (pc) fired boilers are the primary means of raising steam in electricity-generating plants. Very useful descriptions of these systems, their ancillary equipment, and their operation, with abundant details, are available [1,76,77]. Also, detailed overviews of how the properties of low-rank coals affect their performance [78], and of prospects and technologies for upgrading low-rank coals prior to use [79] are available.

Most of the electricity produced in low-rank coal-fired generation is supplied to electric grids. However, some local, specialized uses are also found. For example, electricity generated by burning Victorian brown coal is supplied to the Alcoa-Anglesea aluminum-smelting plant in Port Henry [80].

At the same rate of electricity generation, pc-fired boilers operating on low-rank coals are larger than those using bituminous coals. The increase in linear dimension ranges from a factor of about 1.1 for subbituminous coals to about 1.5 for brown coals [81]. Reasons for this increase in size are thoroughly documented (e.g., [1,81,82,83]). The need for increased size derives from the larger volume of gas to be handled in the combustion of low-rank coals, and from larger heat exchange surface area needed because of lower flame temperatures [81]. Also, a strategy for decreasing the chances of ash deposition inside the boiler is to maintain relatively low flue gas temperatures at the inlet to the secondary superheater [84]; this strategy increases the size of the boiler needed for good combustion efficiency [57].

As mentioned previously, many of the world's low-rank coals have low ash values and low sulfur contents, attractive properties for use in electric power generation. However, these coals often have high moisture contents. High moisture content increases the transportation cost per useful MJ of energy. Up to 7% of the fuel input to the plant can be expended drying the coal [85]. Pre-drying the coal certainly improves the plant efficiency. This has been demonstrated by Great River Energy in the Coal Creek Station (Underwood, North Dakota), where waste heat from the plant is used to dry lignite in a fluidized-bed dryer [85]. The tendency for dried low-rank coals to experience moisture re-absorption and autogenous heating, sometimes leading to ignition, makes it most practical to dry the coals immediately prior to combustion [86]. In addition, drying low-rank coals can increase their friability, leading to production of excessive amounts of fines during various handling and processing operations [78].

Because of the high moisture contents and low calorific values, it is seldom efficient or economic to ship brown coals and lignites long distances. They tend to be consumed in plants located very close to the mines, with short hauls by truck, rail, or conveyors. In contrast, subbituminous coals, with the highest calorific values among the low-rank coals and with low sulfur contents, can be attractive fuels even if considerable transportation is required. In the United States, for example, subbituminous coals from the Powder River Basin are shipped hundreds or thousands of kilometers by rail to meet demands for low-sulfur coals.

Any reduction in the moisture content of the coal will increase the energy efficiency of the boiler. Increased energy efficiency translates into reduced fuel consumption in the boiler for a given load on the plant [87]. Depending on the specific drying technology used, there may be ancillary benefits, such as the possible oxidation of mercury, making it more amenable to downstream capture [86]. Lignite drying technologies up to the mid-1990s have been reviewed previously [57]. Further advances in the following decade are also reviewed [88], and, subsequently, up to the present time [89]. In the United States, the federally funded Clean Coal Power Initiative program resulted in the development of a low-temperature, fluidized-bed drying process that has been employed at the Coal Creek generating station in North Dakota [86]. In Europe, RWE Power AG has also developed a fluidized-bed drying technology [90]. The RWE technology is now employed in plants at Frechen and Niederaußem.

As a rule, low-rank coals are more reactive than bituminous coals in most kinds of processing. Their greater reactivity results in higher reaction rates and/or lower reaction temperatures, which can be an advantage during utilization. However, the high reactivity also can lead to excessive oxidation, or even autogenous heating and "spontaneous" combustion, during storage. Oxidation of the coal will alter its behavior during processing. An additional, general rule-of-thumb is that low-rank coals tend to be low-sulfur coals, although there are certainly some exceptions, as mentioned above. For those low-rank coals that have high concentrations of alkali and alkaline earth elements—especially sodium and calcium—in the ash, some of the sulfur oxides produced during combustion are captured by the ash, helping to reduce emissions.

Unlike coals of higher rank, low-rank coals contain carboxylic acid functional groups as part of the coal structure. These groups can absorb ions of alkali and alkaline earth elements. In some coals, a high percentage, up to about 90%, of elements such as sodium are present on ion-exchange sites rather than being present in grains of minerals. As the coal is heated, the carboxylate groups are cleaved from the coal structure (beginning around 350°C), liberating the associated inorganic ions. This distinction, relative to bituminous coals, is reflected in different ash fusion behaviors and different slag properties, e.g., viscosity. Sodium ions associated with the ion-exchange sites tend to volatize in combustion systems; their subsequent condensation at lower temperatures contributes to the formation of sticky or molten surface layers that allow the formation of other relatively low-melting phases, leading to the build-up of ash deposits. In addition to the issues related to high moisture content and correspondingly low calorific value, the formation of ash deposits on boiler heat-transfer surfaces, i.e., ash fouling, can be another significant problem in combustion of low-rank coals [76,91,92]. The formation of ash fouling deposits often correlates with the sodium content of these coals [57,69].

Carbon dioxide emissions, expressed on a unit of thermal energy produced, can be slightly greater for lignites than for bituminous coals. The lower calorific value of the lignite requires that more lignite be burned to achieve a given energy output than would be required with bituminous coal. Even though lignite has a lower carbon content, the effect of calorific value dominates, leading to higher CO_2 emissions for production of the same amount of energy. However, if the carbon dioxide equivalent for methane emissions from mining is included in the analysis, along with the carbon dioxide produced during combustion, the advantage is with the lignite [93].

1.4.2 Other commercial uses

Though combustion of low-rank coals for electricity generation dominates by far the commercial applications of these coals, other uses in combustion also exist. Small-scale industrial boilers can certainly be fired on low-rank coals. Such boilers are in use in many countries, for producing hot water, low-grade steam, or other applications. Drying the fuel before combustion offers the same advantage as mentioned above, an increased thermal efficiency of the boiler and therefore reduced fuel consumption for a given demand on the boiler. It might be anticipated that the advantages of pre-dried low-rank coals would increase their demand in this market, but at the same time there is likely to be increased competition in the small boiler market from natural gas, especially in those regions in which new gas supplies are becoming available as a result of hydraulic fracturing (fracking) of reservoirs.

The so-called charcoal briquettes, in which the carbonaceous material actually comes from lignite (occasionally subbituminous coals), are produced in numerous countries for domestic combustion, including in outdoor barbecues. Low-rank coal chars can be used to produce formcokes, which have applications as relatively smokeless domestic fuels and as reducing agents in some metallurgical applications [57,94].

The Dakota Gasification plant in Beulah, North Dakota converts lignite to substitute natural gas. By-product tars and oils from the gasifiers are valuable sources of chemical products, such as phenol, catechols, naphthol, anhydrous ammonia and ammonium sulfate fertilizers, and tar-oil for liquid fuel. Useful summaries of gasification technology and its applicability of low-rank coals are available [1,3,57,95]. The Dakota Gasification plant captures carbon dioxide, shipping it about 330 km to Weyburn, Saskatchewan, Canada, where it is used in enhanced oil recovery projects.

The use of by-product gasification tars as a source of chemicals, or even use of the gas for downstream chemical synthesis, provides excellent opportunities for low-rank coals as raw material for an energy complex, which might, for example, produce electricity, synthetic liquid fuel, and a variety of value-added chemicals as co-products. This strategy offers the prospect of increased use of low-rank coals, but also may enhance the commercial viability of these coals because of the multiple revenue streams. It can be speculated that this strategy might be attractive in those regions where regulatory or governmental policy pressure militates against coal-fired plants in which electricity is the sole product.

An extensive report details coal gasification in China [96], including low-rank coal feedstocks. In some facilities, the primary products are chemicals, such as ammonia, urea, and ethylene glycol. It appears, though, that this technology is not a priority for future development [97].

A great deal of research, development, and demonstration effort was put into direct liquefaction of coal, especially during various oil shortages and price spikes in the last quarter of the 20th century, but presently there are no commercial direct liquefaction plants running on low-rank coals. A small 0.8 tonne per day pilot plant for conversion of lignite to gasoline and diesel fuel was scheduled for start-up in Kazakhstan in early 2015 [98].

Lignite is a raw material for the commercial production of activated carbon. Lignite is a raw material for the commercial production of activated carbon, e.g. in Texas and in Saskatchewan. These carbons can be used as adsorbents in water or gas treatment [57]. For example, the Cabot Corporation provides Darco carbon, derived from lignite, for removal of various contaminants, including heavy metals, from incinerator flue gases [99].

Soil conditioners and fertilizers are produced in Australia from brown coal [81,100]. Performance of lignites in this application can be enhanced by oxidation, e.g., with nitric acid [101]. Leonardite, a naturally occurring oxidatively weathered form of lignite, and humalite, an analogous material from subbituminous coals, are also used as soil additives. These materials can be used in reclamation and reconstruction of soils in agricultural areas [102]. Lignites, sometimes modified by various chemical treatments, are used in well drilling fluids (drilling mud) in the oil industry.

Wax is extracted from lignite in China [66], and from German brown coal [21]. Commonly called montan wax, this material occurs in amounts up to about 10% in some Chinese lignites, and finds a variety of commercial applications, including, as examples, floor waxes and polishing waxes, and electric cable insulation [103].

Germanium-enriched lignites from Inner Mongolia might prove to be a useful source of this important element, which has applications in photovoltaics, fiber optics, and infrared optics. Germanium concentrations in the Shengli lignite are an order of magnitude higher than average values for coals [104].

Acknowledgments

It is a pleasure to thank Nita Schobert for her outstanding assistance in searching the literature and obtaining many of the documents cited in this chapter. Thanks also to Michalis Agraniotis for useful comments and information on the European energy scene for improving the preliminary draft of this chapter.

References

[1] B.G. Miller, Clean Coal Engineering Technology, Butterworth-Heinemann, Burlington, MA, 2011 (Chapters 2, 3, 5).

[2] H.H. Schobert, Chemistry of Fossil Fuels and Biofuels, Cambridge University Press, Cambridge, 2013 (Chapters 17, 21).

[3] J.G. Speight, The Chemistry and Technology of Coal, Marcel Dekker, New York, 1994 (Chapters 1, 17).

[4] American Society for Testing and Materials (ASTM), Annual Book of ASTM Standards: Volume 05.05 Gaseous Fuels; Coal and Coke, 388-84, American Society for Testing and Materials, Philadelphia, 1986, pp. 247–252, Standard D.

[5] International Organization for Standardization (ISO). Classification of Coals, ISO Standard 11760.2005(en). <https://www.iso.org/obp/ui/#iso:std:iso:11760:ed-1:v1:en>, 2005.

[6] N. Berkowitz, An Introduction to Coal Technology, Academic Press, New York, 1979 (Chapter 3).
[7] D.W. Van Krevelen, *Coal. Typology—Physics–Chemistry–Constitution*, Elsevier, Amsterdam, 1993 (Chapter 2).
[8] M.S. Rana, F.S. AlHumaidan, Statistical data on worldwide coal reserves, production, consumption, and future demand, in: M.R. Riazi, R. Gupta (Eds.), Coal Production and Processing Technology, CRC Press, Boca Raton, FL, 2016 (Chapter 2).
[9] International Energy Agency (IEA). Coal information 2012. <http://www.iea.org/media/training/presentations/statisticsmarch/coalinformation.pdf>, 2012.
[10] International Commission for Coal and Organic Petrology (ICCP), International Handbook on Coal Petrology: Second Supplement, Centre national de la recherche scientifique, Paris, 1993.
[11] L. Thomas, Coal Geology, Wiley, New York, 2002 (Section 3.4).
[12] S. Pipatmanomai, B. Fungtammasan, S. Bhattacharya, Characteristics and composition of lignites and boiler ashes and their relation to slagging: the case of Mae Moh PCC boilers, Fuel 88 (2009) 116–123.
[13] J.D. Hughes, W.J. McDougall, Geology and coal resources of southern Saskatchewan coalfields, in: D.G. Richardson (Ed.), Investigations Completed by the Saskatchewan Geological Survey and the Geological Survey of Canada under the Geoscience Program of the Canada–Saskatchewan Partnership Agreement on Mineral Development., Geological Survey of Canada Open File Report No. 3119, 1995, pp. 279–302.
[14] G. Alfaro, M.E. Cisternas, Chemical and mineralogical characteristics of the subbituminous coal of the Oligo-Miocene of southern-center Chile, in: J.A. Pajares, J.M.D. Tascon (Eds.), Coal Science, Elsevier, Amsterdam, 1995, pp. 103–106.
[15] K. Birkenmajer, J.A. Frankiewicz, M. Wagner, Tertiary coal from the Lions Cove Formation, King George Island, West Antarctica, Pol. Polar Res. 12 (1991) 229–241.
[16] BP, 2015. BP Statistical Review of World Energy, 64th edition. <http://www.bp.com/statisticalreview>.
[17] H.H. Schobert, Energy and Society, CRC Press, Boca Raton, FL, 2014 (Chapter 30).
[18] BP. BP Statistical Review of World Energy, 65th edition. <http://www.bp.com/content/dam/bp/pdf/energy-economics/statistical-review-2016>, 2016.
[19] World Coal Association. Coal facts 2014. <www.worldcoal.org>, 2014.
[20] World Energy Council (WEC). Coal. <https://www.worldenergy.org/wp-content/uploads/2013/10/WER_2013_1_Coal.pdf>, 2013.
[21] C.S. Gloe, Comparison of Victorian brown coals and other Australian deposits with major overseas brown coal/lignite deposits, in: R.A. Durie (Ed.), The Science of Victorian Brown Coal, Butterworth Heinemann, Oxford, 1991 (Chapter 13).
[22] E.A. Wolfrum, Correlations between petrographical properties, chemical structure, and technological behavior of Rhenish brown coal. The chemical characteristics of Victorian brown coal, in: H.H. Schobert (Ed.), The Chemistry of Low-Rank Coals, American Chemical Society, Washington, 1984 (Chapter 2).
[23] von Bechtolsheim, M.; Kruse, M., 2015. The Future of Lignite Power. Arthur D. Little., Boston <www.adl.com/FutureOfLignitePower>.
[24] Energy Information Agency (EIA), 2016. Germany's renewables electricity generation grows in 2015, but coal still dominant. <http://www.eia.gov/todayinenergy/detail.php?id = 26372>.
[25] Vorotnikov, V., 2013. Kazakhstan prepares to grow coal production, *Coal Age*. **118**(9), 2013, 40–43.

[26] Knoema, 2015. Kazakhstan Coal—Production of Lignite Coal. <http://knoema.com/atlas/Kazakhstan/topics/Energy/Coal/Production-of-Lignite-Coal>.
[27] Ç. Şentorun, H.H. Schobert, Comparison of inorganic constituents in Turkish low-rank coals, Proceedings of 1997 International Ash Utilization Symposium, 1997, pp. 700–706.
[28] H. İnaner, E. Nakoman, Turkish lignite deposits, in: R.A. Gayer, J. Pešek (Eds.), European Coal Geology and Technology, The Geological Society, London, 1997, pp. 77–99.
[29] M. Fotopoulou, G. Siavalas, H. İnaner, K. Katsanou, N. Lambrakis, K. Christanis, Combustion and leaching behavior of trace elements in lignite and combustion by-products from the Muğla Basin, SW Turkey, Bull. Geol. Soc. Greece XLIII (5) (2010) 2218–2228.
[30] G. Scheffknecht, Technologies for efficient utilization of low-rank fuels, in: A.F. Armor, F. Bauer (Eds.), Lignites and Low-rank Coals Conference: Proceedings 2001, Electric Power Research Institute Report No. 1006879, 2002, pp. 5.31–5.48.
[31] Mills, S., 2014. Regional report—Turkey: Turkish delight, World Coal.<www.world-coal.com/special-reports/10112014/World-Coal-Regional-Report-Turkey>.
[32] C. Papanicolaou, T. Kotis, A. Foscolos, F. Goodarzi, Coals of Greece: a review of properties, uses future perspectives, Int. J. Coal Geol. 58 (2004) 147–169.
[33] S. Valčeva, A. Georgakopoulos, K. Markova, The relationship between the petrographic composition and some chemical properties for the Ptolemais basin lignites, Greece, in: J.A. Pajares, J.M.D. Tascon (Eds.), Coal Science, Elsevier, Amsterdam, 1995, pp. 259–262.
[34] D. Pentari, A.E. Foscolos, V. Perdikatsis, Trace elements in the lignite-bearing area of Lofoi, Florina Basin, western Greek Macedonia, Greece, Energy Sources, Part A 30 (2008) 316–324.
[35] D.E. Karageorgiou, A. Metaxas, D. Dimitriou, E. Arapogiannis, G. Varvarousis, Contribution of lignite in the Greek economy, Bull. Geol. Soc. Greece. XLIII (5) (2010) 2229–2235.
[36] N.K. Koukouzas, R. Zeng, V. Perdikatsis, W. Xu, E.K. Kakaras, Mineralogy and geochemistry of Greek and Chinese coal fly ash: research for potential applications, Proceedings, 2005 World Coal Ash Conference, 2005, <http://www.flyash.info/2005/37kou.pdf>
[37] T. Gentzis, F. Goodarzi, C.N. Koukouzas, A.E. Foscolos, Petrology, mineralogy, and geochemistry of lignites from Crete, Greece, Int. J. Coal Geol. 30 (1996) 131–150.
[38] International Energy Agency (IEA), 2011. Energy policies of IEA countries: Greece 2011 review. <https://www.iea.org/publications/freepublications/publication/Greece2011_unsecured.pdf>.
[39] S. Valčeva, M. Yossifova, K. Markova, J. Stefanova, D. Apostolova, M. Chukarski, Petrological characteristics of lignites from Berovo deposit, Fyrom., Annuaire de l'Universite de Sofia. Livre 1—Geologie, vol. 102, 2010, pp. 225–240.
[40] G.D. Šiškov, Bulgarian low rank coals: geology and petrology, in: R.A. Gayer, J. Pešek (Eds.), European Coal Geology and Technology, The Geological Society, London, 1997, pp. 141–148.
[41] V.R. Kotler, D.V. Sosin, A modern thermal power station operating on high-moisture lignite, Therm. Eng. 56 (2009) 526–528.
[42] I. Kostova, J. Kortenski, Inorganic composition of lignite from the Maritza East basin, C.R. Acad. Bulg. Sci. 55 (10) (2002) 49–54.
[43] S.V. Vassilev, C.G. Vassileva, Comparative chemical and mineral characterization of some Bulgarian coals, Fuel Process. Technol. 55 (1998) 55–69.

[44] S.P. Marinov, M. Stefanova, I. Kostova, V. Stamenova, R. Carleer, J. Yperman, Peculiarities of sulphur functionalities in the Thracian coal province, Bulgaria, Bull. Geosci. 80 (2005) 33–38.
[45] S.V. Vassilev, M.G. Yossifova, C.G. Vassileva, Mineralogy and geochemistry of Bobov Dol coals, Bulgaria, Int. J. Coal Geol. 26 (1994) 185–213.
[46] W.G. Steblez, The mineral industries of the northern Balkans: Bulgaria and Romania, 2005 Minerals Yearbook: Northern Balkans, U.S. Geological Survey, Washington, 2007, pp. 3.1–3.14.
[47] N. Ticleanu, D. Scradeanu, M. Popa, S. Milutinovici, R. Popa, et al., The relation between the lithotypes of Pliocene coals from Oltenia and their main quality characteristics, Bull. Czech Geol. Surv. 74 (1999) 169–174.
[48] G. Jankes, O. Cvetković, T. Glumičić, Determination of different forms of sulphur in Yugoslav brown coals, in: R.A. Gayer, J. Pešek (Eds.), European Coal Geology and Technology, The Geological Society, London, 1997, pp. 269–272.
[49] B. Bielowicz, A new technological classification of low-rank coal on the basis of Polish deposits, Fuel 96 (2012) 497–510.
[50] J. Pešek, M. Dopita, Coal production in the Czech Republic, in: R.A. Gayer, J. Pešek (Eds.), European Coal Geology and Technology, The Geological Society, London, 1997, pp. 3–11.
[51] V. Bouška, J. Pešek, K. Žák, Values of $\delta^{34}S$ in iron disulphides of the North Bohemian Lignite Basin, Czech Republic, in: R.A. Gayer, J. Pešek (Eds.), European Coal Geology and Technology, The Geological Society, London, 1997, pp. 261–267.
[52] I. Sýkorová, J. Černý, H. Pavlíková, Z. Weishauptová, Composition and properties of North Bohemian coals, in: R.A. Gayer, J. Pešek (Eds.), European Coal Geology and Technology, The Geological Society, London, 1997, pp. 207–217.
[53] V. Cílek, To Breathe with Birds, University of Pennsylvania Press, Philadelphia, 2015, p. 7.
[54] Joo, N.; Lee-Ashley, M.; Madowitz, M. Fact Sheet: 5 Things You Should Know About Powder River Basin Coal Exports, Center for American Progress. <www.americanprogress.org/issues/green/report/2014/08/19>, 2014.
[55] Luppens, J.A.; Scott, D.C.; Haake, J.E.; Osmonson, L.M.; Pierce, P.E., 2015. Coal Geology and Assessment of Coal Resources and Reserves in the Powder River Basin, Wyoming and Montana, U.S. Geological Survey Professional Paper. No. 1809.
[56] T.F. Edgar, W.R. Kaiser, Resources, properties, and utilization of Texas lignite: a review, in: H.H. Schobert (Ed.), The Chemistry of Low-Rank Coals, American Chemical Society, Washington, 1984 (Chapter 4).
[57] H.H. Schobert, Lignites of North America, Elsevier, Amsterdam, 1995 (Chapters 1, 10, 11, 12).
[58] W.R. Kube, H.H. Schobert, S.A. Benson, F.A. Karner, The structure and reactions of Northern Great Plains lignites, in: H.H. Schobert (Ed.), The Chemistry of Low-Rank Coals, American Chemical Society, Washington, 1984 (Chapter 3).
[59] H.H. Schobert, Coal: The Energy Source of Past and Future, American Chemical Society, Washington, 1987 (Chapter 3).
[60] A.P. Beaton, F. Goodarzi, Geochemical characterization of lignite from southern Saskatchewan, Canada, in: D.G. Richardson (Ed.), Investigations Completed by the Saskatchewan Geological Survey and the Geological Survey of Canada under the Geoscience Program of the Canada–Saskatchewan Partnership Agreement on Mineral Development, Geological Survey of Canada Open File Report No. 3119, 1995, pp. 255–278.

[61] F. Goodarzi, E. van der Flier-Keller, Organic petrology and geochemistry of intermontane coals from British Columbia, Chem. Geol. 75 (1989) 227–247.

[62] British Columbia (BC). British Columbia Coal Industry Overview 2011, Ministry of Energy and Mines. <http://www.mining.bc.ca/sites/default/files/resources/icoal2012-2_1.pdf>, 2011.

[63] T. Lu, G. Laman, A Preliminary Comparison of Coal Classification and Processing Between Canada and China, Tetra Tech, Vancouver, 2011.

[64] Aden, N.; Fridley, D.; Zheng, N., 2009. China's Coal: Demand, Constraints, and Externalities, Lawrence Berkeley National Laboratory Report LBNL-2334E.

[65] A. Pan, China Economy Atlas, Cloud New Zealand, Auckland, 2011.

[66] C. Zhang, X. He, S. Zhu, X. Wang, H. Zhuang, Distribution, character, and utilization of lignite in China, Proceedings, 2011 Asia-Pacific Power and Energy Engineering Conference, 2011, pp. 1–4.

[67] Knoema. China Coal—Production of Lignite Coal. http://knoema.com/atlas/China/topics/Energy/Coal/Production-of-Lignite-Coal, 2015.

[68] C.S. Gloe, G.R. Holdgate, Geology and resources, in: R.A. Durie (Ed.), The Science of Victorian Brown Coal, Butterworth Heinemann, Oxford, 1991 (Chapter 1).

[69] G.J. Perry, D.J. Allardice, L.T. Kiss, The chemical characteristics of Victorian brown coal, in: H.H. Schobert (Ed.), The Chemistry of Low-Rank Coals, American Chemical Society, Washington, 1984 (Chapter 1).

[70] IBIS. Brown coal mining in Australia: Market research report, Report no. ANZSIC B0602. <www.ibisworld.com.au/industry/default.aspx?indid = 59>, 2015.

[71] Joshi, N.C. Development of Lignite Sector in India, Press Information Bureau, Government of India. http://pib.nic.in/newsite/efeatures.aspx?relid = 100924, 2013.

[72] B. Lucarelli, The History and Future of Indonesia's Coal industry: Impact of Politics and Regulatory Framework on Industry Structure and Performance, Program on Energy and Sustainable Development Working Paper. No. 93. Stanford University, Stanford, CA, 2010.

[73] Belkin, H.E.; Tewalt, S.J., 2007. Geochemistry of Selected Coal Samples from Sumatra, Kalimantan, Sulawesi, and Papua, Indonesia, U.S. Geological Survey Open File Report No. 2007-1202.

[74] Sunshine, W.L., 2014. Lignite. About Money. http://energy.about.com/od/Coal/a/Lignite.htm.

[75] Hsueh, M. Thermal Coal: China Import Ban Takes Shape, Deutsche Bank Markets Research. <https://etf.deutscheawm.com/GBR/ENG/Download/Research-Commodities/369a725f-1f7e-4e63-8095-e0db9d586b8c/Special-Report.pdf>, 2014.

[76] J.B. Kitto, S.C. Stultz (Eds.), Steam: Its Generation and Use, Babcock and Wilcox, Barberton, OH, 2005 (Section II).

[77] M. Angleys, W. Kessel, Pulverized fuel technology for lignite and brown coal fired utility boilers, in: A.F. Armor, F. Bauer (Eds.), Lignites and Low-rank Coals Conference: Proceedings 2001, Electric Power Research Institute Report No. 1006879, 2002, pp. 7.37–7.62.

[78] Swanson, M.L.; Sondreal, E.A.; Laudal, D.A.; Hajicek, D.R.; Henderson, A.K.; Pavlish, B.M., 2009. JV Task 129—Advanced Conversion Test—Bulgarian Lignite, University of North Dakota Energy and Environmental Research Center Report. No. 2009-EERC-06-14.

[79] H. Katalambula, R. Gupta, Low-grade coals: a review of some prospective upgrading technologies, Energy Fuels 23 (2009) 3392–3405.

[80] Alcoa Alcoa in Australia: Anglesea power station. <http://www.alcoa.com/australia/en/info_page/anglesea_coal.asp>, 2015.
[81] D.J. Allardice, B.S. Newell, Industrial implications of the properties of brown coals, in: R.A. Durie (Ed.), The Science of Victorian Brown Coal, Butterworth Heinemann, Oxford, 1991 (Chapter 12).
[82] Heil, T.C.; Durant, O.W., 1978. Designing boilers for western coal, in: Proceedings of Joint Power Generation Conference, Paper No. PGTP-78-14.
[83] J.G. Singer, Combustion: Fossil Power Systems, Combustion Engineering, Windsor, CT, 1981 (Chapter 8).
[84] F.I. Honea, E.A. Sondreal, Ash properties and combustion characteristics of lignites, Trans. Soc. Min. Eng. 270 (1982) 1980–1987.
[85] Department of Energy (DOE), 2012. Clean Coal Power initiative. Round 1 Demonstration Projects, Topical Report No. 27.
[86] P.J. Reddy, Clean Coal Technologies for Power Generation, CRC Press, Boca Raton, FL, 2014 (Chapter 6).
[87] Dong, N., 2014. Techno-Economics of Modern Pre-Drying Technologies for Lignite-Fired Power Plants, International Energy Agency Report No. CCC-241.
[88] M. Karthikeyan, W. Zhonghua, A.S. Mujumdar, Low-rank coal drying technologies—current status and new developments, J. Dry. Technol. 27 (2009) 403–415.
[89] N. Nikolopolous, I. Violidakus, E. Karampinis, M. Agraniotis, C. Bergins, P. Grammelis, et al., Report on comparison among current industrial scale lignite drying technologies. (A critical review of current technologies.), Fuel 155 (2015) 86–114.
[90] H.J. Klutz, C. Moser, D. Block, Development status of WTA fluidized-bed drying for lignite at RWE Power AG, Kraftwerkstechnik—Sichere nachhalt. Energie-versorg. 2 (2010) 2–18.
[91] R.W. Bryers, Fireside slagging, fouling, and high-temperature corrosion of heat-transfer surface due to impurities in steam-raising fuels, Prog. Energy Combust. Sci. 22 (1996) 29–120.
[92] E.A. Sondreal, P.H. Tufte, W. Beckering, Ash fouling in the combustion of low rank western U.S. coals, Combust. Sci. Technol. 16 (1977) 95–110.
[93] J. Engelhard, Mining and use of lignite—environmental aspects, in: A.F. Armor, F. Bauer (Eds.), Proceedings: Lignites and Low-rank Coals Conference, 2001, Electric Power Research Institute Report, No. 1006879.
[94] H. Cengizler, M. Kemal, Formcoke production from char fines of hard brown coals by air curing, Miner. Process. Extr. Metall. 115 (2006) 132–138.
[95] C. Higman, M. van der Burgt, Gasification, Gulf Professional Publishing, Burlington, MA, 2008, Chapter 4.
[96] US-China Energy Center, Coal Gasification in China: A Study Report, National Research Center for Coal and Energy, West Virginia University, Morgantown, WV, 2014.
[97] H. Bernton, China rethinking plans to build coal-to-gas plants, Seattle Times (2014), December 25 <http://www.seattletimes.com/nation-world/china-rethinking-plans-to-build-coal-to-gas-plants/>
[98] Caspian Energy. Kazakhstan launches pilot fuel production from lignite coal. <http://www.caspianenergy.net/en/oil-and-gas/18108-kazakhstan-launches-pilot-fuel-production-from-lignite-coal>, 2015.
[99] Cabot Corporation. Powdered activated carbon. <http://www.cabotcorp.com/solutions/products-plus/activated-carbon/powdered>, 2015.

[100] AIMR, 2011. Brown coal, in: Australian Mines Atlas. <www.australianminesatlas.gov.au>.
[101] F. Liu, S. Xing, Z. Du, Nitric acid oxidation for improvement of a Chinese lignite as soil conditioner, Commun. Soil Sci. Plant Anal. 42 (2011) 1782–1790.
[102] A. Bekele, J.L. Roy, M.A. Young, Use of biochar and oxidized lignite for reconstructing functioning agronomic topsoil: effects of soil properties in a greenhouse study, Can. J. Soil Sci. 95 (2015) 269–285.
[103] X. Fu, W. Liang, X. Zhao, J. Tian, A survey of paragenetic and associated resources in lignite of China, Adv. Mater. Res. 524 (2012) 106–110.
[104] H. Qi, R. Hu, Q. Zhang, Concentration and distribution of trace elements in lignite from the Shengli Coalfield, Inner Mongolia, China: implications on origin of the associated Wulantuga Germanium Deposit, Int. J. Coal Geol. 71 (2011) 129–152.

Properties of low rank coals and resulting challenges in their utilization

2

Halina Pawlak-Kruczek
Wrocław University of Technology, Wrocław, Poland

2.1 Introduction: Resources and usefulness of low rank coal

Coal offers an abundant and widely spread fossil energy resource which is used for power production. The degree of alteration (or metamorphism or coalification) that occurs as a coal matures from peat to anthracite is referred to as the "rank" of the coal. Low rank coals (LRCs) include lignite and sub-bituminous coals.

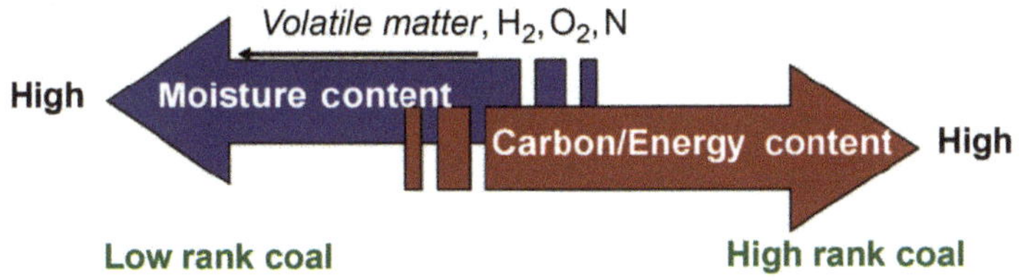

These coals generally characterize a lower energy content than high rank coal (bituminous to anthracite) because they have a low carbon content. They are lighter (earthier) and have high moisture levels.

Sub-bituminous coal is considered black coal, even though its appearance varies from bright black to dull dark brown. Its consistency ranges from hard and strong to soft and brittle due to its intermediate stage—between bituminous and brown coal (lignite). The coal is widely used for generating steam power and industrial purposes. Sometimes called "black lignite," sub-bituminous coal is not stable when exposed to air; it tends to disintegrate.

This type of coal contains more moisture and volatile matter than other bituminous coal types, but lower sulfur levels. Sub-bituminous coal has a heat value of approximately 19–30 MJ/kg, as mined. Sub-bituminous coal is noncoking and contains less sulfur but more moisture—approximately 10%–45%—and volatile matter—up to 45%—than other bituminous coal types. It has 35%–45% carbon content; its ash content ranges up to 10%. The coal's sulfur content is generally under 2 wt %. Nitrogen makes up approximately 0.5%–2%t of the coal's weight.

Low-Rank Coals for Power Generation, Fuel and Chemical Production
DOI: http://dx.doi.org/10.1016/B978-0-08-100895-9.00002-4

Availability of this coal is moderate. Countries with notable resources include the United States (ca. 300,000 million tons), Brazil, Indonesia, and the Ukraine.

Lignite is available in larger degree. Approximately 7% of coal mined in the United States is lignite. Lignite deposits are located relatively near the surface, which means that underground excavation in tunnels is not necessary thus the exploitation is cheaper. According to the *World Coal Association*, the top 10 countries that produce brown coal are (ranked from most to least) Germany, the United States, Russia, Poland, Turkey, Australia, Greece, India, Czech Republic, and Bulgaria.

Currently, LRC in worldwide is mainly used to produce electricity and heat. In Figs. 2.1 and 2.2 coal production in Europe and role of coal in power production in 2014–15 are shown. However these data changed very drastically in England, i.e., the power plants fired with coals are close. Reduction of the power production from coal is caused by negative impacts on the environment and the greenhouse effect. These negative effects are caused by the presence of toxic components in the combustion products of coal (SO_x, NO_x, and dust) and CO_2. Hence a lot of work is carried out for the development of efficient "clean carbon technologies," resulting in the use of LRC for energy production that becomes profitable and expected.

Lately, there has been increasing interest in coals including LRCs chemical processing, which mainly applies to the production of a substitute to natural gas, liquid fuels, and activated carbons. Also the production of humic acid from lignite by extraction methods used for fertilization and regeneration of soils is developed.

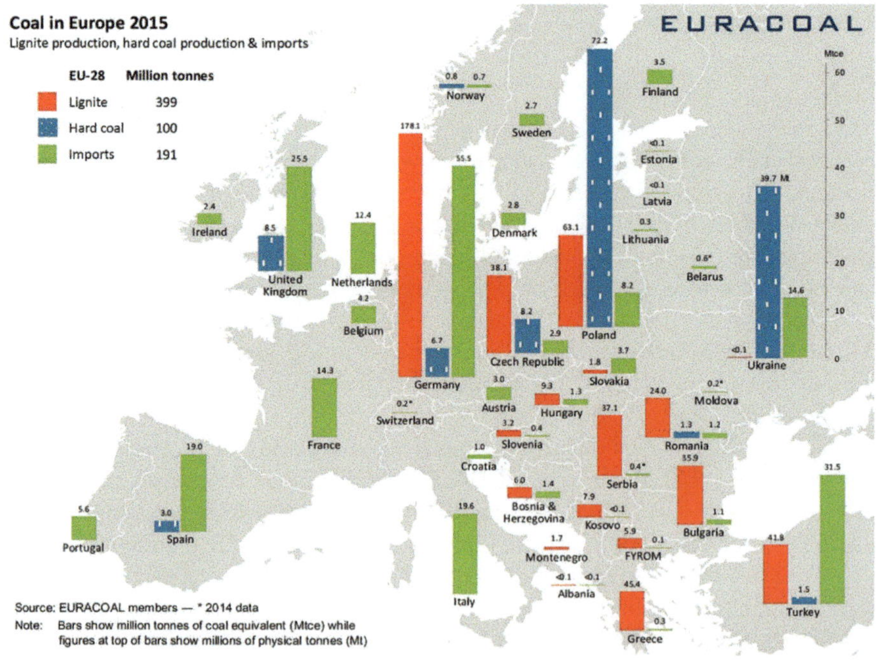

Figure 2.1 Production and import of coal in Europe—Eurocoal data 2015.

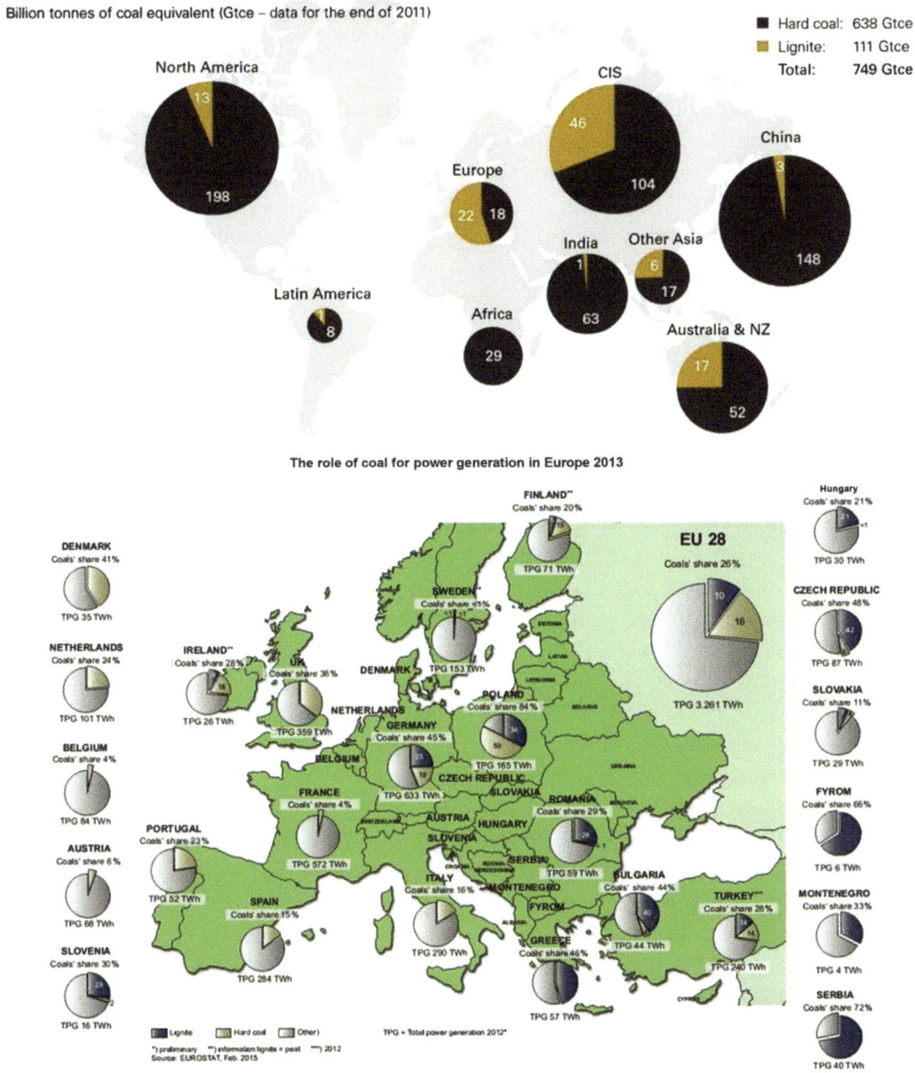

Figure 2.2 Role of coal in power production in Europe—Eurostat source 2015.

According to long-term forecasts, these raw LRCs will continue to be used for energy purposes, but it should be noted that its processing technology would significantly change by applying some upgrading technology.

Among the technological classifications of LRC, which characterize its industrial usefulness, there is a clear division into main parameters, which are present in most studies and characterize the quality of raw material for its possible use, and supplementary parameters, which allow the designation of the specific applicability of a particular processing technology.

2.2 Overview of methods for classification of low rank coals

The classification of LRC requires selected parameters to determine its possible different applications. Three main parameters are considered in classifying coals, namely *type*, which refers to the petrographic composition, *rank*, which refers to the level of coalification, and *grade*, which refers to the amount of inorganic matter content. LRC in the first part of the classification is thus defined on the basis of the degree of coalification, ash content, and petrographic varieties. The problem of selection method and parameters for coal classification was studied by Alpern et al. [1,2], Eremin and Bronowiec [3], Roga and Tomkow [4], Kwiecinska and Wagner [5,6], Wagner [7], and Bielowicz [8]. In the last work the authors study the proposed modification of "UN-ECE International Classification of In-Seam Coals" for LRCs.

Advanced methods of coal utilization at high efficiency of the equipment in the processes of energy production and with near zero-emission require a precise assessment of the quality of coal including method of LRC upgrading. Such a solution would allow the use of resources of lower quality coal which in turn can be solved by the elaboration of classification of LRC particularly aimed for the use of its energy.

Research on the properties of coal and the evaluation of the usefulness of different coal variants is achieved through use results of the technical, physicochemical, and petrographic analysis of coal.

The laboratory tests of the coal samples include the following:

Technical and physical analysis of coal, chemical analysis of coal and its ash, Lithological–petrographic studies, along with the measurement of random reflectance of ulminite B/collotelinite.

Lignite can be classified on the basis of multiple criteria. For industrial uses there are divisions indicating its potential use. There are technological features used —chemical and lithological–petrographic in a possible small number of parameters. Due to great diversity of LRCs' properties, there is no standardized classification particularly of lignite on the international scale until the present day. A distinct lack of uniform international division resulted from the fact that lignite is not suitable for long transportation and is considered as a national energy resource.

2.2.1 Classification of lignite for energy purposes

For the purpose of energy production, the so-called normal coal can be used with a calorific value under the operational condition of at least 6.7 MJ/kg and an ash content in the dry state of less than or equal to 40% is suitable. The total sulfur content recalculated to the dry state (if higher than 2% marks the sulfated coal) and the alkali content based on the sum of $Na_2O + K_2O$ (if the sum is greater than 0.5% coal is considered as saline) [6] are considered as supplementary parameters.

For classification of lignite for energy according to polish standard (PN-91/G-97051/01) the basic parameters are as follows: *ash content in coal* and *calorific value of coal*, while complementary parameters are as follows: content of total sulfur, sand content, fiber xylite content, and melting point of ash.

2.2.2 Classification of technological low rank coal according to chemical and petrographic criteria

Technological type of lignite expresses its degree of coal rank and its technological usefulness. And therefore it should be defined on the basis of parameters which are measures of coal rank and the most important measures of its quality. It is suggested that the technological type of coal should be specified according to its future use and the distinguishing digital features within three types of coal ranks: soft (C), matt (B), and bright (A), assuming, however, the limits of these types according to international classification, corresponding to the calorific value of burning in wet, ash-free condition (Q_S^{maf}). In Fig. 2.3 the classification of LRC based on UN-ECE (Economic Commission for Europe United Nations) in 1995 called Alpern classification or classification in deposit after modification is presented in work [8].

The part of the classification relating to the LRC was modified by Bielowicz [8] adding blocks characterizing the specific requirements for the particular directions of utilization (Fig. 2.3). The presented scheme is limited to the determination of the requirements for normative coal for combustion (normative coal at above LHV = 6.7 MJ/kg), gasification, and direct liquefaction. It is visible that the normative coal for combustion has the most tolerant requirements. The use of LRC for the liquefaction process is limited by its ash content, maceral composition, and random reflectance.

In the work [8] by analyzing the correlations between the random reflectance of ulminite B/collotelinite and gross calorific value in a moisture ash-free basis, it has been found that the suborder of lignite can be determined based on the random reflectance of ulminite B/collotelinite. The proposed classification distinguishes three types of lignite: soft brown coal (ortholignite), dull brown coal (meta-lignite), and bright brown coal (sub-bituminous coal). In the work [8] it is proposed that the boundary between peat and ortholignite should be set at a random reflectance of ulminite B equal to 0.15%. A similar value for this parameter for the lower limit of the LRC was determined by Wagner and Kwiecinska [5] and Kwiecinska and Wagner [6].

The boundary between the soft brown coal (C) and dull brown coal (B) should be set at a random reflectance of ulminite B equal to 0.35%, similar to the suggestion of Kwiecinska and Wagner [5], which is the result of a tendency to complete the jellification of humic components and the resulting rearrangement of the physical and chemical structures of huminite. To separate dull brown coal (B) and bright brown coal (A), a random reflectance of 0.40% is proposed. This value is also used in the classification of ISO 11760. The upper limit of the coalification of bright brown coal is just as in the ISO classification, a random reflectance of collotelinite equal to 0.5%. It should be noted that the introduction of random reflectance as a classifying parameter can take place only for humic coal.

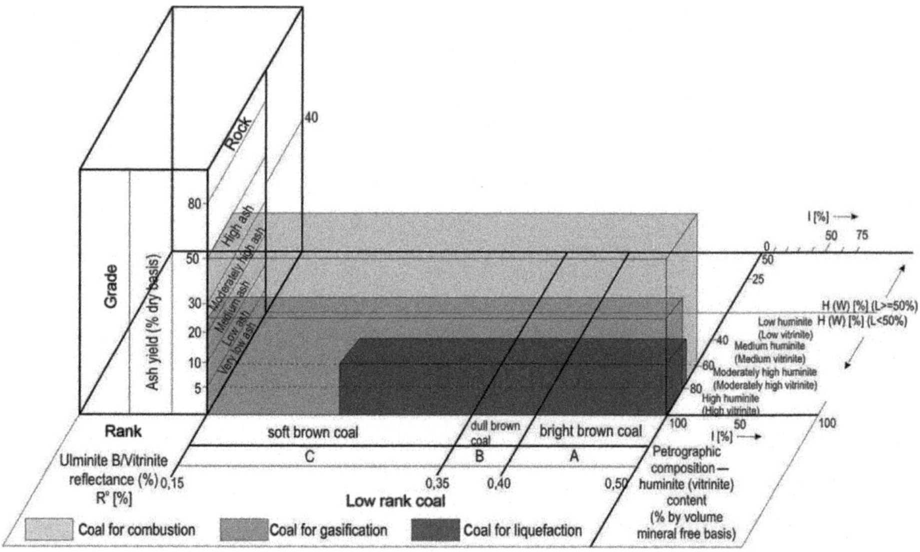

Figure 2.3 Diagram of the international classification of coal, "in coal seam" (ECE-UN, 1993) with block possible utilization of low rank coal with the use of combustion, gasification, and liquefaction processes based on the proposed classification of coal, where R_r^o, ratio of average vitrinite reflectance (ulminite B); Q_S^{maf}, heat of combustion in a dry ash-free condition; *V*, vitrinite content (huminite); *L*, exynite content (liptynite); *I*, inertinite content (*mmf*, non-mineral condition), and A^d, ash content in dry condition.
After permission of Barbara Bielowicz, A new technological classification of low rank coal on the basis of Polish deposits, Fuel 96 (2012) 497–510.

This classification introduces three equal groups of classification: coal rank degree—as a result of action on the organic sediment of geological time (lignite, sub-bituminous coal, the facial type of the sediment expressing the influence of transformations on organic material (aerobic decomposition, anaerobic decomposition in dry or wet conditions)—determined on the basis of macerals' composition and facies, understood as a step change in ash content, i.e., coal grade.

2.3 Physicochemical properties of low rank coals and their effect on the method used to produce power

Typical and limit values of parameters of LRC–lignite—are presented in Table 2.1.

Table 2.2 contains data of proximate and ultimate lignite analyses, i.e., elemental composition of six different lignites come from Polish, German, Bulgarian, two

Table 2.1 Limit values of the selected parameters of lignite at dry ash free [9]

Parameter	Symbol	Unit	Value
Carbon	C^{daf}	%	5878
Hydrogen	H^{daf}	%	4.9–7.0
Oxygen	O^{daf}	%	12–30
Nitrogen	N^{daf}	%	0.6–2.4
Sulfur organic	S_o^{daf}	%	0.2–3.8
Total sulfur	S_t^d	%	0.3–6.0
Sulfur sulfate	S_{SO4}^d	%	0.01–0.4
Sulfide sulfur	S_S^d	%	0.1–3.8
Sulfur combustible	S_{pl}^d	%	0.2–5.0
Volatiles matter	V^{daf}	%	37–68
Heat calorific value	Q_s^{daf}	MJ/kg	23.4–32.6
Low calorific value	Q_i^r	MJ/kg	5.9–23.0
Humic acids		%	5–70
The oxygen functional group	O_{COOH}^{daf}	%	0.2–2.5
	O_{OH}^{daf}	%	5–13
	O_{CO}^{daf}	%	1–2.5
	O_{OCH3}^{daf}	%	0–3
Lignin	L^{daf}	%	3–72
Cellulose	C_{el}^{daf}	%	1–47
Bitumen	B^d	%	2–37
Resins bituminach	Z	%	1080
The content of low temperaturę pyrolisis			
Semicoke	P_k^d	%	40–65
Tars	T_k^d	%	5–27
Decomposed water	W_k^d	%	6.5–14
Low temperature pyrolitic gas	G_k^d	%	8–25
Moisture	W_t^r	%	15–70
Ash	A^d	%	150
The composition of ash	SiO_2	%	2–86
	Al_2O_3	%	0.8–48
	Fe_2O_3	%	0.2–33
	CaO	%	0.5–50
	MgO	%	0.5–11
	SO_3	%	1–40
	$Na_2O + K_2O$	%	0.1–26
	Alkali in the coal $(Na_2O + K_2O)^d$	%	0.01–6
Fusibility of ash	The sintering temperature	°C	830–1050
	Melting temperature	°C	900–1250
	Melt temperature	°C	1050–1550

Table 2.2 **Characteristics of lignites (%)**

Lignites	Proximate analysis			Elemental analysis							
	W^{ar}	A^{ar}	V^{daf}	C^{daf}	H^{daf}	N^{daf}	$(O+S)^{daf}$	Ct^{d}	S^{d}	C_2H_5 (g/cm^3)	A^{a}
Polish Turów raw-I	34.19	15.22	60.69	64.15	6.25	0.84	28.76	0.18	0.54	1.558	4.3
Polish predried FG-I	11	24.5	57.7	70.6	6.3	0.95	22.15	0.46	0.39		
Polish predried CFB I	5.06	9.57	59	68.22	5.95	0.86	24.97		0.22		
German predried CFB-II	13.32	3.54	55.3	63.2	5.17	0.82	30.81	0.1	0.15	1.53	1.2
Bulgarian predried FG-III	13.44	14.37	60.23	62.31	6.28	0.86	30.55	0.08	3.87	1.598	
Bulgarian predried CFB-III	3.08	47.1	67	57.03	5.2	0.9	36.87		1.12		
Romanian predried FG-LT-V	2.12	48.2	61.2	62.44	5.75	2.107	29.703	0.14	1.04		
Romanian raw lignite-V	25.75	36	68.2	51.94	5.89	1.6	40.57	0.21	0.6		
Greek South predried CFB-VI	5.9	23.3	67	63.61	5.4	2.23	28.76	0.3	0.49	1.89	
Greek South predried HP VIS	17	39.7	68	65.5	5.15	1.86	27.49		1.145	2.12	
Greek Kardia 1 CFB VII	8.06	28.5	67.7	33.39	2.78	1.07			0.20		
Greek Kardia 1 predried HPS	17.9	26.088	68	63.98	4.41	1.94	29.67		1.58	1.94	

Lignites		Bulgarian FG	Rhenish CFB	Polish lignite FG	Polish CFB	Bulgarian CFB	Finnish peat HTS	Romanian FG-	Greek South CFB	Greek Kardia 1 CFB
SiO_2	%d	27.51	2.17	36.25	40.11	55.49	35.87	53.51	39.66	41.54
Al_2O_3	%d	16.58	3.78	35.52	14.5	14.59	17.42	17.2	12.62	9.11
Fe_2O_3	%d	14.51	22.07	8.99	9.97	8.97	19.7	8.48	7.51	8.15
Mn_3O_4	%d	0.19	0.42		0.08	0.04		0.04	0.04	0.03
TiO_2	%d	1.22	0.48	1.94	3.24	0.84	0.67	1.19	0.86	1.94
CaO	%d	6.42	31.17	2.92	1.76	0.86	11.6	8.02	28.12	21.16
MgO	%d	3.05	13.06	2.64	7.81	1.85	1.67	1.81	4.13	3.17
Na_2O	%d	2.26	2.30	1.61	5.91	0.44	0.15	0.11	0.43	0.45
K_2O	%d	0.31	0.15	0.2	0.8	0.73	0.12	0.13	0.9	1.01
SO_3	%d	16.71	11.82	0.66	4.94	7.41	67		5.04	8.67
loss of roasting 815°C		14.24	12.58	12.27	10.88	8.78	9.04		0.6	4.57

[a]A is the content of dissolved organic substance in ethanol.

Table 2.3 **Reactivity data obtained by TG technique**

Coal	VM^{daf}	ITR (°C)	E_m^{tg} (kJ/mol)
Polish	57.7	225	92.75
German lignite CFB	55.3	210	34.2
Bulgarian CFB	67	226	87.4
Greek South CFB	67	220	35
Greek Kardia CFB	58	220	49
Greek Kardia	67.7	225	38.35

Greek, and Romanian mines. Also the composition of mineral matter called ash for combustion process is presented.

In Table 2.2 the characteristics (proximate, ultimate, and ash composition) of different raw lignites and after drying are presented. The coals reported in Table 2.2 were dried by different thermal methods at different temperatures and pressures of drying agent. The coals were dried by low steam parameters in fluidized bed—CFB; by high pressure steam (~22 bar) HPS; by flue gas FG at low temperature (LT) and high temperature (300°C and ~1000°C).

The effect of elemental analysis allows for the determination of ratio H:C or O:C which implies decreasing aromaticity of the fuel. Increasing O:C ratio implies increasing hydroxyl, carboxyl, ether, and ketone functional groups in the fuel. Both the aromaticity and oxygen-containing functional groups influence the modes of occurrence of inorganic material in fuels and its transformation during combustion.

For power production typical process is combustion, which depends on base and supplementary properties of coal contained in Table 2.3 and reactivity of coals. The reactivity of coal (of coal combustion rate under certain conditions, expressed in $kg\ m^2\ s^{-1}$) is important for the efficiency of its combustion. Reactivity in turn depends on the type of coal, and particularly on its structure petrographic and thus the degree of coalification. Generally, the reactivity of coal decreases with increasing degree of coalification. The measure of reactivity is also the minimum ignition temperature which increases with the increase of coalification degree. However, the contents of ash and volatile parts also have an effect on the value of the ignition temperature and the reactivity of coals and thus on the combustion process. Generally, with increasing ash content greater than 20% the ignition temperature rapidly increase.

Coal combustion behavior and reactivity can be studied by thermogravimetry, where the mass and temperature history of the fuel sample is continuously monitored throughout the experiment. Full-scale conditions cannot be simulated in this type of reactor, but kinetically controlled conditions are easy to obtain. Thus keeping in mind the need to use simple and rapid tests for fuel characterization, the use of a thermogravimetric analysis (TGA) technique, in which both pyrolysis and char combustion are conducted consecutively, seems to be an appealing possibility for the evaluation of fuel reactivity.

Based on coal burning profile, i.e., the coal mass loss depending on time and temperature at a constant heating rate, the kinetic data as activation energy was determined. The derivation of activation energy values for the coal combustion process in the burning profile test was reported in a paper [10,11] and this was used as a basis for the evaluation of activation energy.

The results of TGA, i.e., characteristic temperatures ITR—initial temperature of reaction starts and activation energies for Polish, German, Bulgarian, and Greek lignite are reported in Table 2.3. It is visible that different lignite differing structure and composition characterize similar temperature of start combustion processes, which strongly depends on volatile matter content but apparent activation energy significantly differs.

Additional parameters characterizing coal for combustion include the ash melting point set in an oxidizing atmosphere (t_B), the content of toxic elements, and the ratio of radioactivity c (f_1) in coal ashes. The content of toxic elements and the ratio of radioactivity $c\ f_1$ have been omitted for coal because there was no exceedance of permissible norms in any of the tested LRC deposits.

Additional parameters include the base/acid ratio, slagging index, silica value, and fouling index. The power production from LRC of high efficiency by advanced technology as IGCC is limited, because it requires special gasification system. The typical properties of LRC, the high moisture content and volatile content, make it difficult to apply these IGCC techniques. Gasification of LRC leads to heavy tar formation and char formation and also it can decrease the degree conversion of carbon, thus reducing the amount of synthesis gas produced.

2.4 Main problems in utilization of low rank coal and effect of valorization on low rank coal by drying

The main problem for utilization of LRC by different planned processes depends on moisture content and ash composition and the low calorific value LHV which results from composition of coal. Up till now the largest market for low quality coals is power generation. Presently LRC based power plants can meet varying load requirements. Lignite competes mainly with hard coal and natural gas. But the utilization of LRC in thermal–mechanical and chemical processes is in progress [12].

Table 2.4 contains the quality parameters of lignite which are important for energy uses. In a competitive market of power production from different sources, lignite can hold its own, when the assumptions were on the reduction of CO_2 by cost effective methods.

Now and for decades to come, lignite and renewables are not opposing, but truly complementary.

However, water in LRCs exerts significant effects in their utilization processes, including combustion, gasification, and liquefaction. The high moisture content

Table 2.4 The names, symbols, and units of quality parameters of lignite for energy uses according to PN-91/G-97051/01

Type of parameters	Parameter	Symbol	Unit
Base	Ash content (as received)	A^{ar}	%
	LHV	Q_i^r	MJ/kg
Supplementary	Total sulfur content in dry condition	S_t^d	%
	Silica content	p^r	
	Content of xylite	k_w	
	Ash melting temperature	t_B	°C

results in lower efficiency, increased transportation cost, and higher carbon emission. On the other hand LRCs have the advantages of low mining cost, high reactivity, high volatile matter content, and low content of pollution forming impurities such as sulfur, nitrogen, and heavy metals. The problem of high CO_2 emission and lower efficiency can be solved by the use of predried LRCs which is one possibility of increasing the efficiencies of LRC fired utility boilers. As a mean value, an increase of 2%–4% in power plant efficiency can be achieved [13–15].

The heat loss in the furnace to evaporate the moisture contained in the coal causes a large increase in fuel consumption with increasing boiler load. These losses can be compensated by heat recovery from condensation of steam containing in flue gas. The relative benefit of heat recovery from the condensation of the steam contained in the exhaust gas referred to the consumption of coal is greater for lower moisture coal at the inlet to the furnace. The main advantages of burning dried coal, what is necessary for oxy-combustion, are as follows:

- external grinding/drying process but usually integrated with power unit to reach highest its efficiency
- lower amount of flue gas, hence, smaller furnace cross section
- flue gas recirculation for sufficient convective heat transfer—especially important for oxy combustion
- low heat loss in furnace for water evaporation
- high heat recovery net from steam condensation from flue gas
- furnace height similar to directly raw coal fired boilers
- no need for fan beater mills and flue gas suction ducts

Additional benefits are its improvement regarding flexibility. Also argument to use of predried lignite in combustion is because stable combustion of raw lignite causes a negative effect on emissions.

The analysis of heat recovery from condensation of steam containing in FG minus heat losses in furnace for evaporization was studied at selected parameters. The calculation of energy balance, i.e., heat for coal moisture evaporation and heat recovery from condensation of steam containing in FG, the following assumption was carried out: fuel lignite at moisture content $M = 54\%$, (Belchatow sample), $Ash^d = 21.6\%$, LHV = 7470 kJ/kg, and reference nominal output 300 MWth.

The conventional combustion process and oxy-combustion process for Oxy25 and Oxy30 atmospheres were analyzed and the last technology requires use of predried coal.

The recovery of net heat from condensation of water contained in FG is reduced to evaporate the coal moisture in the furnace for different inlet moisture coal as shown in Fig. 2.4.

The additional positive effect of combustion predried lignite and heat recovery from condensation of steam containing in FG is lowering of fuel consumption, and the increases of boiler efficiency are presented in Figs. 2.5 and 2.6, respectively.

2.4.1 Effect of lignite drying on coal properties structure reactivity

LRCs, such as lignite, due to high moisture and oxygen content result in many problems including high transportation cost per thermal unit of coal and problems with

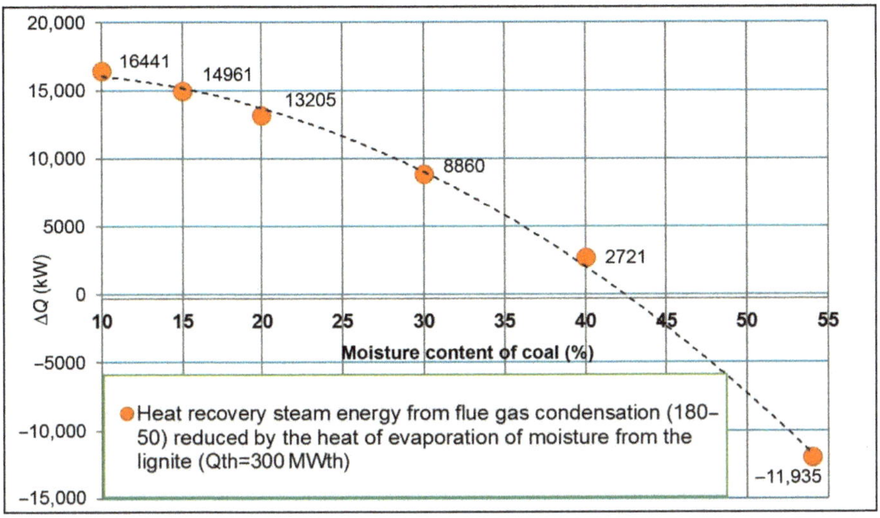

Figure 2.4 Net heat recovery from condensation of steam containing in flue gas including heat losses in furnace for evaporization.

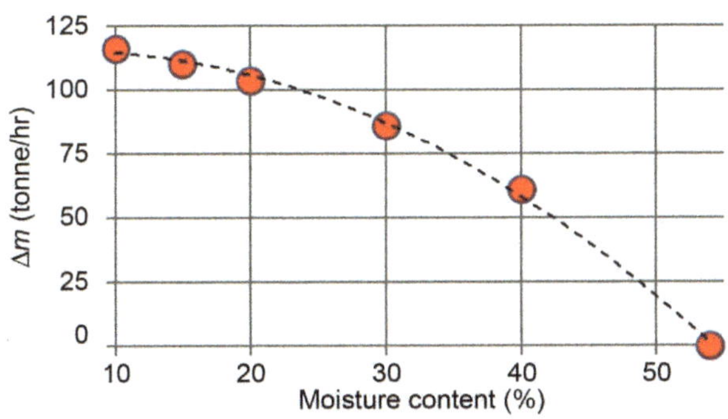

Figure 2.5 The difference of lignite consumption at load 372 MWth at combustion of raw (54% moisture) and predried lignite with heat recovery.

milling. Thus the improved efficiency of lignite utilization for power production using the upgrading by drying is an important method and it is the subject of many studies [13,16–18]. Milling problem of raw lignite needs preliminary drying of the coals, which is one of the principle stages in the preparation of fuel for combustion in PC boilers.

Conventionally, brown coal is dried using recirculated hot FG then fed with all the fumes from drying into the chamber. For the use of LRC with high moisture for power production and/or use for gasification, liquefaction in large and effective way the valorization by drying is the best and required method. The combustion of pre-dried coal without feeding the vapors from drying into the chamber has several

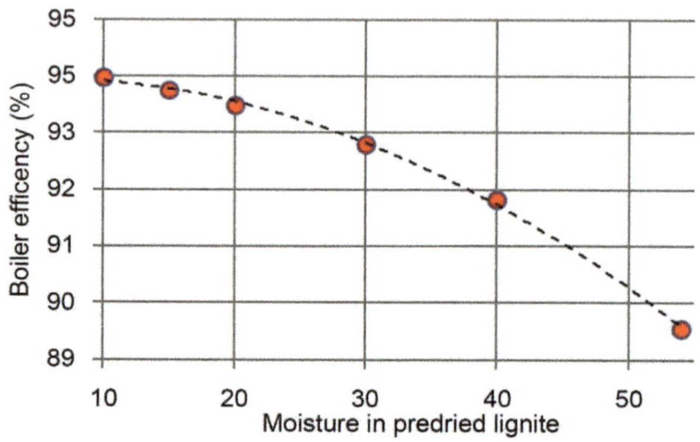

Figure 2.6 The change of boiler efficiency for combustion of predried of lignite and steam condensation.

advantages, particularly economic ones, over the conventional method, provided that the heat from the vapors (WTA) is utilized as a heating medium in the power unit.

However the heating and drying of coal by various methods can initiate a series of physical changes in the fuel particles, e.g., in their dimension, structure, and properties. As a result, the behavior of the fuel particles changes in the course of combustion or gasification.

Besides the effectiveness of drying one must determine to what extent a given drying technology alters coal properties (e.g., reactivity), which affects subsequent coal utilization processes, i.e., coal combustion or gasification and liquefaction.

Presently there are some commercial technology [14,15] for drying which differs from drying agent and its parameters thus the effect on reactivity and composition of final products can be positive but in different degree.

In all the cases, the reactivity of the completely dried coals usually is lower than that of the initial coals, which indirectly indicates structural changes [16,19,20].

During drying low-grade coals show a strong tendency to form cross-links even at low temperatures, which is due to the high concentration of carboxyl and hydroxyl oxide functional groups. The formation of cross-links also produces simple molecules such as CO_2, H_2O, and CH_4. The first two components can form in the air drying process as a result of low temperature oxidation. The results of the author's and other research [16,19,20] have shown that the drying of brown coal by thermal methods causes irreversible changes in the coals' porous structure, to a degree which depends on the drying medium and its parameters. As a result of drying, most of the pores collapse which limits the access of the reagents during combustion.

The shrinking behavior of brown coal particles was investigated in detail for thermal methods.

The calculated mean shrinkage percentages of the different coals are shown in Fig. 2.7.

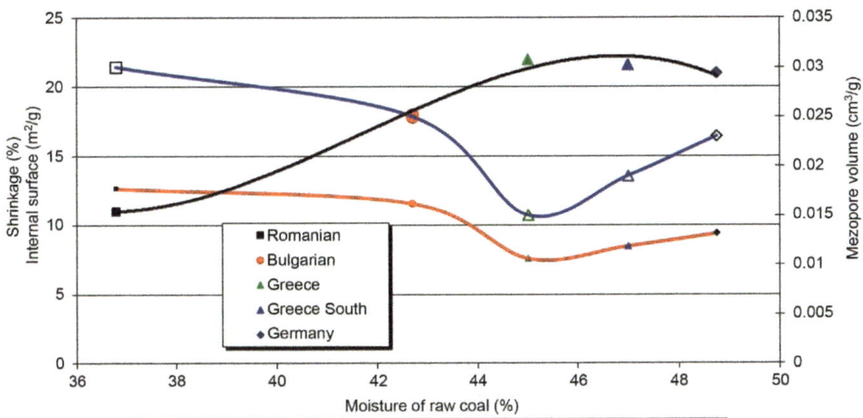

	Romanian	Bulgarian	Greece	Greece South	Germany
	36.8	42.7	45	47	48.75
Internal surface (m^2/g)	11	18	22	21.6	21
	12.7	11.6	7.6	8.5	9.4
Mezopore volume (cm^3/g)	0.03	0.025	0.015	0.019	0.023

Figure 2.7 Effect of initial coal moisture on shrinkage of particles (—) and surface (—) and volume (—) of mesopores after thermal drying in fluidized bed.

The percentage of particle shrinkage is an important parameter regarding the preparation expenditure after drying and a design parameter of the storage containment and the dimensions of the transport system. The thermal drying of lignite by steam lowers the size of lignite particles and simultaneously the surface and volume of capillaries, first of all mesopores.

There were noteworthy correlations between the shrinkage of dried lignite particles and the volume and surface of mesopores (see Fig. 2.7). The shrinkage increases with increasing initial moisture up to value 45% and above this value becomes quasi-constant.

During drying and removal of moisture the coal structure is disrupted and shrinkage or "slacking" is observed, simultaneously the moisture can interact with oxygenated coal species and intermediates in the oxidation process, resulting in change of pore structure and size of coal particle shrinkage. The rate of changes depends on the parameters and agents of drying. For six brown coals Polish, German, Bulgarian, Romania, and Greek South and Kardia the change in internal pore surface depends on drying methods, which is shown in Fig. 2.8 [19].

The internal structure of coals influences on its reactivity.

2.5 The barriers of low rank coal utilization and progress in low rank coal upgrading

The main problem which should be solved for large utilization of rank coal is improving its calorific value by removing moisture and some impurities. Thus an

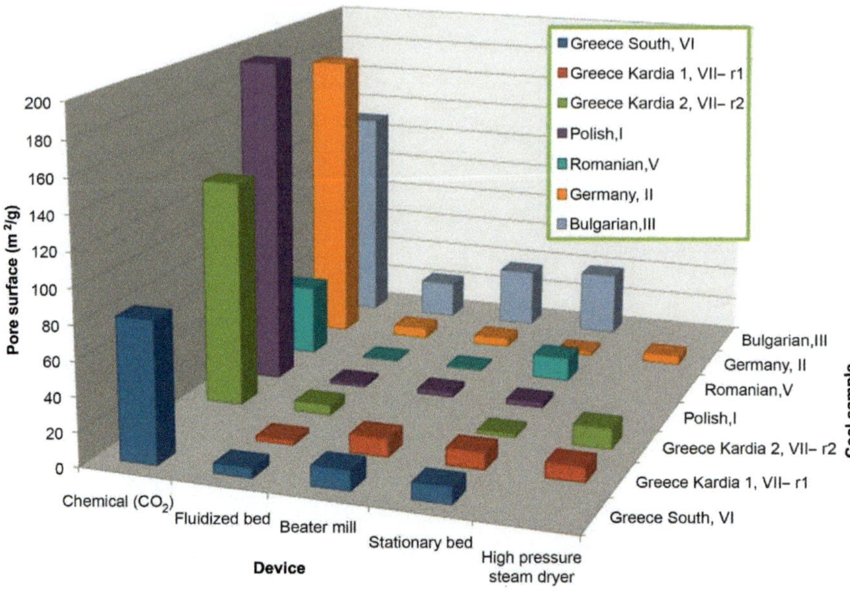

Figure 2.8 The effect of drying methods on internal pores surface.

upgrading of coals technologies permits to improve efficiency of power production and flexibility of this fuel. The LRC drying, cleaning, and upgrading are of great importance for increasing their use, as these make the coals cleaner, safer for transport and storage, and more valuable as an exportable fuel.

The utilization of LRC in very effective scale for advanced power sector and for industrial purpose as well as for production of liquid/gas clean fuel requires its upgrading by advanced technology.

Although drying is an important upgrading method, there are some technologies which transform low quality/rank coal by drying and next pelletizing into high value, dedicated for different types of boiler including small boilers and moving bed gasifiers, and for industrial application purposes.

Recently a number of new advanced methods for upgrading/processing of LRC to liquid and/or gas and "clean solid fuel" have been developed. These technologies have reached either pilot scale or commercial demonstration scale and are intensively analyzed with projects undertaken in many country especially in China, Japan, the United States, Australia, and Indonesia. The review of advanced upgrading technology of LRC to solid crude oil, semicoke and gas, and "clean" solid fuel or special slurry are presented in work [12]. In many countries technology of "coal-to-liquids" (CTL) emerges as an important sector in recent years due to concerns in security of energy supply and high oil/gas price variability.

To this end, a number of high efficiency processes are under development in Australia, Germany, Japan, and the United States. All these developments in power generation, CTL, UCG, drying, and upgrading are likely to ensure a sustainable future for low quality coals.

2.6 Summary remarks

The main problem limiting use of the LRC to produce power results from its properties, in particular with low calorific value which is the result of a high content of the so-called ballast, i.e., water and mineral substance and the variability of the composition within one bed of mine.

However simultaneously LRC especially lignite characterizes high reactivity against the combustion and gasification. Consequently the use of effective method of upgrading is required to get fuel for power production with reduced carbon emission in advanced supercritical boiler or even in IGCC. The upgrading by drying is the most developed technology in commercial scale, drying in condition of low temperature should be favorable due to avoidance of the reduction of reactivity. [21]. Water–coal interactions are very complicated and water exists in LRCs in different forms which behave differently during thermal treatment. Coals undergo chemical and physical changes upon drying resulting in change of pores structure. The change in chemical nature of the coal surface by the destruction of the oxygen-containing polar functional groups makes the coal fewer hygroscopic after drying.

The problem of upgrading of LRC is current and is the subject of research in particular as regards the use of coal in advanced high efficiency technologies of power production and their use for the production of other valuable fuels.

A significant theme is to develop technologies to modify and/or use of ashes from coals.

Advanced lignite coal-fired combined-cycle power systems under development and demonstration have the potential to increase generating efficiency to approach 50%, reduce the cost of electricity, and meet stringent standards on emissions of SO_x, NO_x, fine particulates, and air toxic metals. This advanced coal-fired combined-cycle power system when operated on LRCs depends importantly on the technology of removing water, i.e., upgrading especially lignite, e.g., by use of waste heat from power unit. That scheme *Integrated Drying Gasification Combined Cycle* is under the study by Mitsubishi in Japan and also it was proposed in Australia [22].

References

[1] B. Alpern, M.J. Lemos de Sousa, D. Flores, A progress report on the Alpern coal classification, Int. J. Coal Geol. 13 (1989) 1–19.
[2] B. Alpern, Lemos, M.J. de Sousa, Documented international enquiry on solid sedimentary fossil fuels; coal: definitions, classifications, reserves–resources, and energy potential, Int. J. Coal Geol. 50 (2002) 3–41.
[3] Jeremin I.W., Bronowiec T.M., 1997 Projekt mjedynarożnej klassifikacii ugljej nizkobo, srjednjego i bysokowo rangow, Chimija twjordowo topliwa, no. 2.
[4] B. Roga, K. Tomkow, Chemiczna technologia węgla, WNT, Warszawa, 1971.
[5] Wagner M, Kwiecińska B. Rank of polish coals according to the international classification of coals, in: Anstr in 48th ICCP, Heerlen, 1996.

[6] B. Kwiecińska, M. Wagner, Mozliwosc zastosowania refleksyjności jako metody badawczej w klasyfikowaniu i technologicznej ocenie jakosci wegla brunatnego, Wyd, Andrzej Choczewski, Krakow, 2001.
[7] M. Wagner, Brunatny węgiel bitumiczny ze złóż Turów i Bełchatów w świetle badań petrograficzno-chemicznych i sedymentologicznych, Pr. Geol. Komis. Nauk. Geol. PAN (143) (1996).
[8] Barbara Bielowicz, A new technological classification of low-rank coal on the basis of Polish deposits, Fuel 96 (2012) 497–510.
[9] Tomkow K., 1981 - Surowce mineralne świata, Węgiel Brunatny: Użytkowanie i przetwórstwo. Wyd. Geolog. Warszawa: 131–192.
[10] S.E. Smith, R.C. Neavel, E.J. Hippo, R.N. Miller, DTGA combustion of coal in the Exxon coal library, Fuel 60 (6) (1981) 458–462.
[11] J.W. Cumming, Reactivity assessment of coals via a weighted mean activation energy, Fuel 63 (9) (1984) 1436–1440.
[12] Nigel S. Dong, Utlization of Low Rank Coal, IEA Report CCC/182, April 2011.
[13] Christian Bergins, Alfons Leisse, Sebastian Rehfeldt How to utilize low grade coals below 1000 kcal/kg? in: POWER-GEN Europe, 2014.
[14] Nikolaos Nikolopoulos, Ioannis Violidakis, Emmanouil Karampinis, Michalis Agraniotis, Christian Bergins, Panagiotis Grammelis, Emmanouil Kakaras, Report on comparison among current industrial scale lignite drying technologies (A critical review of current technologies), Fuel 155 (2015) 86–114.
[15] Nenad Sarunac, Edward K. Levy, Mark Ness, Charles W. Bullinger, Jonathan P. Mathews, Philip M. Halleck, A novel fluidized bed drying and density segregation process for upgrading low-rank coals, Int. J. Coal Prep. Util. 29 (6) (2009) 317–332. Available from: http://dx.doi.org/10.1080/19392691003666387.
[16] Final Report JOF3-CT97-0037, ERBIC20-CT97-0046, Institute of Process Engineering and Power Plant Technology, Universität Stuttgart WUT.
[17] Kruczek H. et al. (1999). Thermodynamic analysis of modified power plant cycle utilizing condensation heat from predried brown coal. Pre-drying processes for the efficient and clean utilization of brown coals in the enlarged EU market. Workshop proceedings. University of Stuttgart, National Technical University of Athens, Athens, 2 April.
[18] Kruczek H., Grochowski A. (2000). Low ran coal effect, its predrying and Ca, Na additives on NO_x and SO_2 emission, in: 28th International Symposium on Combustion, Edinburgh, Scotland (WIP), Report PRE 32/2000, TUW Poland.
[19] Halina Kruczek, Klaus R.G. Hein, Joerg Maier, The Effect of Different Drying Methods on Brown Coals Structure and Reactivity, Report PRE, Wrocław University of Technology, Poland.
[20] Jianglong Yu, Arash Tahmasebi, Yanna Han, Fengkui Yin, Xianchun Li, A review on water in low rank coals: the existence, interaction with coal structure and effects on coal utilization, Fuel Process. Technol. 106 (2013) 9–20.
[21] Pawlak-Kruczek Halina, Plutecki Zbigniew, Ł. Michalski Marcin, Brown coal drying in fluidized bed applying low-temperature gaseous medium, Drying Technol. 32 (11) (2014).
[22] Terry Johnson, HRL Developments Pty Ltd., Gasification Technology for Brown Coal Power Generation, APP Brown Coal Best Practice Workshop, Melbourne, June 2008.

Critical review of current industrial scale lignite drying technologies

3

Ioannis Violidakis[1], *Panagiotis Drosatos*[1,2] *and Nikolaos Nikolopoulos*[1]
[1]Centre for Research & Technology Hellas, Ptolemais, Greece, [2]School of Mechanical Engineering, NTUA, Athens, Greece

3.1 Introduction

Lignite is characterized as the lowest rank of coal due to its high moisture and ash content and its low carbon content. From the total global coal reserves, a significant part of about 45% is low-rank coals (LRCs) [1]. The electricity production by lignite combustion in power plants is widespread in many countries, such as Australia, the United States, Canada, India, Germany, Greece, Poland, Serbia, Russia, and many other parts of Europe. The significance of lignite in the energy production mix can be seen in Fig. 3.1 [2].

The extended use of lignite, which is expected to further increase in the future, is basically driven by its advantages, i.e., low mining cost, high reactivity, high amount of volatiles, and low impurities (sulfur, nitrogen, and heavy metals). However, a further parameter that constitutes a major issue in terms of its commercial utilization is the high moisture content of lignite, since it lowers the heating value of the fuel and consequently its energy density. Therefore, the high moisture content of lignite lowers the plant efficiency, leads to higher CO_2 emissions per unit of energy output, and increases the capital costs due to the necessary larger size of the lignite boilers compared with the current state-of-the-art hard coal fired boilers. In order to enhance the energy efficiency of the unit, lower the pollutant emissions, and assure stable combustion, it is necessary to implement the drying and pulverization processes in the large-scale applications. In addition, the presence of moisture in the inserted fuel deteriorates its friability and imposes further difficulties in its blending and pneumatic transportation. The impact of moisture content on the efficiency of coal fired power plants is depicted in Fig. 3.2.

The lignite predrying concept aims toward the improvement of the lignite utilization. The decrease in the amount of moisture in LRC leads to lower energy losses during combustion, lower flue gas mass flow, higher plant efficiency, and lower transportation costs. However, this system demands the increase of the safety measures for the transportation and storage of lignite [1], since the risk for self-ignition gets higher.

Low-Rank Coals for Power Generation, Fuel and Chemical Production
DOI: http://dx.doi.org/10.1016/B978-0-08-100895-9.00003-6

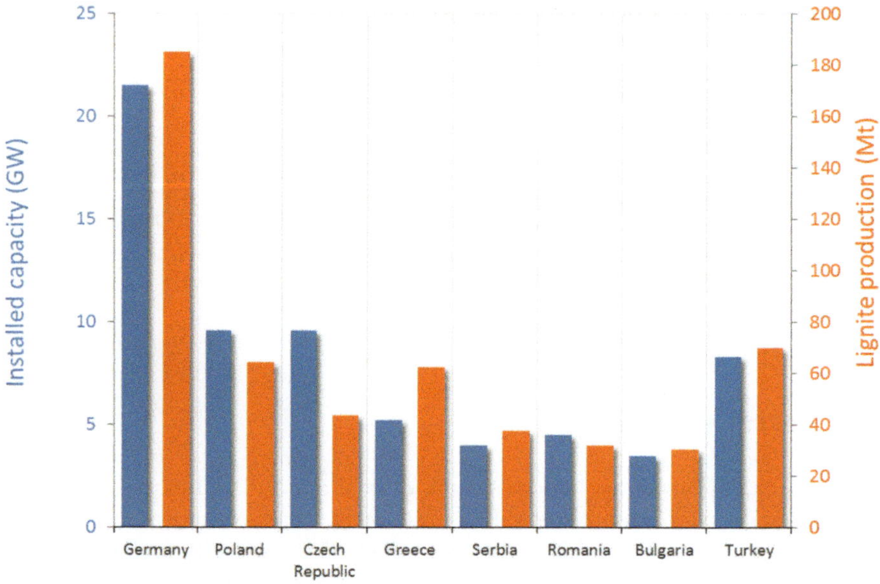

Figure 3.1 The role of coal for power production in Europe [2].

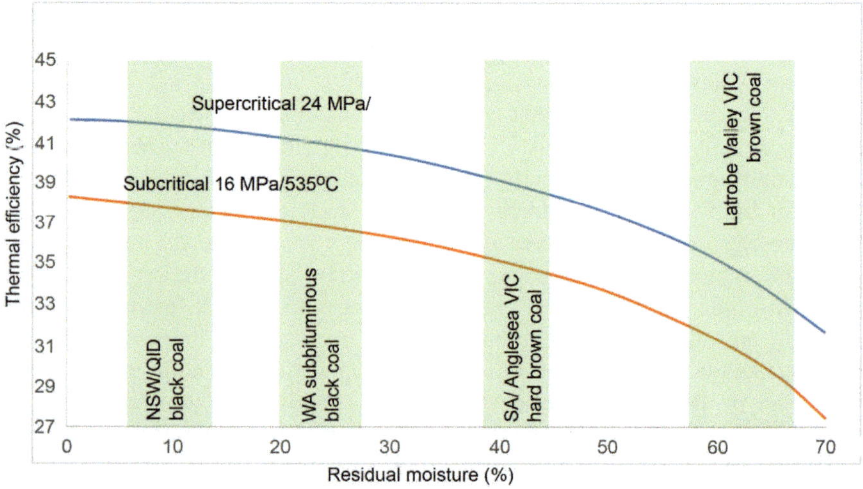

Figure 3.2 Power plant thermal efficiency as a function of moisture content of coal [3].

3.2 Drying process

Each material is characterized by a drying curve which plots the rate of moisture loss against time for a sample of the investigated material exposed to the drying medium. The exact curve varies with (1) the velocity of the drying medium, (2) the temperature of the heating medium, (3) the particle size distribution (PSD), and (4) the particles' dimension and pore size.

At the beginning of the drying process, the material heats up and the drying rate gradually increases. This period is succeeded by the constant drying rate phase, in which the diffusion of water through solid is sufficiently rapid to maintain saturated conditions at its surface and keep the drying rate constant. However, in the final phase, the internal diffusion of water is not capable of maintaining the saturation conditions at surface, so the rate slope constantly decreases. One key variable that needs to be determined is *the critical moisture content*, i.e., the moisture content at which the constant rate period ends and the respective falling rate drying phase starts. This parameter is dependent on the material structure, thickness, and initial moisture.

Clearly, the drying of the fuel to low moisture levels implies progressively longer drying time (and therefore larger size of equipment) to achieve the desired final moisture level, and therefore drying beyond a predefined limit should be generally avoided, given the condition that a longer drying time results in too large and high energy consuming industrial sized facilities [4].

3.3 Dryer classification

Dryers can be classified into different categories based on the drying method, the controlling heat transfer mechanism, the drying or heating medium, the heat source, and the applied pressure.

The different dryer types are presented in Table 3.1.

3.3.1 Drying method (evaporative–nonevaporative)

Based on the first categorization the dryers can be distinguished into evaporative and nonevaporative dryers. The operating principle of the former is based on the drying of the material through the provision of the necessary latent heat increased by a specific amount to cover heat losses. On the other hand, the nonevaporative

Table 3.1 Classification categories of dryers

Parameter	Alternatives
Drying method	Evaporative or nonevaporative
Heat transfer mechanism	Direct or indirect
Drying medium (i.e., the stream passing through the material to be dried)	Flue gas, hot air, or superheated steam
Heating medium	Flue gas, hot air, steam, or hot water
Heat source	Dryer burners, boiler (flue gas or steam), recovered waste heat from facility processes
Pressure	Atmospheric, vacuum, or high pressure.

dryers use other than conventional methods, such as mechanical thermal dewatering and hydrothermal dewatering (HTD).

3.3.2 Heat transfer mechanism (direct–indirect)

In terms of heat and mass transfer, the available evaporative dryers can be classified into (1) direct drying systems and (2) indirect drying systems.

- *Direct* (*or convective*) dryers operation is based on the direct contact and the heat transfer via convection mechanism between the material and the heating medium, which in this case is also the drying medium. Direct drying has two main variants, in terms of the respective flows of (1) the material to be dried and (2) the heating medium; parallel or counter-current flow (Fig. 3.3).
- *Indirect* (*or conduction*) dryers operation is based on the heat transfer via conduction and radiation mechanisms, which do not require the contact between the material and the heating medium, but, instead, they are wall separated. This type of systems is more suitable for thin products or very wet solids. The necessary heat of evaporation is supplied through stationary or moving heated surfaces, which simultaneously support, convey, and/or confine the solids. The evaporated moisture is carried away by vacuum operation or by a stream of gas that is mainly a carrier of moisture. Vacuum operation is recommended for heat-sensitive solids [5] (Fig. 3.4).

Between the direct and the indirect-fired dryers, the former are more efficient, since the latter present an inefficiency linked to the heat transfer from the steam tubes to the material. The efficiency of an indirect dryer is increased if (1) no air is injected and (2) the moisture that is vented away from the dryer in the form of steam is recovered to serve any heating process needs [7]. Table 3.2 compares and

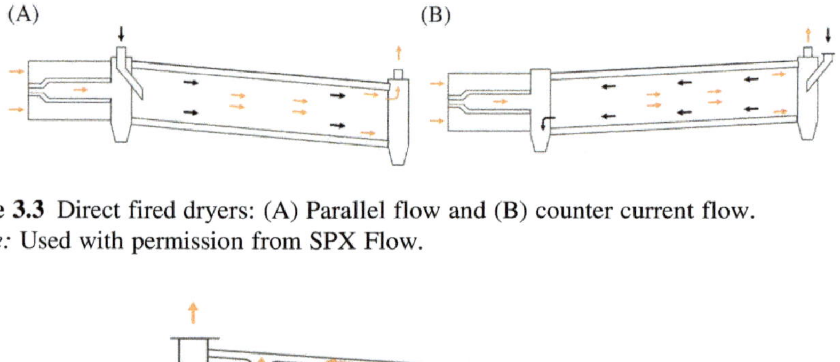

Figure 3.3 Direct fired dryers: (A) Parallel flow and (B) counter current flow.
Source: Used with permission from SPX Flow.

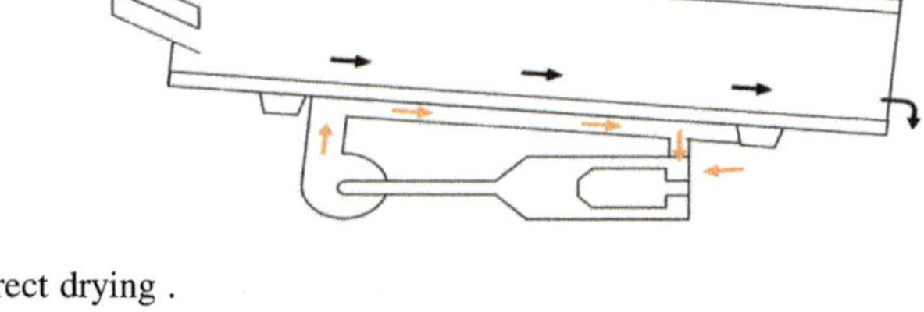

Figure 3.4 Indirect drying .
Source: Used with permission from SPX Flow.

Table 3.2 Comparison of characteristics between direct and indirect dryers

Dryer type	Capital & operating cost	O&M requirements	Environmental emissions	Energy efficiency & heat recovery	Fire hazard
Direct dryer	Lower	Lower	Heating medium treatment required	High efficiency	Relatively high
Indirect tube dryer	Higher	Higher	Less emissions	Low efficiency in general equally efficient with direct dryers when heat recovery option from steam is available	Decreased fire hazard —absence of oxygen

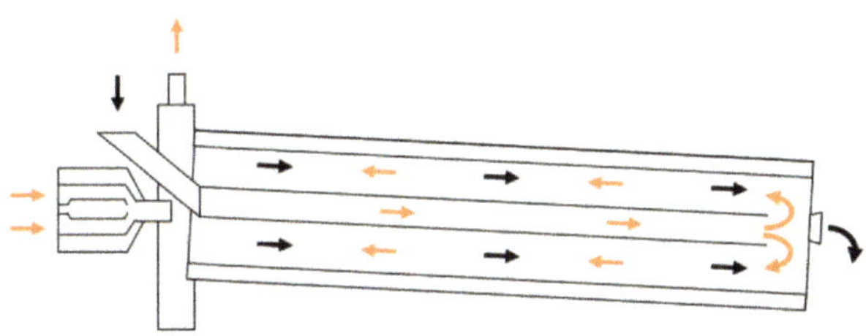

Figure 3.5 Indirect/direct drying .
Source: Used with permission from SPX Flow.

presents the main properties of these two types of dryers. Nowadays, there are efficient hybrid systems that combine the characteristics and the advantages of both direct and indirect drying methods (Fig. 3.5), such as fluidized-bed dryers (FBDs) that use immersed heating tubes or coils to dry especially heat-sensitive materials. This dryer can be as small as the one-third of the size of a purely convective FBD for the same duty [5].

3.3.3 Heating medium (flue gas–air–steam)

Depending on the drying demands and the availability of streams, the utilized heating medium can vary. The most common heating mediums are (1) flue gas, (2) air, or (3) steam.

1. Flue gas dryer:

   The flue gas dryer uses the waste heat of the flue gas to dry the injected fuel, thus improving the gross energy efficiency of the unit. However, a basic drawback of this system is the demand for large flow of flue gas if its temperature is low, which subsequently mandates a large-scale dryer with considerable electricity consumption. A typical energy consumption value for a flue gas dryer is about 3200 kJ/kg H_2O.

2. Air dryer:
   Recently, low temperature air dryers have received significant interest for drying systems because of low grade heat and low gas emissions. The heat demand for a low temperature air dryer is estimated to be around 2700 kJ/kg H_2O (including heat recovery) [8].
3. Steam dryer:
   Steam dryers have the advantage of being able to recover a large amount of input heat through the condensation of the evaporated moisture at a later stage in the process. Due to this energy recovery step, the net heat demand for a steam dryer can be as low as 600 kJ/kg H_2O [9].

3.4 Types of coal dryers

Coal drying is carried out in various types of drying and dewatering equipment [5], e.g.,

- Rotary dryers
- FBDs
- Pneumatic dryers
- Mill-type dryers
- HTD
- Mechanical thermal expression (MTE)

There are also other types of dryers emerging, using superheated steam, while other types of novel drying technologies (e.g., using high velocity air flow stream) are also currently being developed.

The main types of dryers are described in the following.

3.4.1 Rotary dryers

3.4.1.1 Direct rotary dryers

Direct rotary dryers are commonly used in many industrial plants for the drying of coal and coal muds. In order to avoid the possibility of fuel self-ignition, the co-current mode is preferred. The combustion of coal or natural gas produces the necessary flue gas that will be used as the drying medium. The layout of a typical direct rotary dryer is shown schematically in Fig. 3.6. The main component of this dryer is the steel drum (3) lined with a refractory lining and set up on rollers (7) by means of bandages (hoops, 10) located on the drum. A toothed gear (5) is typically used to rotate the drum. The drum is set up obliquely with a slope of 2–5° to the horizontal.

Inside the drum, there are lifters attached to the inner surface of the drum. During operation, as the drum rotates, these lifters lift the coal granules and shower them gradually inside the drum in the stream of the flowing heating gas [5].

Various lifters are shown in Fig. 3.7.

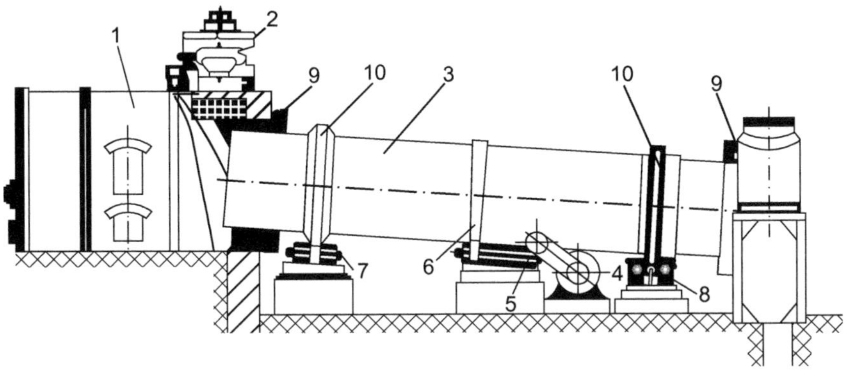

Figure 3.6 Schematic of a rotary dryer [5].

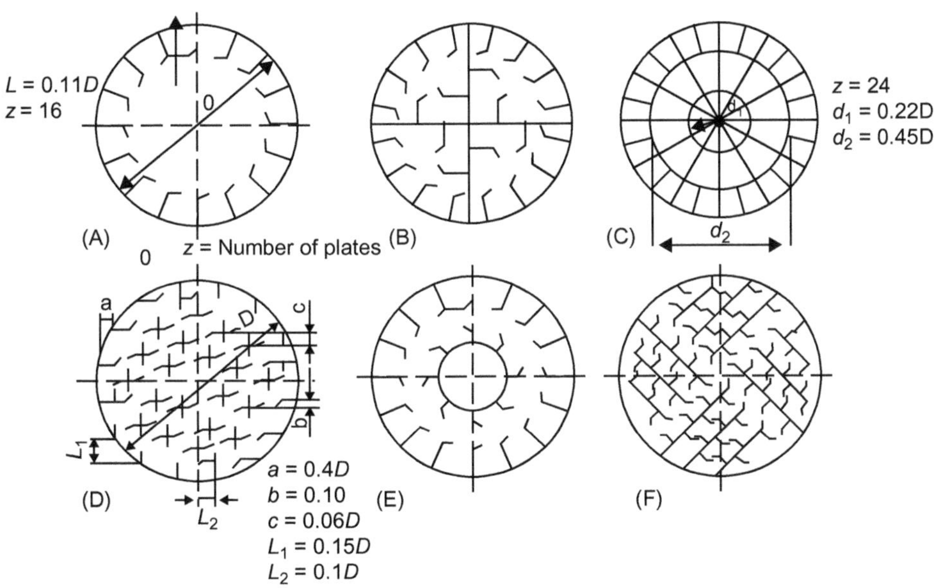

Figure 3.7 Scheme of the lifting–mixing devices for rotary dryers [5].

3.4.1.2 Rotary-tube (indirect) dryers

The second type of rotary dryers is the so-called indirect or rotary-tube dryer. This kind of dryers is widely used for drying of lignite in briquetting plants and drying of hard coal. The drying medium used is saturated steam, whose pressure ranges between 0.15 MPa and 0.55 MPa. In the most common configuration of indirect dryers the system consists of a jacketed drum through which steam or other heating medium flows. However, the fraction of solids that is exposed to the heated wall at any time is very small, resulting in low heat transfer rates and low drying efficiency. In order to increase the drying efficiency and the heat transfer rate, it is necessary to further increase the contact area between the heated walls and the particles. This is achieved by introducing a number of tubes inside the rotary shell and passing steam through them. In the steam-tube dryer, wet solids are lifted and showered within the rotary shell in the usual sense, and heated by radiant heat and contact with the outer surfaces of the tubes [10].

In the second type of indirect dryers, the tubes contain coal and not steam. The wet material enters the inclined tubes and moves under the influence of gravity, while the vessel rotates. The tubes diameter is at least 150 mm and they house helical wires serving as stirrers, regulating the flow of the inserted material and ensuring sufficient residence time. Normally, this type of dryer is able to reduce the moisture content of 3 mm coal particles from 20% to 4% by maintaining the shell temperature above 120°C. In general, the coal-in-tube dryer should produce less dust compared to the steam-tube dryer since the tumbling action in the former is not as intense as in the case of the steam-tube dryer. The general configuration of this type of dryer is presented in Fig. 3.8. It consists of a sloping drum (1), whose perforated walls have seamless tubes fastened to them (2). The dryer drum rolls on special rings (6) and rollers (7) and is driven by power transmitted (8) by means of a gear wheel (9). The heating steam enters the drum through the pin (3). The moist coal is fed in the heating tubes.

Due to slope and rotation, the coal is gradually transferred from the inlet to the outlet of the heating tubes. The displacement of coal is controlled by the

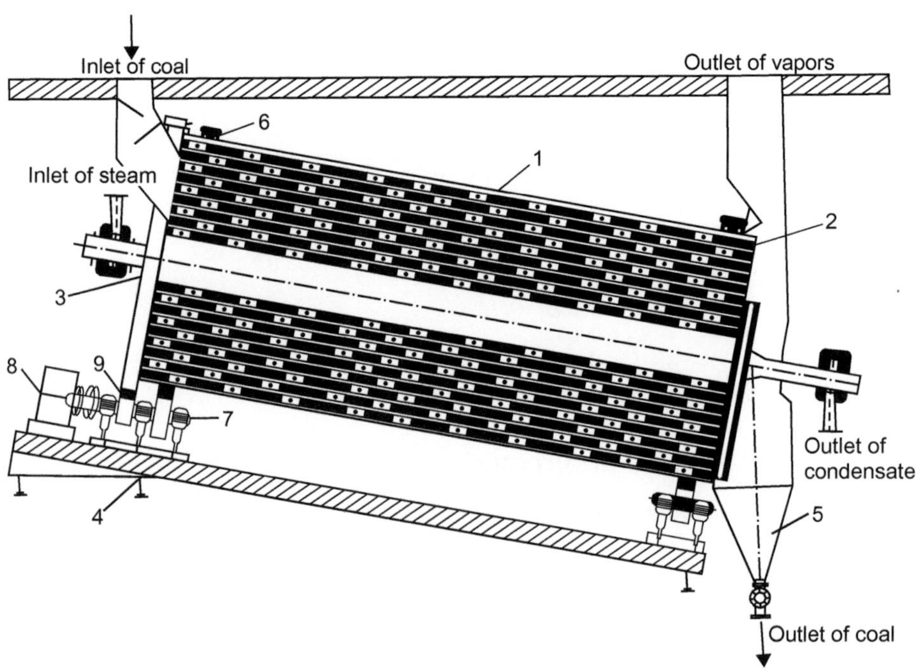

Figure 3.8 Rotary-tube dryer layout [5].

appropriate screw guides that are installed at the heating tubes. The dried coal is collected at the lower part of the tubes, at the outlet (5). The steam condensates are drained through a pin placed in the lower end of the drum. In this type of dryer, the convective heat transfer coefficient from steam to coal is about 25 W/m^2 K [5].

3.4.2 Fluidized-bed dryer (FBD)

Another type of dryers that is majorly implemented on industrial scale is the FBD, because of its simple design formulation and high capacity. This type of dryer can achieve high intensity of the drying process due to the good mixing of the coal with the drying medium, the high temperature values developed, and the easy controlling of the process. The operation of this dryer is achieved by the provision of gas through a coal layer. When the gas flow reaches a suitable velocity, then a fluidized state of the coal layer can be achieved. The gas flow velocity at which the packed bed is converted into a fluidized bed is known as the minimum fluidization velocity. Fig. 3.9 schematically shows a characteristic example of this type, i.e., the Fluo-Solids dryer with the roller chamber designed by Dorr-Oliver for coal drying.

Wet coal from the bin (1) is proportioned by feeder (2) to the drying chamber (3). Combustion gases from the combustion chamber (4) are mixed with air to obtain the desired temperature (300–700°C). The heat consumption is typically in the range of 3100–4000 kJ/kg H_2O. The hot gases pass through a screen and the coal layer, causing fluidization of coal. The exhaust gases leaving the dryer are cleaned in a cyclone (5) and cloth filter (6) [5].

3.4.2.1 Fluidized-bed dryer with fountain bed (spouted bed)

Recently, the FBDs with a fountain bed (spouted bed) have become popular for coal drying processes. Fig. 3.10 presents the conical–cylindrical apparatus, i.e., the characteristic component of this specific dryer type, which develops the fountain bed. The particles supplied by screw (2) are moving upward drifted by the hot flue gas provided by tube (3) and then move peripherally falling downward, forming the

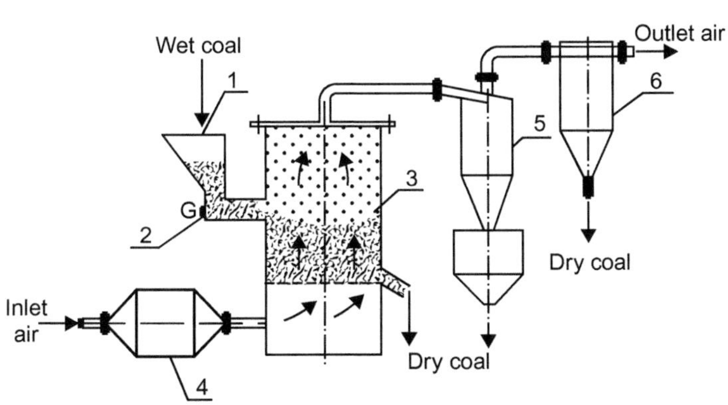

Figure 3.9 Schematic of Fluo-Solids-type fluidized-bed dryer [5].

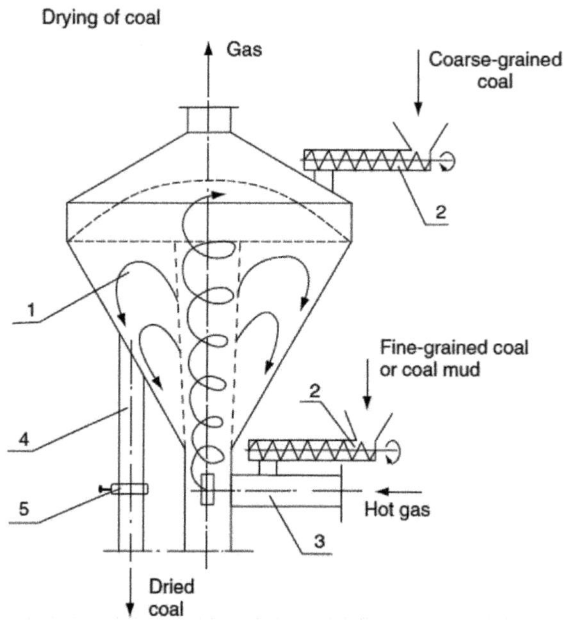

Figure 3.10 Schematic of dryer with spouted bed [5].

fountain bed state. For this type of dryer, it is a prerequisite to estimate the values of the velocities that define the start and the end of the spouting process. The critical velocities are dependent on the bed height, the physical properties of the coal and gas, and the geometrical characteristics of the dryer (shape and dimensions). For coal drying, the temperature of the drying medium (e.g., combustion gases or air) is usually not more than 200°C, unlike the case of conventional FBDs, which operate under much higher inlet temperatures [5].

Fig. 3.11 shows a schematic depiction of a spouted bed dryer for coal as designed by Fiodorov and Michailov [5]. Wet coal is fed to the conical drying chamber, and it is carried by the flue gas upflow. The coarser coal grains fall downward due to the gradual increase of the cross section and the corresponding decrease of the flue gas velocity in the conical chamber. The finest grains are entrained by the gas stream and then separated in the cyclone (3). The intermediate fractions are circulated in the drying chamber (1), after which they pass through a vertical slit in the chamber wall for discharge. Control of the slit opening allows control over the dryer holdup. The coarse coal particles fall into the bin (5) from which they are recycled to the dryer by a bucket elevator (6).

3.4.2.2 Fluidized-bed dryer with immersed heater

Even though the FBDs assure high heat and mass transfer rates, as well as high drying rate and also prevent individual coal particles from overheating, further increase of the thermal efficiency is possible. Under this scope, a new type of FBD has been

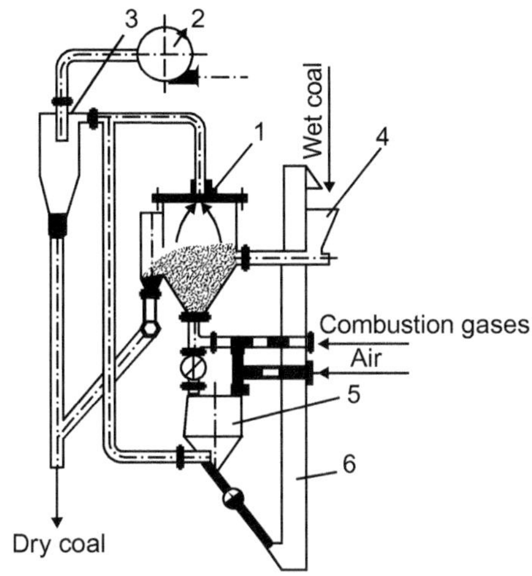

Figure 3.11 Schematic of fluidized system with spouted bed [5].

developed, immersing heat exchangers in the particle bed. Based on this modification, the drying process can be achieved by both convective (gas flow) and conduction (immersed heaters) mechanisms, permitting increased temperature and vapor capacity [11]. It is to be noted that the heat transfer coefficients are strongly affected by the local hydrodynamic conditions, which differ depending on the position, geometry, and properties of the internal element, the gas flow characteristics, and the form of the bed [12].

3.4.2.3 Fluidized-bed dryer (WTA)

RWE Power in Germany has also developed a fluidized-bed drying technology for lignite. The technology concept developed by RWE is called Wirbelschichttrocknung mit interner Abwärmenutzung (WTA) technology (English: Fluidized-bed drying with internal waste heat utilization), which is arguably the most advanced superheated steam drying technique [13].

In the scheme of this very sophisticated system (Fig. 3.12), the lignite is first milled to a fine size by hammer mills that are placed in series with a two-stage FBD. The dried material exiting the bed is separated from the continuous phase and mixed with coarser particles from the bed bottom and directly injected into the boiler. The heat demands are provided by external steam originating from the turbine and transferred to the fuel particles inside the bed through tube bundles.

A slightly modified design of the WTA process (Fig. 3.13) includes (1) an FBD using superheated steam, (2) a vapor compression step for recovering the latent heat from the process, and (3) the supply of energy to the drying solids. It is estimated that this specific system can provide drying of the raw material by reducing the moisture content by 48% (from 60% to 12%) using steam at 110°C and

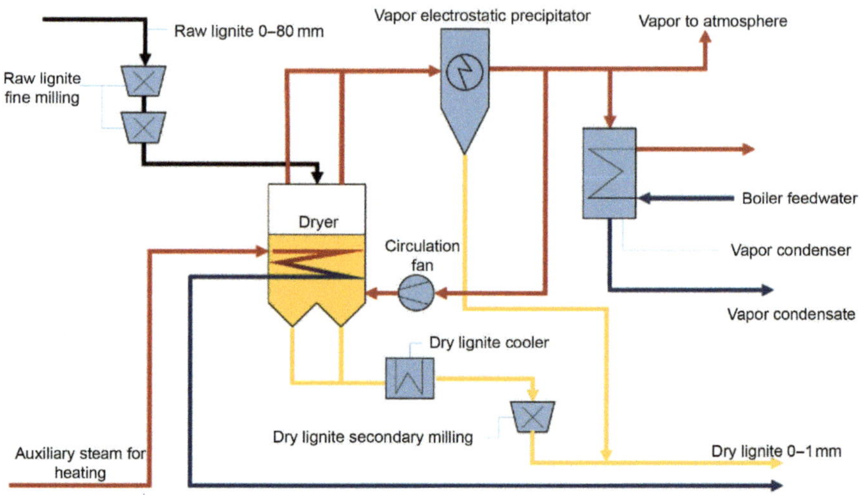

Figure 3.12 WTA concept with vapor condensation in RWE Power - Niederaussem power station
Source: Used with permission from RWE Power, WTA Technology, www.rwe.com.

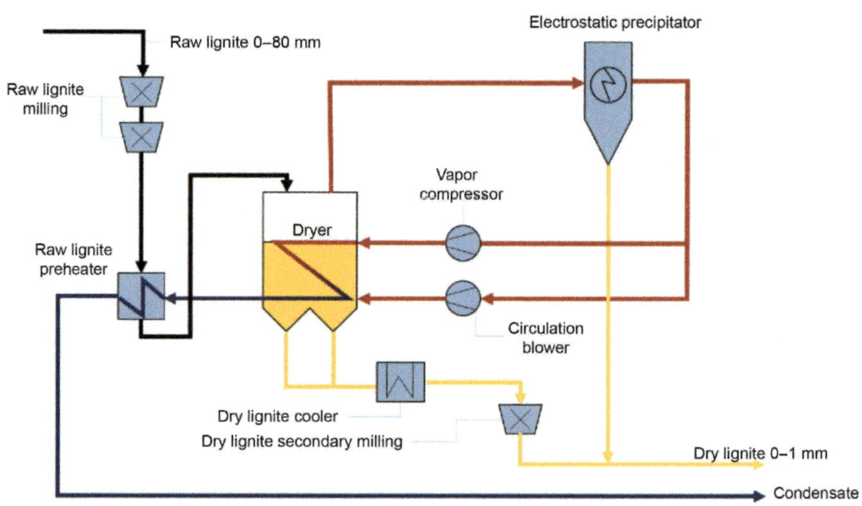

Figure 3.13 Process principle of WTA fine-grain drying with vapour recompression.
Source: Used with permission from RWE Power, WTA Technology, www.rwe.com.

50 mbar. A part of the steam at higher temperature is used for indirect heating of the fluidizing bed through submerged tube bundles [15].

3.4.2.4 Pressurized fluidized-bed dryer

Another recent scientific step, which is schematically presented in Fig. 3.14, has been conducted by Lechner et al. [16], who have successfully performed drying of

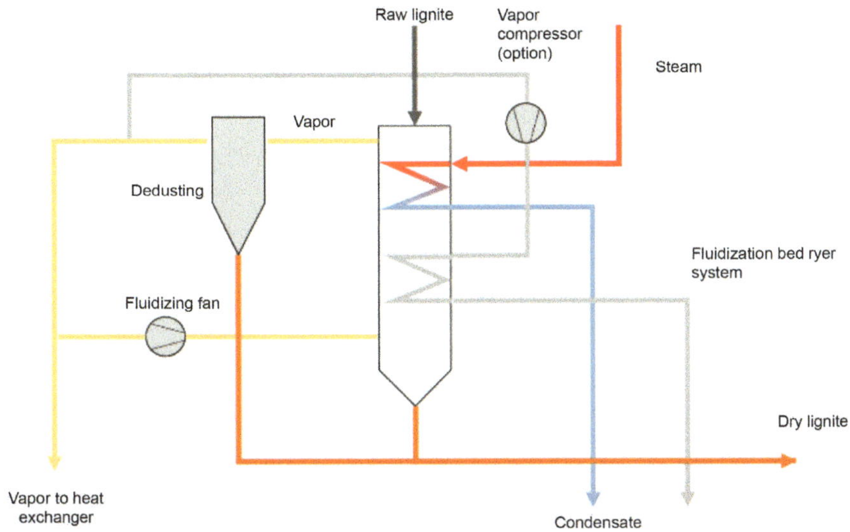

Figure 3.14 Schematic view of the pressurized fluidized-bed drying (PFBD) process [17].

lignite using high-pressure superheated steam in a pilot scale reactor. Experiments were carried out for a flow rate of 240–500 kg/h dry lignite starting in on a raw basis from 50%–60% moisture and ending to a 5%–30% moisture content (mass basis). Raw lignite is injected into the pressure vessel containing the fluidized bed. Steam condenses in a heat exchanger in the fluidized bed, while releasing its heat. The transferred heat vaporizes the water in the lignite, which then exits the dryer at its top, where subsequently is de-dusted. The heat of the vaporized lignite moisture can be recovered and used either internally for the heat requirements of drying with a vapor compressor or elsewhere besides the dryer.

The basic difference of this process compared to the standard WTA scheme is the high pressure of steam that can reach up to 10 bar.

3.4.3 Pneumatic (flash) dryers

The basic element of this type of dryers, presented in Fig. 3.15, is a vertical tube of 650–1100 mm diameter and 14–35 m length. The hot drying medium flows from the bottom to top carrying the coal particles, which are heated and dried. The velocity of the drying medium is specified by the grain size, but generally it varies from 10 to 40 m/s. More specifically, the velocity increases with the increase of the particles diameter. The mixing of coal with hot flue gas/air occurs in the disintegrator which is installed at the bottom of the tube. Due to the fact that the mixing is intense, the drying process is highly accelerated managing to reduce the initial moisture of coal by usually 50%, while the final moisture content is about 4%–6%. The heating medium temperature at the dryer inlet is 550–700°C and 70–170°C at the dryer outlet. The most intensive drying occurs during the first 2–3 m of the

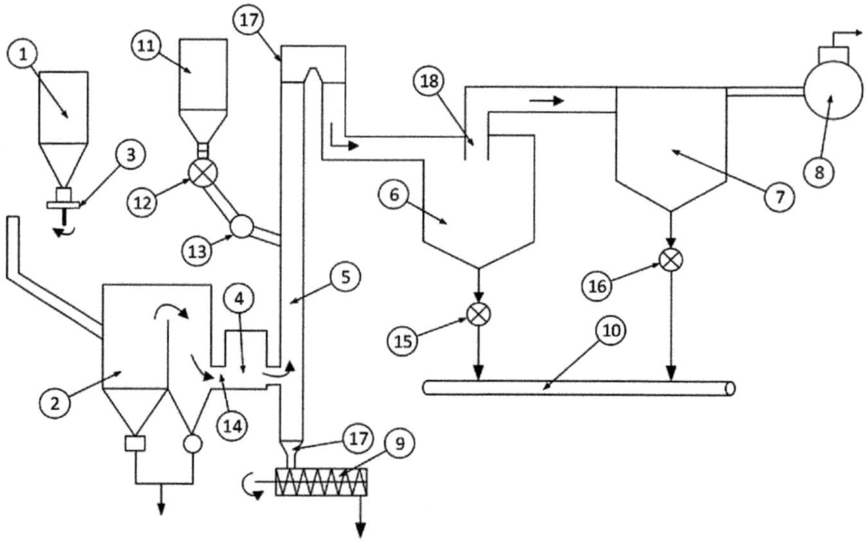

Figure 3.15 Schematic of pneumatic dryer for coal drying [5].

dryer. Since the velocity increases with the grain size, the coarser particles reside a longer time in the heating medium than in the case in which moderate and lower grains occur. On the other hand, the large contact surface of the finer grains allows rapid drying.

Wet coal is stored in the bin (11) and proportionally distributed through the feeder (12) to the lifting tube (5), where it is carried away by the combustion gas stream derived from fuel combustion in the furnace (2). The drying process occurs in the tube, after which the coal is separated from combustion gases in the cyclone (6) and then in a bank of cyclones (7). Negative pressure in the system is generated by an induced draft fan (8). The dried and separated coal is directed to the conveyor (10) through the shutter closures (15, 16), which are the lower seals of the cyclones (6, 7). The heavier coal particles which fall down in the lifting tube (5) are sent to the dried coal conveyor (10) through the screw (9). The system is equipped with explosion flaps (17) [5].

3.4.4 Mill-type dryers

The mill-type dryers are suitable in the cases in which the grinding and drying operations take place together. The mills vary significantly in terms of fundamental properties, shape, and operation depending on the fuel that is going to be dried. Therefore lignite milling differs fundamentally from milling bituminous and hard coals. The mill-type dryers are majorly utilized in steam power plants. However, it is essential to take into consideration the risk of fire. In order to minimize this possibility and simultaneously attain adequate drying, a mix of flue gas and air is used.

This provides the heat required and reduces the oxygen concentration to a safe level.

The beater wheel mills are suitable for the soft nature of lignite and achieve the predefined prerequisites of the fuel particles as regards their fineness and moisture content in order for a stable and efficient combustion to be attained. The fineness of pulverized coal is usually assessed on the basis of the residues on the 1 mm sieve. The typical values for pulverized lignite are well below 10% (usually between 3% and 6%), depending on the quality of coal, the combustion system, and the boiler size [18]. Also, for lignite combustion there are no special demands on the fineness of the pulverized fuel in terms of primary NO_x emissions control. However, finer fuel particles tend to reduce deposition and slagging problems [18]. Finally, the number of mills in operation in a unit is dependent on the energy input (calorific value) of the fuel and the maintenance requirements.

3.4.4.1 Beater mills

For drying of lignite, beater mills are typically used. Several types of beater mills are shown in Fig. 3.16.

Normally, the mill of this type consists of articulated rotating beaters, while sometimes vapor separation, particle classification, or staged grinding is included as well. However, it does not use a fan for the drying and grinding processes.

The primary materials of the beaters are manganese steel, manganese cast steel, cast iron, or carbon steel. Each one of them weighs 4–10 kg depending on the size. Due to wear and erosion problems they must be periodically replaced. It is estimated that a lifetime span for this type of dryer is in between 200 hours and 400 hours.

One important operational parameter that plays a vital role to the drying of the particles is the rotational speed of the beaters which is 40–60 m/s (720–1420 rpm).

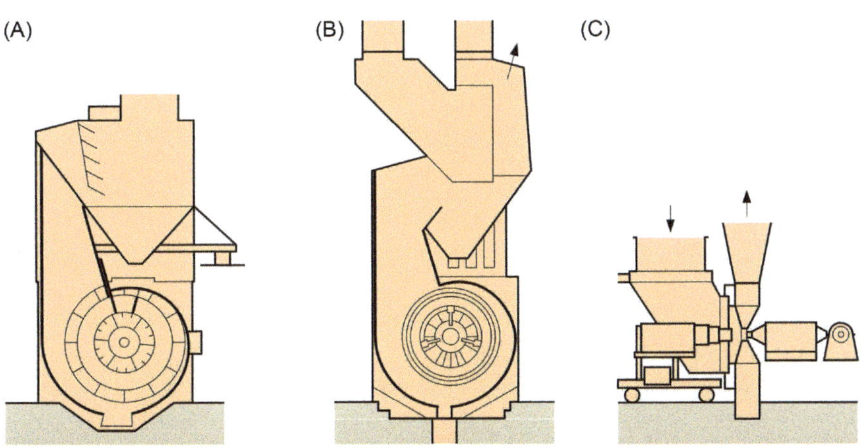

Figure 3.16 (A) Mill with classifier [18], (B) mill with vapor separator classifier [18], and (C) mill with staged grinding [18].

More specifically, the wet coal is driven by gravity into the mill through the feeder, while the combustion gases are supplied by stub pipes. The rotation of the beaters bounces the wet particles up to a certain height of the shaft, resulting in their grinding and drying. Then, the finer grades are supplied directly to the boiler furnace, while the coarser ones are recycled for further grinding.

The power consumption for grinding is typically 15–20 kWh/tonne for hard coal and 4–14 kWh/tonne for lignite. The metal consumption is 60–160 g/tonne for hard coal, 20–80 g/tonne for lignite, and 2000 g/tonne for quick coke. For coals at moisture levels of 10%–14%, air at 300–400°C may be used as the heating medium. When the moisture content is above 15%–25%, combustion gases at temperature of 700–1000°C should be used. Coal at a high moisture content (25%–35%) should be predried partially at the chute to the mill by means of hot combustion gases [5,18]. Several arrangements of beater mill dryers are described more thoroughly in the following.

Beater wheel mill with classifier

In this arrangement of mills, classifiers are also included as shown in Fig. 3.16A. The use of the classifiers ensures that the residue on 1 mm sieve does not exceed 3%–5%. The fraction of $<63\ \mu m$ is equivalent to about 70%–90%.

The used deflection surfaces separate the coarse dust fraction from the continuous phase and the finer fuel dust and recycle it to be further grinded. However, the classifiers do not fulfil the current demands of the large lignite-fired power plants, since there is a lack of vapor classifier. Thus, a high concentration of vapor is injected within the boiler along with the fuel particles [18].

Beater wheel mill with vapor separation classifier

The basic drawback of the previous type of dryer, i.e., the absence of vapor classifier, has been overcome in the next mill type (Fig. 3.16B). The beater wheel mill with vapor separation classifier performs two actions. First, through the implemented vapor separation classifier, it separates the coarse grains from the mix of the continuous phase and the finer fuel grades and second, it separates the mixture into fuel-rich and fuel-lean vapor streams. The degree of vapor separation can be set as 90% fuel dust with 70% of the initial carrier gas in the fuel-rich stream, and a 10% fuel dust with a remaining 30% carrier gas in the fuel-lean stream. The fuel-rich stream is injected through the lower main burners, while the fuel-lean stream is injected through the upper main burners or the vapor burners, if the latter are indeed developed in the unit [18].

Beater wheel mill with staged grinding

In the case of lignite with high hardness and poor grindability, the beater wheel mills with staged pulverizing/grinding can be used to achieve the required fineness. These mills are designed without classifiers. However, due to this modification, there are greater fluctuations in the grinding of the mills and the residue on the 1 mm sieve is, on average, 6%–9%.

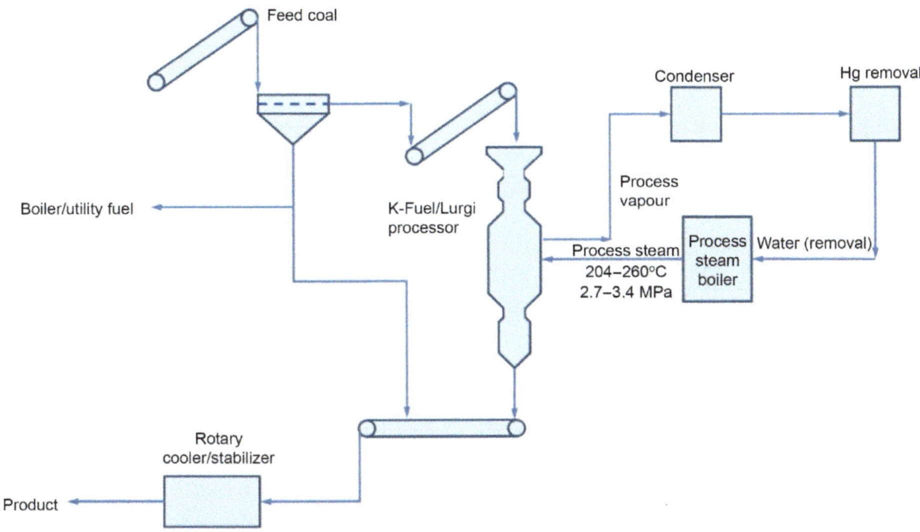

Figure 3.17 Flowsheet of the K-Fuel process [19].

3.4.5 Hydrothermal dewatering (HTD)

The operation of this process is based on the heating of coal under pressure until it reaches a temperature value ranging from 250°C to 310°C. The operational conditions drive the coal to break down and shrink while the water is released as liquid.

3.4.5.1 K-Fuel

A patented technology that follows the previous operational scheme is the K-Fuel process developed by Evergreen Energy Inc. which is presented in Fig. 3.17. It heats the low value coals in high temperatures (204–260°C) and pressurizes up to 2.7–3.4 MPa removing irreversibly the water content by more than 50% and improving the energy input of the fuel. An important co-benefit of this technology is that it can remove significant amounts of mercury and other impurities reducing overall emissions. The first experiments on industrial scale have confirmed that the utilization of such a system does improve the efficiency and reduce the emissions.

3.4.5.2 Continuous hydrothermal dewatering (CHTD)

The continuous HTD, presented in Fig. 3.18, is a process developed by Exergen and demonstrated at pilot scale. The central part of this method is a vertical autoclave that uses gravitational head pressure and a small amount of energy to transforms the molecular structure of lignite for the removal up to 80% of its moisture content and for its decarboxylation using less than 2% of the coal's energy. The sophisticated autoclave design ensures heat recovery of the fuel greater than 90%. The operational conditions (10 MPa and 300°C) imposed on the coal slurry inside

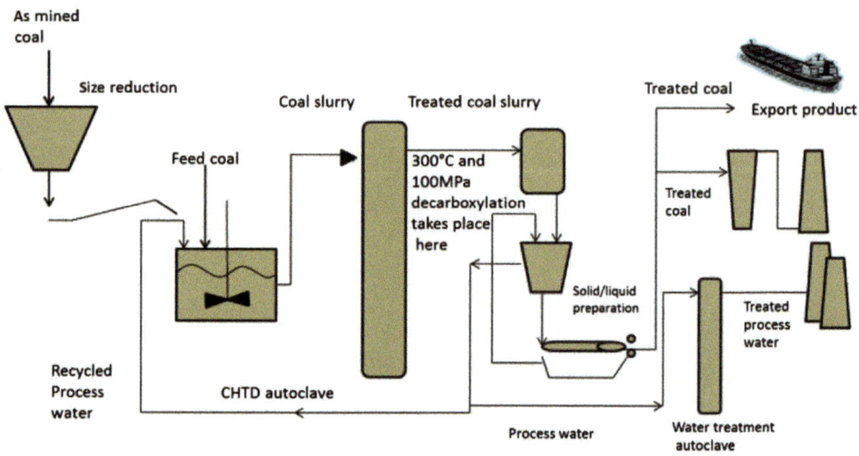

Figure 3.18 CHTD process [20].

the autoclave alter the molecular structure of the coal and collapse the pores making the particles hydrophobic (from hydrophilic) and unable to hold high moisture. The extracted water is in a liquid state, producing a coal with higher energy density and fulfilling at least 40% of the make-up requirements for power station cooling. Finally, this process is able to remove some impurities, resulting in a coal with improved combustion characteristics. Based on the findings of the pilot scale operation it can be concluded that this technology has small footprint and can be easily scaled-up [20].

3.4.5.3 Hot water drying (HWD)

The hot water drying process, presented in Fig. 3.19, dries coal in a water medium under high temperature and pressure values. More specifically, the wet coal experiences high temperatures, beginning from 240°C, under the saturated steam pressure for less than 10 minutes. The released CO_2 due to the decarboxylation process expands the pores of the particles and expulses the moisture from coal. The devolatilized tars/oils that remain on the surface develop a hydrophobic coating that prevents the particles from reabsorbing moisture. Since the coating prevents moisture reabsorption and retains the volatile matter, an excellent combustion performance can be obtained. Also, it is indicated that the alkali cations, which are majorly responsible for boiler fouling, are released in the aqueous phase and removed during the final mechanical dewatering step.

3.4.5.4 The catalytic hydrothermal reactor technology

The most recent development in the field of HTD dryers is the catalytic hydrothermal reactor technology (Cat-HTR), whose scheme is graphically presented in Fig. 3.20. This technology uses water at or near supercritical temperatures and

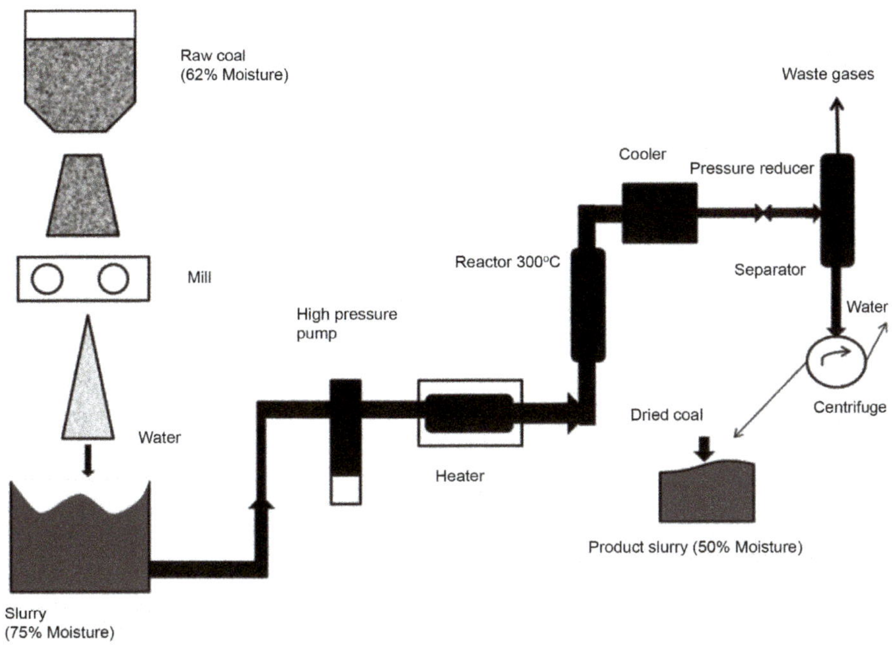

Figure 3.19 Hot water drying concept [21].

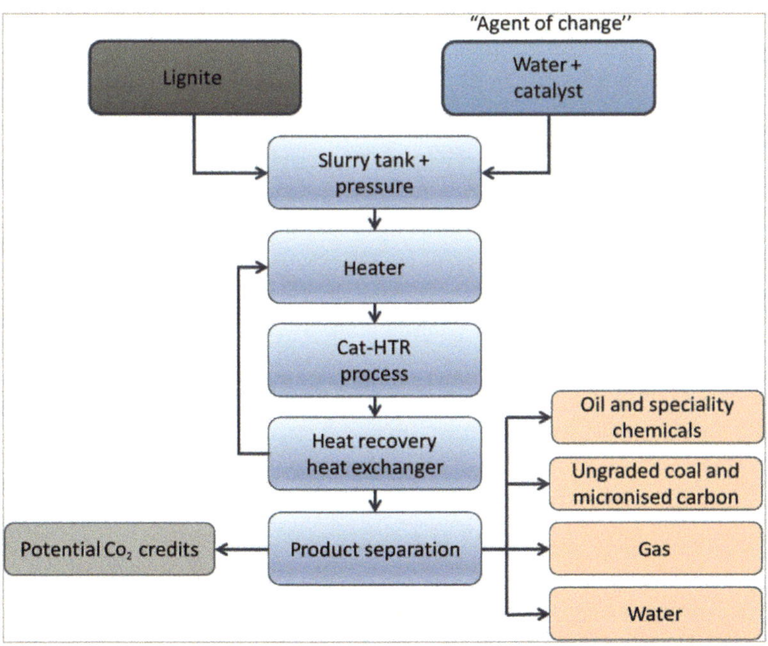

Figure 3.20 Cat-HTR process diagram [22].

pressures along with an appropriate catalyst to selectively de-polymerize and de-oxygenate lignite and convert it into various higher density energy fuels and high grade clean coal products.

3.4.6 Mechanical thermal expression (MTE)

The MTE technology, presented in Fig. 3.21, was initially developed in the mid-1990s and it suitably combines pressure (<12 MPa) and temperature (<200°C) to effectively remove the moisture of lignite. Initially, the raw lignite is treated in saturated conditions (0.5–2 MPa) and gradually heated at a level of 150–200°C. A mechanical pressure of around 6 MPa is then applied to squeeze the water out of the lignite. The design has been tested at laboratory and pilot scales investigating various parameters such as (1) the feeding of the MTE press with coal under quasi-continuous conditions, (2) the treatment of the raw lignite, and (3) the subsequent dry lignite treatment.

A further development of MTE technology was performed by Cooperative Research Centre for Clean Power from Lignite (CRC Lignite) in Australia. The CRC's MTE technology uses a different configuration compared to the conventional one. In this advanced methodology the coal is fed as a preheated slurry. The latter is further heated by saturated steam under pressure in a heating chamber to the required process temperature. Then, it is directly driven under the influence of gravity toward a compression cylinder. The dewatering processes occur through both axial and circumferential filter surfaces. This feature improves the capability of drying of this specific process compared to the axial surfaces, since it becomes independent of PSD. The available reports indicate that this technology can be a low cost, energy efficient, and CO_2-saving option for partial drying (down to

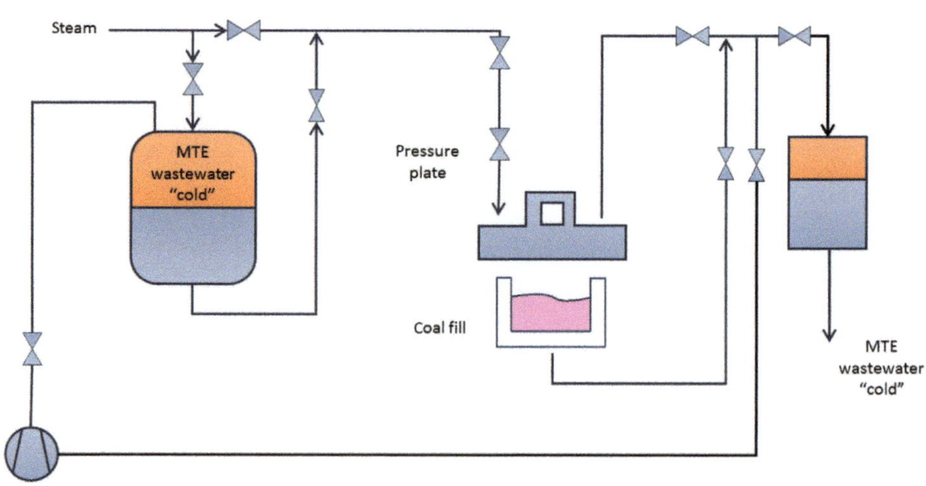

Figure 3.21 MTE process design [3].

around 30% moisture) at the large scale required for power plants [3,23]. However, an important task that needs to be faced is the formation of a cost-effective treatment of the wastewater, since it presents both organic and inorganic matter [18,24,25].

3.4.7 Microwave drying

Microwave (MW) drying is well known for its advantages, such as volumetric heating, faster drying rates, and possibility of using intermittent exposure of wet solids to MWs. Recently, MW drying has been applied for drying of low-rank coal. The CoalTek process developed by CoalTek Inc. and the Drycol Process developed by DBAGlobal (Milton, Queensland, Australia) are based on the use of MW for coal quality improvement.

3.4.7.1 CoalTek process

The CoalTek process is a MW-based process developed by CoalTek Inc. which removes moisture, ash, sulfur, and mercury from LRCs and transforms it into cleaner one with an increase of 50% energy content. Typically, using this method, 40%–50% of moisture can be removed in low temperature values. The fact that CoalTek is capable of removing moisture while preserving the key metallurgical properties of coking coals makes it an ideal option for both thermal and metallurgical coal.

3.4.7.2 Drycol process

Another process that uses the controlled application of MW radiation is the so-called Drycol process developed by DBAGlobal Australia Pty, Ltd. (Fig. 3.22). The stock of coal is first separated depending on the fineness of the particles. The finest particles are directed toward a conveyor where they get dried through a MW-energized heating chamber. The dehydration of these particles is so sufficient that

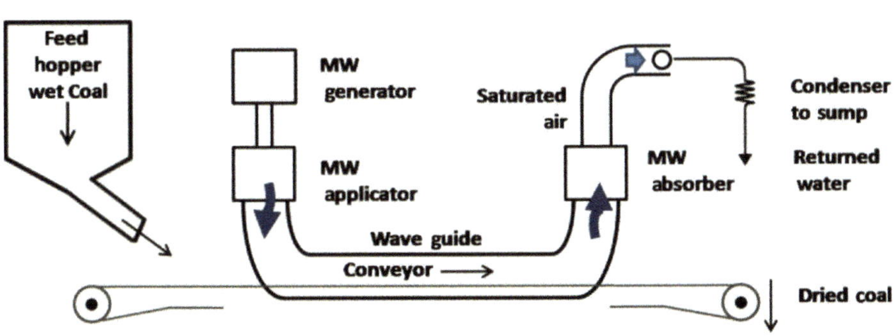

Figure 3.22 Drycol process [22].

when they are recombined with the larger grade coals, the moisture content of the aggregate coal is within a target moisture content range. It is reported that the operation of such system can reduce the vapor content from 24% to 12% [26] in much faster rate than the conventional ways of drying. Further, it is able to reduce impurities such as sulfur, potassium, and phosphorous. However, the presence of impurities can result in hot spots, and the high dielectric losses for coal can also result in fire hazards during drying [1].

3.4.8 Superheated steam dryer

This type of dryer, presented in Fig. 3.23, uses superheated steam as drying medium. This factor is directly fed into the dryer and not through tubes or other heat exchangers as in indirect-fired dryers. The steam evaporates the moisture content of the fuel particles transferring only sensible heat with no possibility of condensation, as long as the operational conditions remain above the saturation point. With this configuration the dryer can operate in a closed-loop operation in which the low-pressure steam can recirculate after being reheated. The heat of the released

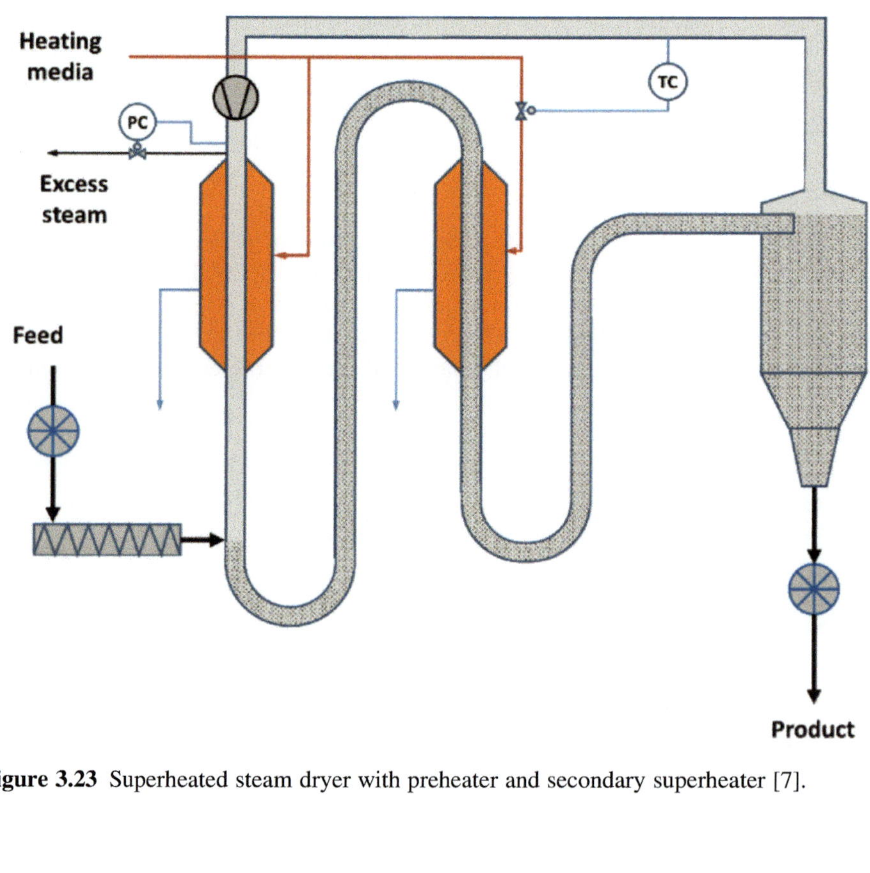

Figure 3.23 Superheated steam dryer with preheater and secondary superheater [7].

Table 3.3 Advantages and limitations of superheated steam dryers [30]

Advantages	Limitations
Energy is easily recovered at high temperature by condensation of steam. This reduces operating costs through substantial net energy saving	Need for tightly closed system to prevent high pressure steam from escaping and to prevent air from entering the system. This adds to capital costs
Emissions mainly appear in the steam condensate and can be easily removed, reducing the need for expensive accessories like after-burners, scrubbers, or filters. This reduces capital costs	Need for stainless steel parts to prevent corrosion since condensation at start-up and shut-down is unavoidable. This adds to capital cost
High drying rates are easily achievable due to high heat capacity of steam and high drying temperatures. Reduced drying time enables the dryer to be built smaller thereby reducing capital costs	Higher product temperatures compared to hot air drying during constant drying rate period. Too high a temperature may cause unwanted melting, lumping, and stickiness
Solvents and volatiles are easily recovered from the condensate by separation or distillation	Condensation of incoming material causes rapid heating of product but also temporarily wets the product
Explosion and fire risks are eliminated due to the absence of air, enabling high temperature drying if necessary	Cleaning of heat exchangers required due to deposition of fine particles during condensation of steam

steam can be used in other processes. To conclude, this method presents many benefits regarding the reduced risk of spontaneous combustion [27,28], the increased drying rates, the better energy efficiency, and finally the improved grindability [19,29]. Table 3.3 summarizes the benefits and the limitations of SSD.

3.5 Dryer comparison

It is generally accepted that there is no universal dryer type suitable for every case but the choice of the dryer depends on the particular conditions and the intended application. Various drying methods have been discussed so far which are essentially used for coal drying. A major objective for LRC upgrade and optimization is to minimize the overall energy consumption in LRC drying, while making the drying process cost effective. It should be noted that each drying method can still be further improved from a technical and economic point of view. The main characteristics and limitations which pose possibilities for various commercial drying improvements are presented in Table 3.4 [31].

Table 3.4 **Advantages and limitations of dryers [1]**

Dryer type	Advantages	Limitations
Rotary dryer	Drying along with disintegration, internal heating with coils, flue gas with low O_2 as drying medium to eliminate fire hazard	High maintenance
Rotary-tube dryers	Indirect heating, no fire hazard, good efficiency	Capital intensive
Fluidized-bed dryer	Intensive drying due to good mixing	High pressure drop, attrition
Spouted bed dryer	Very good heat and mass transfer rates	Scale-up issues, limited particle size
Pneumatic dryers	Simple construction	Attrition
Microwave drying	volumetric heating, faster drying rates, reduction of impurities	Hot spots resulting from the presence of impurities, fire hazards due to high dielectric losses, high cost involved for handling large amounts
Various types using superheated steam	High thermal efficiency, no danger of fire or explosion, energy efficient	Suited for high capacity continuous operation, energy in exhaust should be usable elsewhere in plant

3.5.1 *Technical characteristics*

Conventional evaporative dryers (rotary, fluidized bed, flash, etc.) use air or combustion gases as the drying media, with temperatures of about 700–900°C at the dryer inlet and 60–120°C at the outlet [32]. For coal drying, such high temperatures may lead to spontaneous ignition and loss of volatiles. As a result the use of such high temperature media is typically avoided unless the coal is used directly in a combustion process (e.g., mill drying in pulverized fuel boilers). Hence, in current practice indirectly heated rotary dryers have been a more common choice. Rotary dryers were also reported to have higher energy efficiency and lower energy consumption per unit mass of coal dried compared to other conventional dryers [1,32,33].

On the other hand, although the use of superheated steam reduces the odds for spontaneous self-ignition because of the absence of oxygen, the cost involved in such processes is high due to the complexity in design and operation of such dryers for very high throughput. At lower production rates steam drying is thought to be an expensive operation [1].

Typical values of the technical characteristics and specifications for the main types of dryers used for lignite, which play a major role at the dryer performance, are (1) the specific energy consumption, (2) the dryer efficiency, (3) the operating

Table 3.5 Typical range of design and performance specifications for various types of dryers

	Rotary	Pneumatic (flash)	Fluidized bed/ spouted
Specific consumption (MJ/kg-evaporation)	3.0–4.0	2.7–2.8	2.2[a]–3.5
Drying efficiency (%)	50–75	75–95[a]	70–90[a]
Evaporation (t/h)	3–23	4.8–17	2–41
Drying medium temperature (°C)	200–600	150–280	160–280
Capacity (t/h)	3–45	4.4–16	1.5–25
Feed moisture (%)	45–65	45–65	45–65
Moisture discharge (%)	10–45	10–45	10–45
Typical residence time within dryer	5–10 min	0–10 s	5–10 min

[a]With heat recovery.

Table 3.6 Economic comparison of conventional and SSDs

	Conventional	Superheated steam dryer (SSD)
Capacity (kg/h)	15,000	15,000
Air/steam flow (kg/h)	335,000	241,000
Fan volume (m^3/h)	415,000	290,000
Motor (kW)	900	630
Heat demand (MW)	15.5	9
Heat recovery (MW)	0	5.5
Capital cost (Euro)	1,350,000	2,850,000
Energy cost (Euro)		
w/o heat recovery	2,900,000	1,400,000
w/ heat recovery	–	560,000

range (drying temperature and moisture), and (4) the residence times achieved. These values are presented in Table 3.5.

Osman et al. [31] have performed a comparison of the capital and energy cost between conventional and superheated steam dryers (SSDs), presented in Table 3.6.

The capital investment according to Mujumdar [5] can be calculated roughly from the following equation:

$$C = (\text{MS Index})\left[30 + 36.74\frac{W}{(17.77 + T)}\right] \tag{3.1}$$

where C is the installed cost (US$ in the year of the M&S Index) covering the dryer and auxiliaries, including foundations and erection, but no secondary dust collectors or building; M&S Index is the Marshall and Swift All-Industry Equipment Cost Index; *W* is the evaporation capacity (kg moisture/h); and *T* is the inlet-gas temperature (°C).

Considering $(\text{M\&S index})_{2011} = 1536$, the following diagram (Fig. 3.24) can be drawn, presenting the capital investment as a function of evaporation capacity and with the drying temperature as a parameter ranging from $T = 200°C$ to 600°C.

Bruce and Sinclair [34] have investigated the capital cost and operating consumptions for the major types of dryers (Table 3.7).

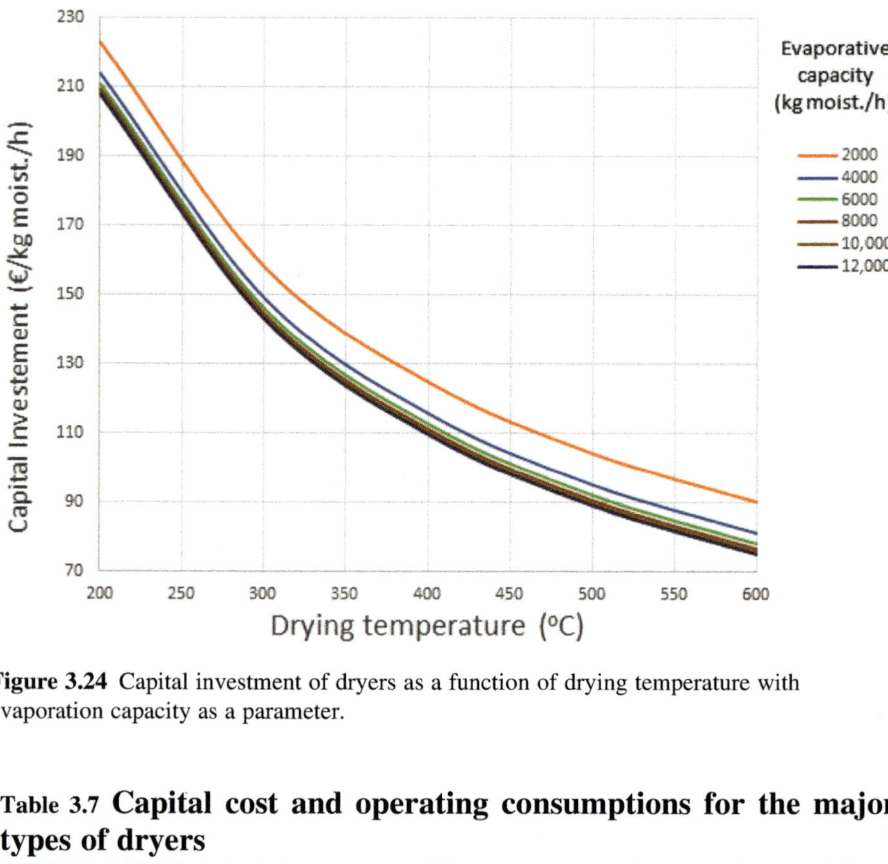

Figure 3.24 Capital investment of dryers as a function of drying temperature with evaporation capacity as a parameter.

Table 3.7 Capital cost and operating consumptions for the major types of dryers

	Rotary dryer	Pneumatic (flash) dryer	Fluidized bed/spouted bed dryer	Superheated steam dryer
CAPEX	120–290 €/dry kg/h	260–680 €/dry kg/h	160–285 €/dry kg/h	235–315 €/kg water/h
Operating electrical consumption	8–14 kWh/dry t	16–38 kWh/dry t	15–20 kWh/dry t	20–45 kWh/dry t

Regarding the maintenance cost, this is assumed to be around 5% of the capital investment for simple systems and closer to 10% for complex systems with significant number of rotating or vibrating parts [34].

Recently, a technoeconomic comparison has also been performed between predrying technologies on retrofit cases of lignite-fried power plants as well as on new build power plants [35].

Comparing the fluidized-bed drying concepts the pressurized variant operated by Vattenfal (pressurized fluidized-bed drying, PFBD) could be 7% − 10% cheaper potentially as compared to the atmospheric variant in the retrofitting case of an existing power plant. From the pilot plant results, the power plant efficiency has been proved to potentially increase by several percentage points and the CO_2 emissions to decrease by up to 5% − 10% [35].

In the case of the WTA concept, a total investment of €50 million was required for the erection and operation of the prototype WTA dryer at Niederaussem as reported in 2009 [35]. Significant reduction in the dryer size and the investment costs of the WTA fine-grain drying concept have appeared in comparison with the coarse grain drying concept. The specific investment costs of the entire WTA predrying system have been reduced by 60% to 70 €/kW, compared to those for WTA coarse grain drying, due to size reduction of the dryer, ESP and fluidization blower, vapor recompressor, and other supplementary equipment optimization (Fig. 3.25) [35].

When comparing the WTA and PFBD concepts, the investment costs of the PFBD concept which uses coarse grains were found higher than those of the WTA fine-grain concept. In particular they were 3.2 times higher than those for the Frechen WTA-2 plant and 1.9 times higher than those for the Niederaussem WTA-2 plant [35].

Finally, an evaluation of all the described dryers by using some key factors with a scale from 1 to 5 for each factor is presented in Table 3.8.

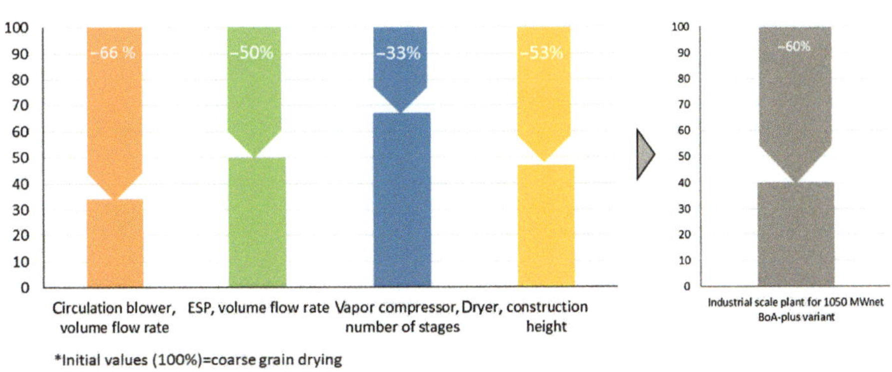

Figure 3.25 Significant reduction in the dryer size and investment costs of the WTA fine-grain drying concept [32].

Table 3.8 **Evaluation of dryer types**

	CAPEX (1 for highest, 5 for lowest cost)	OPEX (1 for highest 5 for lowest cost)	Capacity range (1 for shortest, 5 for widest range)	Final moisture content (1 for highest, 5 for lowest)	Heat recovery opportunity (1 for highest, 5 for lowest opportunity)	Fire Hazard Avoidance (1 for lowest, 5 for greatest avoidance)
1. Rotary-tube dryers						
a. Direct	5	5	5	4	5	4
b. Indirect	4	5	5	4	5	5
2. Fluidized-bed dryers (FBD)	4	3	4	4	5	5
3. Pneumatic (flash) dryers	2	4	3	5	5	4[a]–5[a]
4. Mill-type dryers	N/A[b]	5	5	N/A	N/A	4
5. Hydrothermal Dewatering						
a. K-Fuel	N/A	N/A	N/A	N/A	N/A	N/A
b. Continuous hydrothermal dewatering (CHTD)	N/A	N/A	1[c]	N/A	5	N/A
c. Hot water drying (HWD)	N/A	N/A	1[c]	N/A	5	5
d. Catalytic hydrothermal reactor technology (Cat-HTR)	N/A	N/A	1[c]	N/A	N/A	N/A

6. Mechanical thermal expression (MTE)	5	N/A	4	3	N/A	5
7. Microwave drying						
a. CoalTek process	N/A	N/A	2	N/A	N/A	N/A
b. Drycol	N/A	N/A	2	4	N/A	3
8. Superheated steam dryer	3	2	N/A	5	5	5

[a]Depending on the drying medium (5 for steam and 4 for any other medium containing oxidizing agent).
[b]Although there are currently no available data for the CAPEX of mill drying, it has to be noted that, unlike the other predrying technologies, mill drying cannot be considered to impose an extra cost in pulverized lignite power plants, since mills have to be included anyway for pulverization of the fuel.
[c]Currently the capacity of the pilot plant is low, potentially it could reach high values if future plants with higher capacity are implemented.

References

[1] S.V. Jangam, M. Karthikeyan, A.S. Mujumdar, A critical assessment of industrial coal drying technologies: Role of energy, emissions, risk and sustainability, Drying Technol. 29 (4) (2011) 395–407.

[2] The role of coal in EU Power generation, 2011, European Association for Coal and Lignite (EURACOAL) (14.05.2013).

[3] Wibberley, L., et al., Techno-Economic Assessment of Power Generation Options for Australia, Technology Assessment Report 52, 2006, Queensland, Australia, p. 252.

[4] J. Brammer, A. Bridgwater, Drying technologies for an integrated gasification bio-energy plant, Renew. Sustain. Energy Rev. 3 (4) (1999) 243–289.

[5] A.S. Mujumdar, Handbook of Industrial Drying, CRC Press, Boca Raton, 2006.

[6] *Dryer Handbook*, 1980, APV Company, APV Crepaco, <https://books.google.gr/books/about/Dryer_Handbook.html?id = sxJcNwAACAAJ&redir_esc = y>.

[7] C.J. Roos, C. Northwest, A. Center, Biomass Drying and Dewatering for Clean Heat & Power, Northwest CHP Application Center, 2008.

[8] H. Holmberg, P. Ahtila, Evaluation of energy efficiency in biofuel drying by means of energy and exergy analyses, Appl. Therm. Eng. 25 (17–18) (2005) 3115–3128.

[9] P. Rofouieeraghi, Biomass Gasification Integrated into a Reference Canadian Kraft Mill, Ecole Polytechnique, Montreal, Canada, 2012, p. 109.

[10] C.E. Bill, Rotary steam tube dryer, Ind. Eng. Chem. 30 (9) (1938) 997–999.

[11] H. Groenewold, E. Tsotsas, Drying in fluidized beds with immersed heating elements, Chem. Eng. Sci. 62 (1–2) (2007) 481–502.

[12] C. Gutfinger, N. Abuaf, Heat transfer in fluidized beds, in: P.H. James, F.I. Thomas (Eds.), Advances in Heat Transfer, Academic Press, San Diego, 1974, pp. 167–218.

[13] H.J. Klutz, C. Moser, D. Block, WTA Fine Grain Drying—Module for Lignite-Fired Power Plants of the Future, VGB Power Tech, Essen, Germany, 2006.

[14] E. Nshimiyimana, *Application of Windhexe Dehydration Technology for Producing β-Carotene Rich Flours from Swe*etpotatoes, Food Science Department, North Carolina State University, Raleigh, North Carolina, 2012.

[15] F. Buschsieweke, Dampfwirbelschichttrocknung von Braunkohle, Brown Coal Drying in a Steam Fluidized Bed, Universität Stuttgart, Germany, 2006, in German.

[16] Lechner, S., O. Höhne, and H.J. Krautz. Pressurized Steam Fluidized Bed Drying (PSFBD) of Lignite: Constructional and Process Optimization at the BTU Test Facility and Experimental Results.

[17] Godfrey, B., Recent developments in innovative drying technologies, in: International Symposium on the Sustainable Use of Low Rank Coals, 27–30 April 2010, The Department of Primary Industries (DPI), Melbourne, Australia, p. 31.

[18] Q. Zhu, Update on Lignite Firing, IEA Clean Coal Centre, 2012.

[19] A. Suwono, Hamdani, Upgrading the Indonesian's low rank coal by superheated steam drying with tar coating process and its application for preparation of CWM, Coal Prep. 21 (1) (1999) 149–159.

[20] Exergen, A Major Innovative in Clean Coal Technology, Exergen Pty Ltd, Melbourne, Victoria, Australia, 2009, p. 7.

[21] C.Z. Li, Advances in the Science of Victorian Brown Coal, Elsevier Science, Oxford, 2004.

[22] B. Godfrey, Recent Developments in Innovative Drying Technologies. Power Point Presentation, Brown Coal Innovation Australia, Low-Rank Coal International Symposium, Melbourne, 2010.
[23] H. Katalambula, R. Gupta, Low-grade coals: a review of some prospective upgrading technologies, Energy Fuels 23 (7) (2009) 3392–3405.
[24] C.J. Butler, A.M. Green, A.L. Chaffee, Remediation of mechanical thermal expression product waters using raw Latrobe Valley brown coals as adsorbents, Fuel 86 (7–8) (2007) 1130–1138.
[25] H. Nakagawa, et al., Hydrothermal dewatering of brown coal and catalytic hydrothermal gasification of the organic compounds dissolving in the water using a novel Ni/carbon catalyst, Fuel 83 (6) (2004) 719–725.
[26] G. James, Microwave for Coal Quality Improvement: The Drycol Project, DBAGlobal, Milton, Queensland, Australia, 2008.
[27] E.F. Faber, et al., A technoeconomic compression of air and steam drying, Drying 86 (1986) 588–594.
[28] Mujumdar, A.S., Superheated Steam Drying: Principles Practice and Potential for Use of Electricity, 1990.
[29] G.D. Bongers, W.R. Jackson, F. Woskoboenko, Pressurised steam drying of Australian low-rank coals: Part 1. Equilibrium moisture contents, Fuel Process. Technol. 57 (1) (1998) 41–54.
[30] Van Deventer, H., Industrial Superheated Steam Drying, in TNO Report, 2004, p. 239.
[31] H. Osman, et al., Drying of low-rank coal (LRC)—a review of recent patents and innovations, Drying Technol. 29 (15) (2011) 1763–1783.
[32] M. Karthikeyan, W. Zhonghua, A.S. Mujumdar, Low-rank coal drying technologies—current status and new developments, Drying Technol. 27 (3) (2009) 403–415.
[33] W.G. Willson, D.A.N. Walsh, W. Irwinc, Overview of low-rank coal (LRC) drying, Coal Prep. 18 (1-2) (1997) 1–15.
[34] D.M. Bruce, M.S. Sinclair, Thermal Drying of Wet Fuels: Opportunities and Technology, H.A. Simons Ltd, 1996.
[35] N.S. Dong, Techno-Economics of Modern Pre-Drying Technologies for Lignite-Fired Power Plants, IEA Clean Coal Centre, 2014.

Upgrading and advanced cleaning technologies for low-rank coals

4

Sihyun Lee, Sangdo Kim, Donghyuk Chun, Hokyung Choi and Jiho Yoo
Korea Institute of Energy Research, Daejeon, South Korea

4.1 Introduction

Low-rank coals—lignite and subbituminous coals—refer to the lowest and second lowest rank of coals. Unlike high-rank coals deposited in specific areas, low-rank coals are evenly distributed around the world and their reserves are approximately 500 billion tons, which is similar with that of high-rank coals.

Demands for coal are expected to keep rising due to recent rapid economic growth in China, India, and ASEAN countries. But this leads to supply limit of high-rank coals. Therefore, it is important to find ways to efficiently utilize the abundant and unused low-rank coals.

Low-rank coals are high in moisture or ash content and susceptible to spontaneous combustion. Since low-rank coals are high in moisture, using them as fuel in energy-producing facilities like power plants increases coal consumption, emitting more CO_2, and lowering power generation efficiency. Ash causes slagging and fouling while it is burned, interrupting heat exchange and thereby lowering power generation efficiency and causing overload and operational problems in power plant. In addition, fuel loss occurs due to spontaneous combustion in the process of stockpiling or transporting of coal. The risk of spontaneous combustion is increased after coal is dried. Therefore, technologies should be developed to better utilize low-rank coals [1,2].

There are largely three types of low-rank coal utilization technologies: a high efficiency and low emission (HELE) combustion technology, an upgrading technology, and conversion technologies such as liquefaction and gasification. The HELE combustion technology enhances efficiency by running a boiler steam condition of a power plant at an ultra-supercritical level. This technology lowers pollutants by using facilities removing pollutants in flue gas such as SOx, NOx, and PM. Adding a new facility removing mercury in flue gas is now considered [3].

Conversion technologies are gasification of coal to produce power and liquefaction of synthetic gas with catalyst to produce liquefied fuels or chemical products. Conversion technologies include IGCC (integrated gasification combined cycle) and CTL (coal to liquid), which are already commercialized and used. These technologies are expected to reduce carbon emissions. But that does not mean technologies specialized for low-rank coals are developed.

This chapter introduces upgrading and advanced cleaning technologies, which are low-carbon utilization technologies. Upgrading is necessary to use as a fuel for

Low-Rank Coals for Power Generation, Fuel and Chemical Production
DOI: http://dx.doi.org/10.1016/B978-0-08-100895-9.00004-8

HELE combustion and conversion technologies. Upgrading is a process of removing moisture or ash from low-rank coals before using the coals to enhance facility efficiency. As such, we would like to introduce existing coal drying technologies and stabilization technologies as well as new drying/stabilization technologies, which are under development. This chapter also introduces the production technology of ash-free coals (AFCs), which is expected to be used for various purposes in the future.

4.2 Upgrading: drying and stabilization of low-rank coals

4.2.1 Self-heating property of low-rank coals

The spontaneous combustion of coal is caused by low-temperature oxidation [4]. While mining, crushing, or transporting coal, unexpected explosions like dust explosion sometimes occur. Spontaneous combustion can also occur while stockpiling coal in a yard as carbon reacts with oxygen due to a rapid temperature increase on the coal particle surface (Fig. 4.1).

It is well known that the low-temperature oxidation of coal causes spontaneous combustion. So far, the main focus of studies on low-temperature oxidation has been on the dynamics between coal, oxygen, and moisture. In this regard, studies have been conducted on the relationship between coal and oxygen. However, few studies have been conducted on the impacts of moisture, which greatly influences the spontaneous combustion of coal. It was not until recently that it began to receive attention [5].

On the one hand, moisture in coal inhibits low-temperature oxidation, while on the other it increases the temperature of coal particles and eventually causes combustion as moisture in the air is attached to the coal, generating heat of adsorption. While moisture is adsorbed into the hydrophilic functional group of coal, oxygen is

Figure 4.1 Spontaneous combustion phenomena happened in coal yard.

adsorbed into the hydrophobic functional group. Therefore, no competitive adsorption between moisture and oxygen takes place on the surface of coal. The adsorption of oxygen increases the temperature of particle surface, causing moisture desorption. The adsorption of oxygen increases particle temperature, desorbing moisture, which in turn leads to the readsorption of oxygen, using moisture as a catalyst for low-temperature oxidation.

4.2.2 Analysis of self-heating properties

Numerous studies have been conducted to analyze the characteristics of spontaneous combustion of coal, and methods such as crossing-point temperature (CPT) and isothermal oxidation have been presented. This chapter introduces some of the methods that analyze the characteristics of spontaneous combustion.

Measuring CPT is one of the most widely adopted ways to analyze the characteristics of spontaneous combustion of coal. CPT is measured by putting a certain amount of coal into a cell and steadily increasing the oven temperature, injecting air or oxygen in the process (Fig. 4.2). The coal temperature in the cell increases proportionately with the surrounding temperature until it begins to rise rapidly due to the heat of reaction with oxygen to a point where the oven temperature crosses the coal temperature, and we call the crossing-point CPT.

The more self-heating ability of coal is, the higher coal temperature increases, lowering CPT. CPT can be changed depending on the amount of moisture in coal, particle size, how fast surrounding temperature increases, and the flow of oxygen or air. Therefore, to compare the characteristics of spontaneous combustion of coal by types, CPT should be measured under the same measurement conditions. In

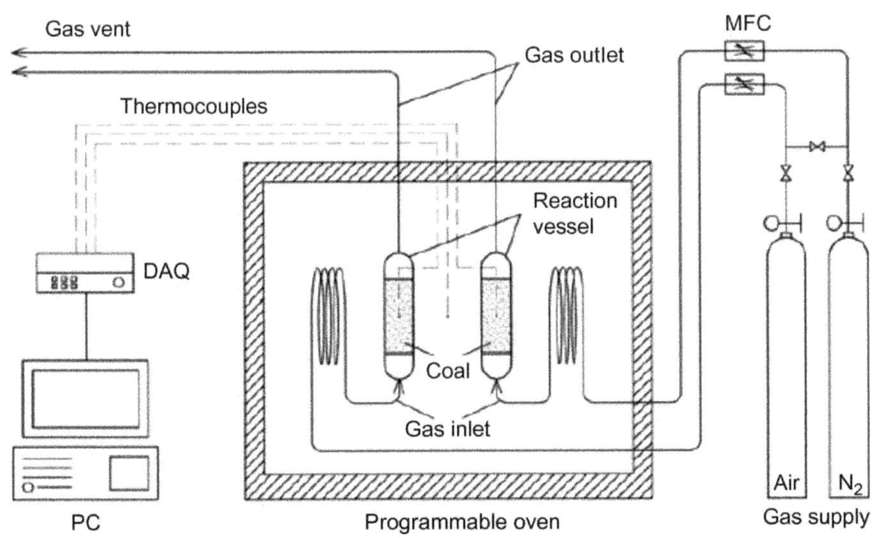

Figure 4.2 Schematic diagram of the CPT testing system [5].

particular, moisture requires heat to be evaporated. When moisture absorbs heat, the rate of coal temperature increase begins to slow down (Fig. 4.3). This results in a very high CPT and disrupts accurate measurement. To prevent this, the coal should be dried sufficiently in a nitrogen environment before measuring CPT [6,7].

Since CPT is susceptible to measurement conditions, instead of using CPT, liability index (LI) is used to analyze the nature of spontaneous combustion. Together with CPT, LI was suggested by Feng et al. [7] as a way to consider the speed of temperature increase.

$$\text{LI} = (\text{average heating rate between 110 and } 220°C/\text{CPT}) \times 100. \tag{4.1}$$

Umar et al. [8] calculated LI by considering the rate of temperature increase 10 minutes before and after CPT without restricting the temperature range between 110 and 220°C. Umar et al. [8] classified the tendency of spontaneous combustion into three categories: low, where LI is between 0 and 5; medium, 10; and high, over 10.

Low-temperature oxidation is a process where temperature is maintained at a certain level while oxygen or air is injected to analyze gases emitted from oxidized carbon (Fig. 4.4). In doing so, the concentration of CO and CO_2, emitted from the oxygen and carbon reaction, is measured and reactivity is compared. The more CO and CO_2 are generated, the higher the risk of spontaneous combustion gets (Fig. 4.5). Low-temperature oxidation enables evaluating the tendency of spontaneous combustion of carbon at different temperatures. It is particularly useful in figuring out the characteristics of spontaneous combustion at low temperatures [5].

Adiabatic oxidation method analyzes the tendency of spontaneous combustion by insulating coal at a certain temperature (60 or 80°C), injecting a sufficient amount of oxygen with no energy being injected from outside and measuring the rate of temperature increases from carbon–oxygen reaction [9]. The tendency of spontaneous combustion is classified with low (0–8), medium (5–10), considerable (8–18), and extraordinary (over 18).

Wang and Luo [10] used thermo-gravimetric analysis method to analyze the tendency of spontaneous combustion. In an early stage of the process, coal weight

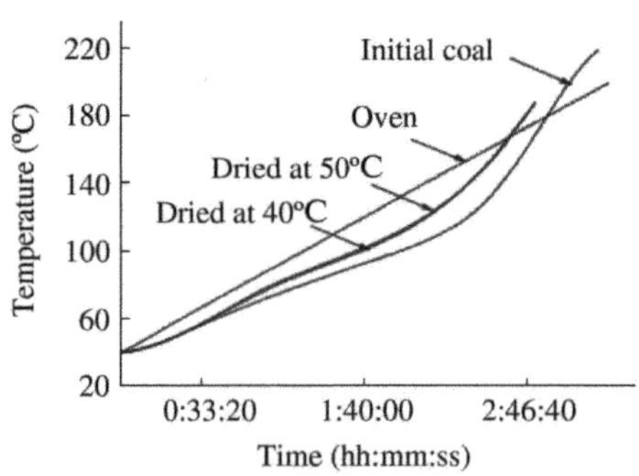

Figure 4.3 Temperature trend curves for CPT measuring method [6].

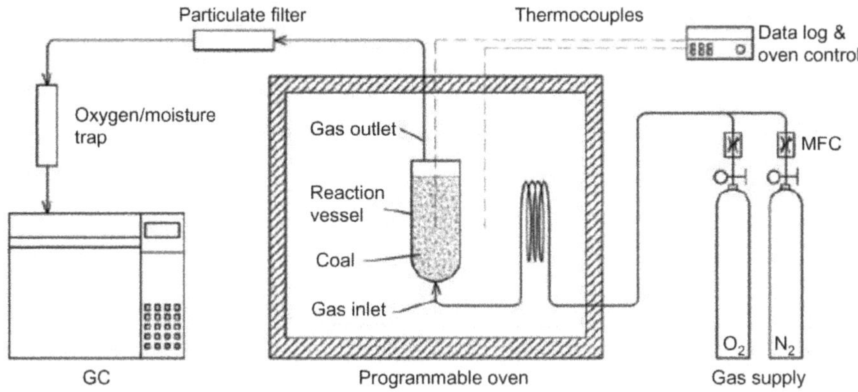

Figure 4.4 Experimental apparatus of low-temperature oxidation [5].

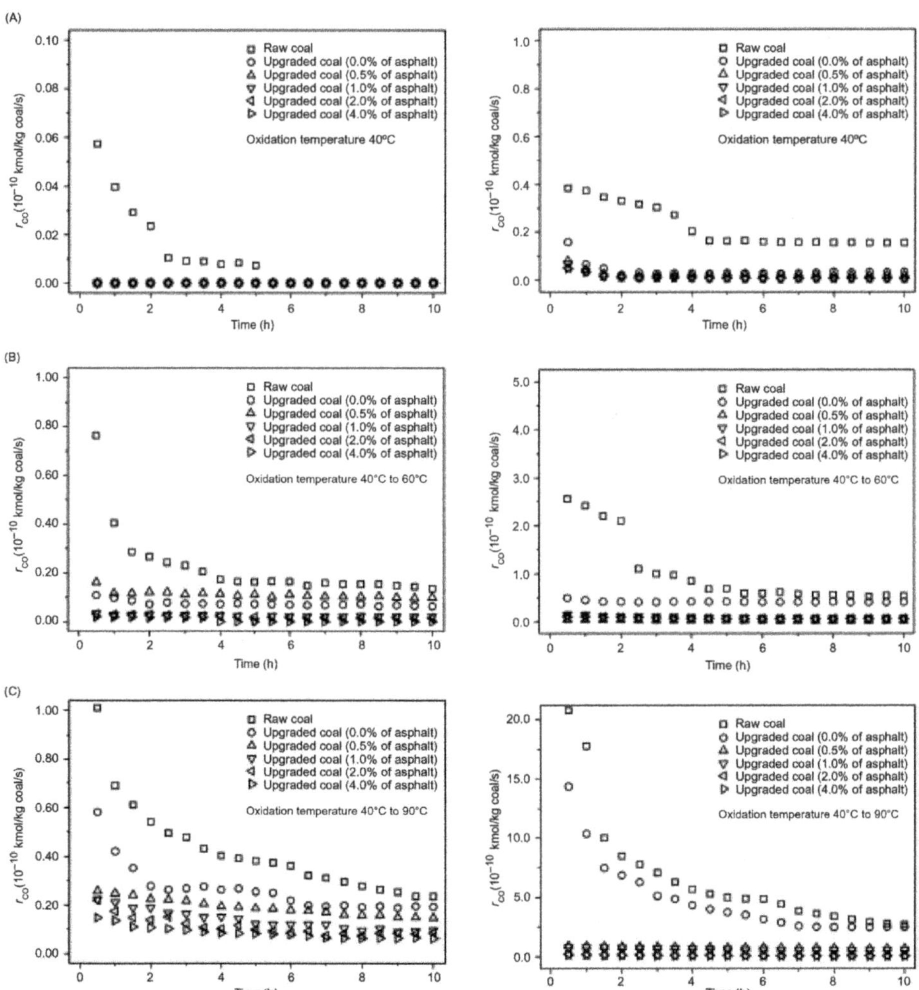

Figure 4.5 Low temperature oxidation results: effect of temperature on the formation of CO_2 (right) and CO (left); (a) 40°C, (b) 60°C, (c) 90°C [5].

increases as it adsorbs oxygen. The tendency of spontaneous combustion is calculated by analyzing activation energy in this stage. The lower the activation energy is, the higher reactivity is, causing spontaneous combustion. This method allows quick analysis with a small quantity of coal.

The above-mentioned methods are some of the ways to analyze the tendency of spontaneous combustion of coal. Since each method has its own characteristics, it is important to understand them and use them properly. To better evaluate the tendency of spontaneous combustion, it is recommended to use several methods together. For instance, CPT is observed at high temperatures over 100°C, so this method cannot analyze the characteristics of spontaneous combustion at low temperatures. In this case, low-temperature oxidation method can be used to get complementary results.

However, since coal is usually piled up out in the open while transported and stored, laboratory analysis alone is not enough to completely prevent or predict the spontaneous combustion of coal. The tendency of spontaneous combustion is affected by various conditions such as surrounding temperature, humidity, air flow rate, and how they are piled up. Therefore, it is important to check the temperature of piled coal to predict or prevent the spontaneous combustion of coal. A study was conducted in a real coal yard to simulate and measure an air flow in tires of piled coal. The aim of the study was to find an area exposed to oxygen and block an air flow. The results of the research were released and they are expected to be widely adopted [11].

4.2.3 Technologies reducing moisture and self-heating properties of low-rank coals

Much attention has been made to upgrading technology of low-rank coals. This technology is largely divided into drying and stabilization technologies. The drying technology uses unused energy such as combustion flue gases or steam from on-site power plants to dry coal. The stabilization technology prevents spontaneous combustion from occurring in the process of stockpiling or transporting of coal after drying it.

Many countries around the world have recently developed coal drying technologies by using waste heat. Among them, DryFining technology and WTA (Wirbelschichttrocknung mit interner Abwärmenutzung) technology are commercialized. What they have in common is that they both use fluidized-bed technology. The DryFining technology uses air fluidized-bed drying technology, and the WTA technology uses steam fluidized-bed drying technology (Fig. 4.6 [12]).

The DryFining technology uses air fluidized-bed drying technology. It collects heat from a cooling tower and increases air temperature to 43°C and uses it as a fluidizing medium. It is reported that this technology has reduced moisture in coal by 9% from 37% to 28%, the use of coal to 14%, increased generating efficiency to 2%–4%, and significantly reduced SOx, NOx, and CO_2.

The WTA technology uses steam fluidized-bed drying technology. It uses waste steam from power plants. This technology reduced moisture in lignite from 50% to

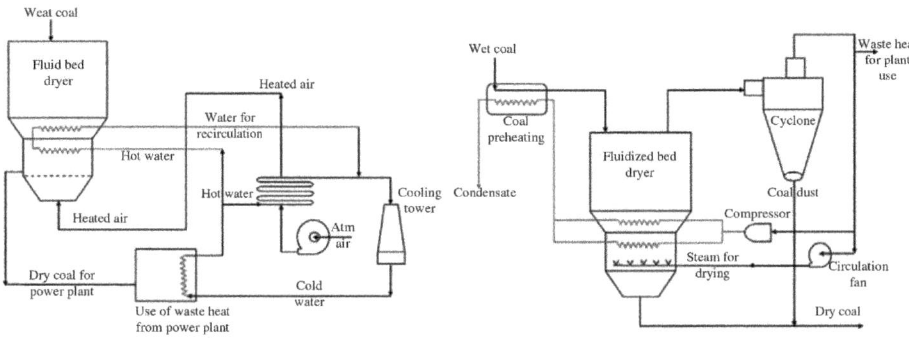

Figure 4.6 Use of plant waste heat for drying coal [12].

12%, increasing lignite combustion boiler efficiency by 4%, from 43% to 47%. When it is used for a 1000 MW power plant, it can reduce approximately 1 million tons of CO_2 annually.

In addition, technologies such as steam tube dryer (STD), pyrolysis technology (GEO Coal), Coldry, GTLE, LiMax, and JCF are under development [12].

COMBDry technology is a hybrid technology combining the advantages of fluidized-bed drying technology and flash drying technology. It uses a vertical column to bring into contact coal and a drying medium through counter current. Several baffle plates attached within the column disrupt drying medium flow, forming turbulent flow, and increasing contact efficiency (Fig. 4.7). It is reported that this technology has drying efficiency of more than 70% with relatively short residence time (3–5 minutes) at a relatively low temperature (<150°C).

The reason why the COMBDry technology has high efficiency with relatively short residence time at a relatively low temperature is that it keeps driving force during drying period (Fig. 4.8). Low-temperature coal particles are injected from the upper part of the column while high-temperature drying media are injected from the lower part of the column. The temperature of coal particles increases gradually as the particles move down through the column while the temperature of the drying media decreases as it moves up through the column. As a result, surface moisture, which is removed at low temperatures, is removed at the upper part of the column and inherent moisture, which is removed at high temperatures, is removed at the lower part of the column.

When dried low-rank coals are directly used in a boiler, they do not require stabilization. However, stockpiling or transporting them requires stabilization due to the risk of spontaneous combustion. Spontaneous combustion refers to a condition in which oxygen in the air is absorbed into coal during mining and transporting of coal, reacting to some of the hydrocarbon in the coal and being oxidized. Low-rank coals include functional groups, most of which are volatile, susceptible to oxygen and highly likely to combust spontaneously.

Stabilization technology reduces the risk of spontaneous combustion in the process of drying, stockpiling, and transporting of coal by using various polymer

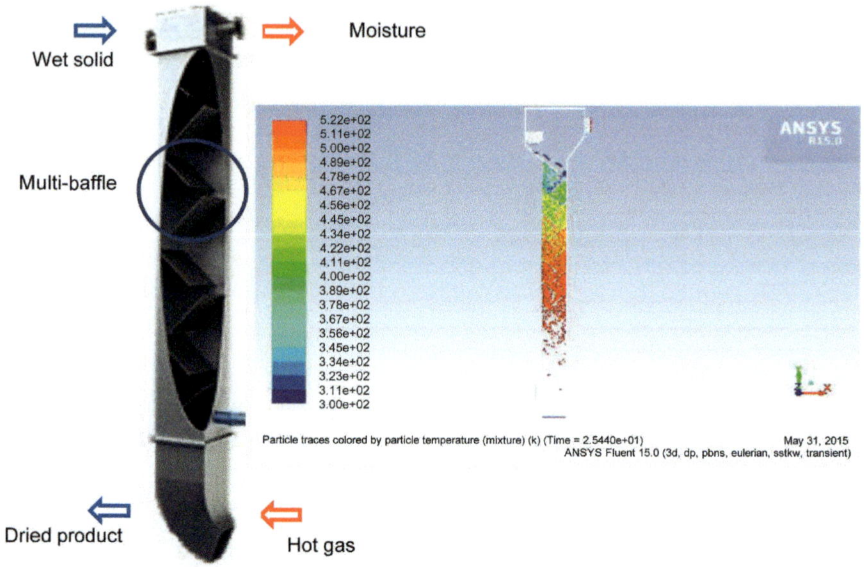

Figure 4.7 Schematic view of COMB technology and CFD results of gas flow [13].

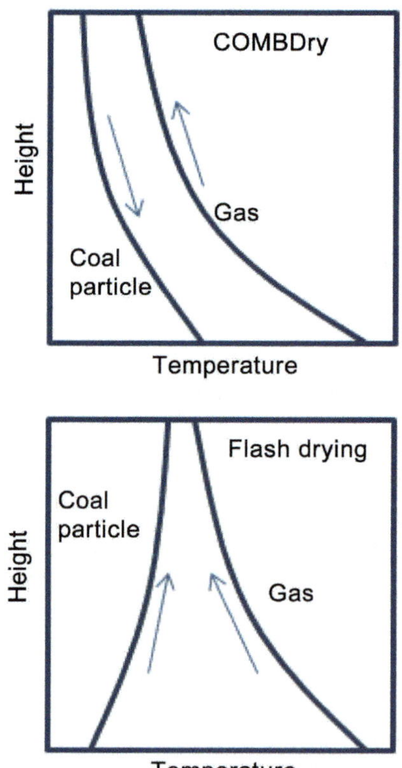

Figure 4.8 Variation of particle and gas temperature during drying with column length [13].

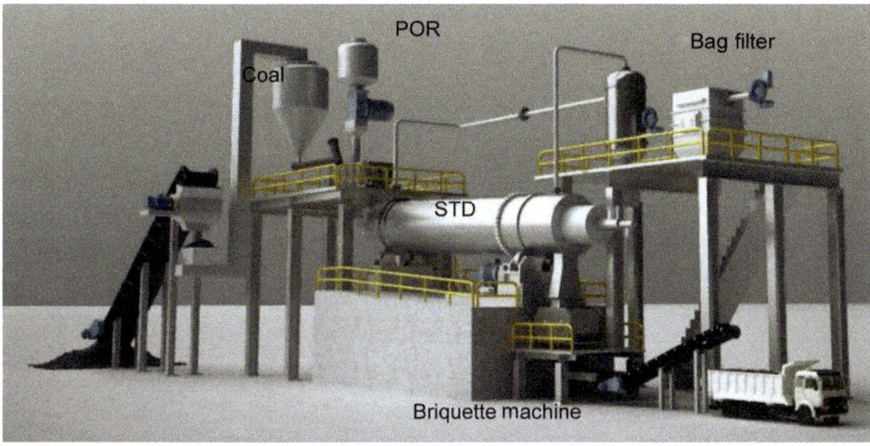

Figure 4.9 Schematic process of coal upgrading by palm oil residue (CUPO) [13].

materials that coat the surface of dried coals. A case in point is UBC (upgraded brown coal) process, developed by a Japanese manufacturer Kobe Steel [14]. It dissolves asphalt into solvent, mixes it with coal, and vaporizes it. About 0.5% of asphalt is used to coat the surface of coal. Indonesia and India are currently working to commercialize the technology.

STD technology uses latent heat and condensation heat in steam at the same time. CUPO (Coal Upgrading by Palm Oil residue) technology is developed by a Korean research team (Fig. 4.9). It uses STD as a drying device and palm residues such as PFAD and PAO as coating materials [13]. The palm residues are solid at room temperature, liquid around 60°C and have low viscosity. They have a high affinity with coal, so they are easily mixed with coal during drying and become solid after drying, serving as a binder. Indonesia plans to commercialize this technology for low-rank coal mines.

Since coal is a major greenhouse gas emitter, many people argue that we should limit the usage of coal. However, coal has been the driving force behind industrialization and contributes to 30% of the world's primary energy, 40% of global power production, and 60% of steel production in the world [15]. Therefore, instead of blaming it for emitting greenhouse gas emissions and trying to limit the usage, we should develop technologies to use it in an efficient way so it can contribute to sustainable development of industries.

4.3 Advanced cleaning of low-rank coals

4.3.1 Advanced cleaning technologies

Coal consists of various kinds of components and impurities that take various forms. For example, N is usually in the form of pyrrolic nitrogen, with some of them in the form of ammonium nitrogen. S is mostly in the form of organic sulfur, with some of them are in the form of iron pyrite, metal sulfides or sulfates, which are mineral types, while the others are elemental sulfur. Cl is mostly in the form of inorganic matter, but some of them are organically bound chloride [16,17].

Metal ingredients in coal take even more various forms. Some of them are in ion form, either being chemically combined with organic matters or being dissolved into inherent moisture in coal structure. Others are in the form of mineral grains, which are interconnected, in the form of silicates, oxides, carbonates, sulfides, sulfates, and phosphates [18].

Ash accounts for the second largest share of impurities following moisture. For this reason, ash is a major target of coal cleaning. Ash in coal is dissolved or volatized in the process of combustion. Some of them are fused within the combustion system, causing slagging, or fouling and others are emitted into the air in the form of fly ash, causing environmental problems. Ash is an obstacle to recently developed coal utilization technologies such as direct carbon fuel cell (DCFC) and direct injection carbon engine (DICE). For this reason, various technologies have been developed to remove ash from coal.

Ash removal methods are largely divided into physical and chemical ones. Physical method uses physical characteristics of coal such as density, electromagnetic property, and surface property to separate ash from organic materials of coal. The physical removal methods are divided into dry process and wet process. Examples of physical ash removal methods include gravity separation, froth flotation, oil agglomeration, magnetic separation, and electro-static separation. While physical ash removal methods are relatively easy to use, they are not that efficient because they cannot completely remove ash, which is chemically combined with organic materials of coal or inside coal matrix [19,20].

Chemical ash removal methods use chemical agents such as aid, alkaline solutions, or organic solvents to selectively remove necessary or unnecessary components from coal. The chemical removal methods are divided into two categories, which are leaching/washing and extraction. Leaching/washing method dissolves ash and removes it from coal. Extraction methods dissolve organic materials and collect them. This chapter introduces some of the highly efficient chemical ash removal processes.

Leaching/washing method was developed long ago. It uses acid or alkaline solutions to directly dissolve ash into coal. This method combines acid solutions such as HNO_3, HCl, H_2SO_4 and HF with alkaline solutions such as NaOH and KOH.

Examples include Bettelle hydrothermal coal process and TRW Gravimelt process. Researchers originally developed these processes to remove sulfur in coal, and later found that this process removed ash, too. CENfuel process removes ash from coal by using HF. CENfuel was designed by Australian researchers. The research began in 1981, and performance tests have been conducted in United States, Europe, and Japan for the past two decades [21–24].

Ultra clean coal (UCC) process was codeveloped by CSIRO and CCUJ. The process consists of caustic digestion and acid soaking. To increase ash removal rate, hydrothermal washing step was added at a high temperature [25]. For filtrate liquor, slaked lime is added to recycle and reuse NaOH. The UCC process has repeated improvements and was able to remove ash content to around 0.1% level in 2008. This process does not require the pulverizing of coal. It increased the efficiency of alkali recycling, which was regarded as the weakest point of alkali leaching process. This process is now almost ready to be commercialized.

Extraction method uses organic solvents to extract organic materials while leaving ash, an inorganic substance. The advantage of using solvent extraction is that it can significantly lower ash content in coal compared to coals produced by acid washing or alkali leaching method. This means, not just extracted coals but also residue coals, which are unextracted, can be used as fuels in power plants. Examples of this process include Hyper-coal process and SinExt process.

Hyper-coal process was codeveloped by AIST, ETRI, Kobe Steel, and JCOAL. The Hyper-coal process development was conducted from 1990 to 2007, resulting in building a facility capable of producing 0.1 ton/day. A validity study to build a production facility for 6000 tons/day was conducted. It is reported that AFCs, where ash content is reduced to 0.02% and sodium content is reduced to 0.5 ppm level, can be produced [26,27].

Since AFCs almost have no particle components and nor emit pollutants after combustion, they can be used for a gas turbine. They are expected to increase power generation efficiency to 48% while reducing CO_2 to 15%. In the meantime, when lignite is used as a raw material for Hyper-coal manufacturing process, the residue coal can be used as a fuel for power generation, with quality similar to that of bituminous coal.

While the price of Hyper-coal, produced at a coal mine, increases by about 2.4 times compared to that of raw coal, its caloric value will be 8000–8700 kcal/kg, 10%–20% higher than that of raw coal. This means that transportation costs and insurance fees become lower based on the same calorific value, which is economically feasible. For coal importers such as Korea and Japan, producing and importing coals from a coal mine can increase generation efficiency and reduce expenses such as ash processing costs and carbon tax. In Japan, Hyper-coal is expected to be used not just as a cofiring fuel and firing fuel at pulverized coal power plants but as a fuel for direct fired gas turbine or additive for coking.

According to the evaluation of a power production system using AFC conducted by CSIRO, energy efficiency was 45% for Hyper-coal-CCGT and 44% for UCC-CCGT system, 40% higher than that of existing supercritical PF [28]. For greenhouse gases, using Hyper-coal as fuel showed about 9% of reduction compared with existing supercritical PF method.

Another ash removal process by using solvent extraction is SinExt process, which has been developed by Korea Institute of Energy Research (KIER) since 2006. This process reduced ash in coal to 0.02% level and used reforming reaction and adsorption mechanism to remove sulfur in coal by more than 90%.

SinExt process refers to a solvent extraction process using a single column extraction reactor (SinExtor) (Fig. 4.10). It conducts solvent extraction and solid liquid separation in the same reactor to simplify the process and reduce heat loss. This process allowed the production of AFC, whose caloric value was over 8000 kcal/kg, and residual coal, whose caloric value was over 6000 kcal/kg, as products from lignite, whose caloric value was less than 4500 kcal/kg, as a raw material (Table 4.1). Performance test and operability tests for this process were conducted in 2016 for a pilot system capable of 1 ton/day for feed coal. Economic evaluation for a production facility capable of 10,000 tons/day was conducted.

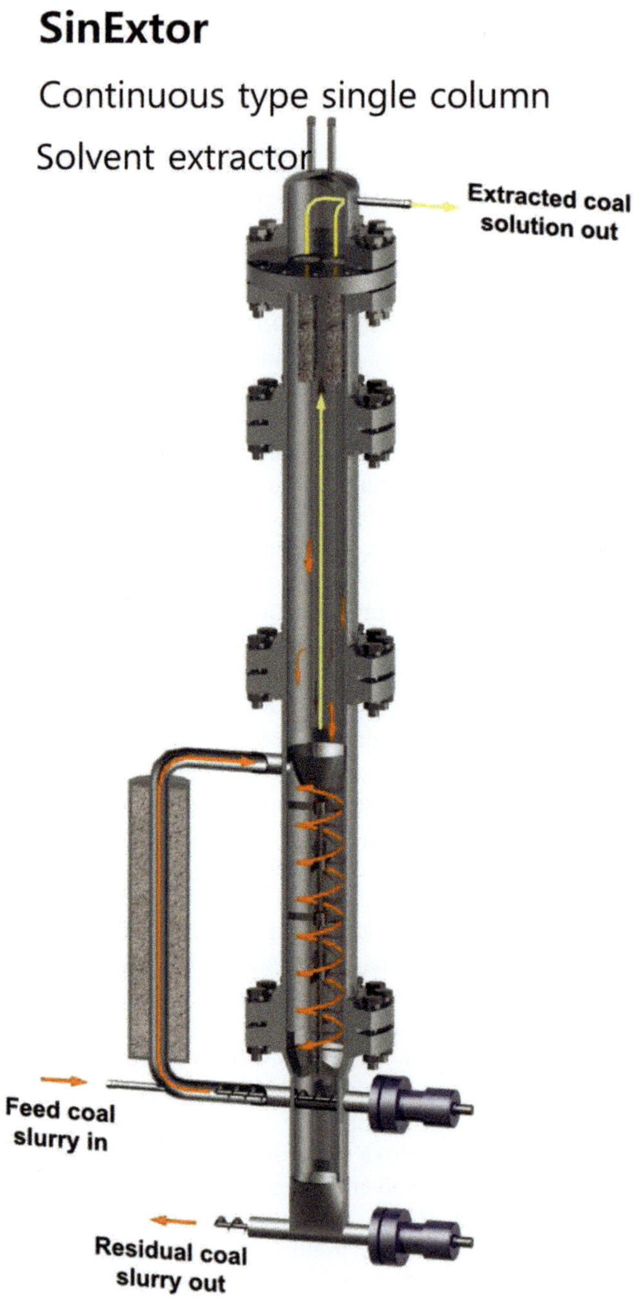

Figure 4.10 SinExtor and SinExt process (1 ton/day).

Table 4.1 Properties of coal produced by SinExt process

Sample	Proximate analysis (wt.%)				Heating value (kcal/kg)
	M	VM	A	FC	
Feed coal	31.25	38.22	3.47	27.06	4240
Extracted coal (Solvent: NMP)	4.45	58.70	0.75	36.10	7228
Residue coal (Solvent: NMP)	6.76	55.61	4.40	33.23	6788
Extracted coal (Solvent: 1MN)	0.65	68.26	0.02	31.07	8280
Residue coal (Solvent: 1MN)	4.84	35.70	9.52	49.94	6010

AFC produced through this process can be used as fuel not only for combustion process but also for gasification, DCFC, and DICE [29–31]. Studies are conducted to use it with a coke binder or carbon materials, which will be discussed in the next session.

4.3.2 Utilization of cleaned coal: application and utilization of AFC

Ash, an inorganic substances found in coal, reduces the efficiency of power generation, and is emitted into the air as pollutants. In addition, ash prevents coal from being used as a raw material of material chemistry. For this reason, researchers have made efforts to remove ash from coal [32–34]. If AFCs are commercialized, they can be used for various purposes, including as a feedstock of catalytic steam gasification, fuel for DCFC, a binder for coking coal manufacturing, fuel for DICE, and an anode substance of electrochemical device designed to smelt aluminum.

Coals account for more than 60% of electricity generation in the world, emitting pollutants which are harmful to the environment and human health. They account for 35% of total GHG emissions, causing global warming. IGCC is based on the gasification of coal. It boasts high efficiency with lower CO_2 emissions. It will help us switch from coal to pollution-free new renewable energy. Switching to syngas (H_2 and CO) is one of the most important processes for the IGCC system. It is high in hydrogen and environmentally friendly, so it is regarded as a next generation fuel source. Most gasification processes are run around 1400°C and 20–70 atm due to the rate of gasification and slagging issue. Such high-temperature and high-pressure conditions require huge capital investments. In addition, oxygen, which is injected to supply a source of heat, lowers conversion efficiency.

While gasification is feasible by adding catalysts under mild conditions, catalytic gasification is not yet commercialized. This is because catalysts react with ash in an irreversible way and are deactivated. In this regard, it is advised to produce AFCs and use them as a feedstock to continuously use catalysts. Many studies have been conducted on this.

Catalytic coal gasification has been studied since the 1970s [35,36]. The kinetics of coal to gas conversion can be accelerated with the introduction of alkali, alkali earth, and Ni/Fe-based catalysts. Potassium carbonate shows the greatest performance when it is applied to solid carbonaceous resources. Its catalytic activity is strong, and it has fluidity when it is mixed with solid coals. This means that there are no mass transport limitations [37].

AFCs were produced for various ranks of coals including lignite, subbituminous, bituminous, and coking coal to conduct steam gasification and compare their activity with parent raw coals. For AFCs, which was produced with thermal extraction method by using 1-methylnaphthalene, a nonpolar solvent, there was no big difference depending on the types of parent raw coals. All had similar reactivity as shown in Fig. 4.11. The steam gasification reaction of AFCs and CO to CO_2 conversion were increased significantly with K_2CO_3 [38].

Researchers from AIST also produced Hyper-coal (ashless coal), whose ash content was less than 0.05 wt%, to conduct the gasification of catalyst by using K_2CO_3. Its gasification rate was increased by almost four times compared to that of raw coals. The catalyst was recycled four times, without noticeable loss in activity [39–41].

Efforts have been made to ensure efficient use of carbon fuels. Among them, DCFC has been the focus of studies for the past decade [42–44]. DCFC uses solid carbon as fuel instead of gas-phase fuels such as H_2 and CO. It is drawing a lot of attention as a next generation power generation method due to its high energy efficiency and easy CO_2 capture system based on the occurrence of pure CO_2. It allows the utilization of various carbon materials such as coal, biomass, solid waste as well as production of syngas (H_2 and CO) through the internal reforming of these materials. It shows these materials can be used as fuel. However, since most carbon fuels

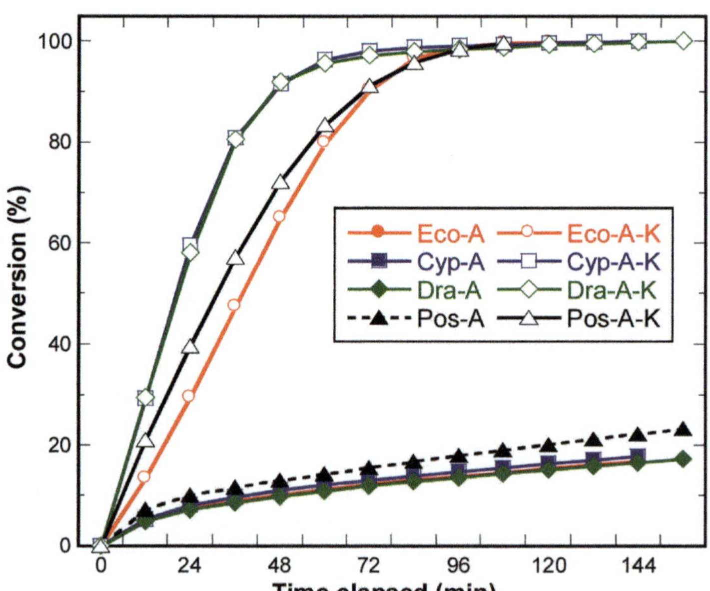

Figure 4.11 Effect of K_2CO_3 (K) catalyst on steam gasification of various AFCs at 750°C.

include ash and do not engage in electrochemical reactions, they remain in electrodes after reaction and disrupting the contact between carbon fuels and electrodes [45].

The comparison of DCFC performance between AFCs and coal including ash showed that although low ash content had little impact on power density, long-term accumulation of the ash content will increase the content and end up having fatal impacts on the power density (Fig. 4.12 [46]). Therefore, using coals including ash as fuel requires a very complex process where ash should be removed in the middle of DCFC operation to ensure the sustainable use of coal. Using ash-free fuel can remove the inconvenience. In particular, AFCs, which are inexpensive and widely available, can bring about many benefits.

Many studies have been conducted to use AFCs as fuel [47–51]. A research team in Poland used Hyper-coal, AFCs produced by a Japanese manufacturer Kobe steel, as fuel to operate DCFC. They confirmed that the maximum power density was over 80 mW/cm^2 at 800°C. While its performance was only 30% compared to that of the existing hydrogen fuel-based SOFC, the performance is expected to be increased by developing efficient electrode contact methods and optimizing solid fuels.

Using coal for Iron smelting accounts for the second largest coal consumption following thermal power generation. Recently, the price of coking coal is on the rise as its reserves decrease and demands for steel increase. In addition, it is currently in short supply due to peripheral production of the resource. Against the backdrop, efforts have been made to mix it with inexpensive raw materials like low-rank coals or use a binder with relatively inexpensive raw materials to ensure coke quality [52,53]. It is expected that if a binder with reasonable quality and price is developed, it will replace huge amounts of coking coals with relatively inexpensive ones.

Many researchers currently conduct studies to use thermally extracted AFCs as a binder to produce coking coals [54–58]. It is expected that AFCs can be produced

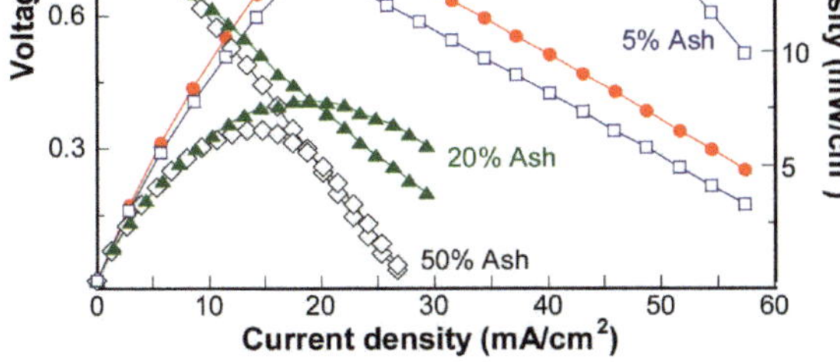

Figure 4.12 Effect of 5%, 20%, and 50% fly ash on the performance of AFC-powered DCFC at 950°C under N_2.

by all ranks of coals under reasonable price, which can be further lowered when inexpensive low-lank coals are used.

The above-mentioned coal extraction processes usually use organic solvents at high temperature over 300°C to dissolve carbonaceous ingredients in coal. Then, AFCs are produced through separation and drying. The quality of AFCs can be changed depending on the polarity of solvents or the character of raw coals. It can have high-quality thermoplasticity and fluidity as a binder.

Efforts to develop engine operation technology by using coal have been made since the early 1990s with coal dusts [59,60]. Such efforts have been made continuously with coal–water mixture and coal upgrading technologies. These efforts have been made thanks to the abundance of coal and anticipated financial gains, especially when oil prices were high. Recently, CSIRO formed a research-oriented consortium DICE-net, where researchers from Australia, Japan, Indonesia, and China joined to develop DICE technology. They focus on gaining economic benefits as well as minimizing environmental impacts by promoting the efficient use of coal with the relatively high efficiency of DICE [61]. To use coal as DICE fuel, using resources, which are low in ash content, is must. This addresses plugging, which is caused by accumulated corrosion and ash in engine room. Considering ash content affects MRC (micronized refined carbon) production costs, AFCs, whose ash content is less than 200 ppm, can be the best option as fuel. While DICE test, where AFCs are used as fuel, has not been conducted, necessary jobs are done to prepare the test.

CSIRO started the DICE project in 2011 and currently works on it to mix solid carbon fuel with water to directly spray it to a diesel engine to run the engine. It provides high thermal efficiency over 50% when applied to a low-speed diesel engine, and limits CO_2 emissions by 20%–50% compared to coal thermal power generation using low-rank coals. Another advantage of the DICE technology is that it can save time and money for commercialization because it uses already proven diesel engine technology, allowing users to promptly respond to power demand.

Aluminum metals are smelted by electrical smelting process using Hall-Heroult electrochemical cells. Anode is made by carbonizing petroleum coke, where small amount of coal tar pitch is added as a binder. Anode coke consumes huge amounts of energy. There has been short supply of petroleum coke, increasing the need for alternative substance. Impurities containing ash in anode serve as a catalyst for oxidation and CO_2, shortening its life [62]. In addition, durability should be ensured against thermal shock, fatigue, and mechanical damage considering a high temperature above 1000°C.

AFCs, where ash was removed, are expected to be very resistant to oxidation response. In addition, it is believed that the use of coal including specific raw materials will allow the production of anode coke with reasonable quality. Researchers from University of Kentucky use mild solvent extraction process to produce AFCs and apply it with an anode substance. The extracted coal satisfies the specification of existing anode coke. Although there are not enough study results regarding this, it is believed that the development of anode coke using AFCs will be activated as demands for aluminum anode coke increase.

4.4 Conclusion

Since coal is blamed for a large part of greenhouse gas emissions, many argue that we should reduce the use of coal. We can reduce the coal consumption by upgrading coal, achieving sustainable industrial development. To this end, this chapter contemplated the efficient use of low-rank coals, which are abundant but has been untapped due to its high moisture content and the risk of spontaneous combustion. In particular, moisture drying technology and spontaneous combustion reduction technology were examined as utilization methods of low-rank coals. For coal utilization technologies, recent trends in technological development on the production of AFC and its efficient use were introduced.

References

[1] Z. Rao, Y. Zhao, C. Huang, C. Duan, J. He, Recent developments in drying and dewatering for low rank coals, Progress Energy Combust. Sci. 46 (2015) 1–11.
[2] W. Xia, G. Xie, Y. Peng, Recent advances in beneficiation for low rank coals, Powder Technol. 277 (2015) 206–221.
[3] Y. Liu, T.M. Bisson, H. Yang, Z. Xu, Recent developments in novel sorbents for flue gas clean up, Fuel Process. Technol. 91 (2010) 1175–1197.
[4] G.B. Stracher, A. Prakash, E.V. Sokol, Spontaneous combustion in open-cut coal mines; Australian Experience and Research, Chapter 1 Coal and Peat Fires; A Global Perspective, vol. 3, Elsevier, Amsterdam, 2015, pp. 1–36
[5] H. Choi, C. Thiruppathiraja, S. Kim, Y. Rhim, J. Lim, S. Lee, Moisture readsorption and low temperature oxidation characteristics of upgraded low rank coal, Fuel Process. Technol. 92 (2011) 2005–2010.
[6] Q. Xuyao, D. Wang, J.A. Milke, X. Zhong, Crossing point temperature of coal, Mining Sci. Technol. (China) 21 (2011) 255–260.
[7] K.K. Feng, R.N. Chakravorty and T.S. Cochrane, Spontaneous combustion a coal mining hazard, in: The Canadian Mining and Metallurgical (CIM) Bulletin, UK, 75-84, October, 1973, pp. 75–84.
[8] D.T. Umar, B. Daulay, H. Usui, T. Deguchi, S. Sugita, Characterization of upgraded brown coal (UBC), Coal Preparation 25 (2005) 31–45.
[9] V. Zubicek, Assessment of susceptibility of coal to spontaneous combustion in OKR, Geo Sci. Eng. LIV (2008) 1–9.
[10] X.Y. Wang and Y. Luo, Experimental studies of spontaneous combustion propensity of coal, in: 2011 International Pittsburgh Coal Conference, 12–15 September, 2011.
[11] B. Taraba, Z. Michalec, V. Michalcova, T. Blejchar, M. Bojko, M. Kozubkova, CFD simulation of the effect of wind on the spontaneous heating of coal stockpiles, Fuel 118 (2014) 107–112.
[12] N. Nikolopoulos, I. Violidakis, E. Karampinis, M. Agraniotis, C. Bergins, P. Grammelis, et al., Report on comparison among current industrial scale lignite drying technologies (a critical review of current technologies), Fuel 155 (2015) 86–114.
[13] S. Lee, S. Kim, D. Chun, Y. Rhim, J. Yoo, H. Choi, et al., Low rank coal drying and upgrading technologies: COMBDry™ and CUPO™, in: 10th Korea-China Clean Energy Workshop, 31 August–3 September, Seoul Korea 2014.

[14] S. Kinoshita, S. Yamamoto, T. Deguchi, T. Shigehisa, Demonstration of upgraded brown coal (UBC) process by 600 tonnes/day plant, Kobelco Technol. Rev. 29 (2010) 93–98.

[15] B. Sporton, Coal-building on strength, in: World Clean Coal Week China Focus, 23–24 October, 2013, Beijing, China.

[16] W.H. Calkins, The chemical forms of sulfur in coal: a review, Fuel 73 (4) (1994) 475–484.

[17] G. Domazetis, M. Raoarun, B.D. James, J. Liesegang, P.J. Pigram, N. Brack, et al., Analytical and Characterization Studies of Organic and Inorganic Species in Brown Coal, Energy Fuels 20 (2006) 1556–1564.

[18] N. Wijaya, A. Zhang, Critical review of coal demineralization and its implication on understanding the speciation of organically bound metals and submicrometer mineral grains in coal, Energy Fuels 25 (2011) 1–16.

[19] H. Katalambula, R. Gupta, Low-grade coals: a review of some prospective upgrading technologies, Energy Fuels 23 (2009) 3392–3405.

[20] P. Meshram, B.K. Purohit, M.K. Sinha, S.K. Sahu, B.D. Pandey, Demineralization of low grade coal—a review, Renew Sustain Energy Rev 41 (2015) 745–761.

[21] S.K. Kawatra, T.C. Eisele, Aqueous Base Processing in Coal Desulfurization: High-Efficiency Preparation Methods, 325p, Taylor and Francic, Inc., London, 2001.

[22] E.P. Stambaugh, J.F. Miller, H.N. Conkle, E.J. Mezey, R.K. Smith, Project summary. Process Improvement Studies on the Battelle Hydrothermal Coal Process, EPA/600/S7-85/023, 1985.

[23] R.A. Meyers, L.C. Mc Clanathan, W.D. Hart, TRW Gravimelt Process for desulfurization and demineralization of coal, in: Proceedings of American Institute of Chemical Engineers Annual Winter Meeting, Chicago, IL, USA, 10 November 1985.

[24] A.R. Ansari, Lignite upgradation and utilization, in: International Seminar on Mineral Processing Technology (MPT-2005), 6–8 January, 2005.

[25] L. Wibberley, D. Palfreyman, P. Scaife, Efficient use of coal water fuels, Technology Assessment Report 74, CSIRO, 2008.

[26] I. Saito, Current status and perspective of research and development on coal utilization technology in Japan, in: 20th International Pittsburgh Coal Conference.

[27] N. Okuyama, A. Furuya, N. Komatsu, T. Shigehisa, Development of a hyper-coal process to produce ash-free coal (in Japanese), Kobe Steel Engineering Reports, Vol. 56, No. 2, 2006.

[28] CSIRO, System assessment of Hypercoal for electricity supply in Japan, Technology Assessment Report 65, 2007.

[29] Y.J. Kong, J.K. Kim, D.H. Chun, S.H. Lee, Y.J. Rhim, J.H. Lim, et al., Comparative studies on steam gasification of ash-free coals and their original raw coals, Int. J. Hydrogen Energy 39 (2014) 9212–9220.

[30] S.H. Park, S.W. Chung, S.K. Lee, H.K. Choi, S.H. Lee, Thermo-economic evaluation of 300 MW class integrated gasification combined cycle with ash free coal (AFC) process, Appl. Thermal Eng. 89 (2015) 843–852.

[31] J.P. Kim, H.K. Choi, Y.J. Chang, C.H. Jeon, Feasibility of using ash-free coal in a solid-oxide-electrolyte direct carbon fuel cell, Int. J. Hydrogen Energy 37 (2012) 11401–11408.

[32] K. Miura, M. Shimada, K. Mae, H.Y. Sock, Extraction of coal below 350 ºC in flowing non-polar solvent, Fuel 80 (2001) 1573–1582.

[33] H. Shui, Z. Wang, G. Wang, Effect of hydrothermal treatment on the extraction of coal in the CS2/NMP mixed solvent, Fuel 85 (12–13) (2006) 1798–1802.

[34] C. Li, T. Takanohashi, T. Yoshida, I. Saito, H. Aoki, K. Mashimo, Effect of acid treatment on thermal extraction yield in ashless coal production, Fuel 83 (6) (2004) 727–732.
[35] Y. Ohtsuka, K. Asarni, Highly active catalysts from inexpensive raw materials for coal gasification, Catal. Today 39 (1997) 111–125.
[36] J. Corella, J. Toledo, G. Molina, Steam gasification of coal at low-medium (600-800°C) temperature with simultaneous CO2 capture in fluidized bed at atmospheric pressure, Ind. Eng. Chem. Res. 45 (2006) 6137–6146.
[37] Y. Nishiyama, Catalytic gasification of coals: features and possibilities, Fuel Process Technol. 29 (1991) 31–42.
[38] Y.J. Kong, J.K. Kim, D.H. Chun, S.H. Lee, Y.J. Rhim, J.H. Lim, et al., Comparative studies on steam gasification of ash-free coals and their original raw coals, Int. J. Hydrogen Energy 39 (2014) 9212–9220.
[39] A. Sharma, T. Takanohashi, K. Morishita, T. Takarada, I. Saito, Low temperature catalytic steam gasification of HyperCoal to produce H2 and synthesis gas, Fuel 87 (2008) 491–497.
[40] A. Sharma, T. Takanohashi, I. Saito, Effect of catalyst addition on gasification reactivity of HyperCoal and coal with steam at 775–700°C, Fuel 87 (2008) 2686–2690.
[41] J. Wang, K. Sakanishi, I. Saito, High-yield hydrogen production by steam gasification of HyperCoal (ash-free coal extract) with potassium carbonate: comparison with raw coal, Energy Fuels 19 (2005) 2114–2120.
[42] S. Giddey, S. Badwal, A. Kulkarni, C. Munnings, A comprehensive review of direct carbon fuel cell technology, Progress Energy Combust. Sci. 38 (2012) 360–399.
[43] X. Li, Z. Zhu, R. De Marco, J. Bradley, A. Dicks, Modification of coal as a fuel for the direct carbon fuel cell, J. Phys. Chem. A 114 (2010) 3855–3862.
[44] B. Yu, Y. Zhao, Y. Li, A SnO2-samarium doped ceria additional anode layer in a direct carbon fuel cell, J. Power Sources 306 (2016) 387–393.
[45] T. Gür, M. Homel, A. Virkar, High performance solid oxide fuel cell operating on dry gasified coal, J. Power Sources 195 (2010) 1085–1090.
[46] I. Lee, S. Jin, D. Chun, H. Choi, S. Lee, K. Lee, et al., Ash-free coal as fuel for direct carbon fuel cell, Sci. China Chem. 57 (2014) 1010–1018.
[47] H. Ju, J. Eom, J. Lee, H. Choi, T. Lim, R. Song, et al., Durable power performance of a direct ash-free coal fuel cell, Electrochim. Acta 115 (2014) 511–517.
[48] M. Dudek, P. Tomczyk, R. Socha, M. Hamaguchi, Use of ash-free "Hyper-coal" as a fuel for a direct carbon fuel cell with solid oxide electrolyte, Int J. Hydrogen Energy 39 (2014) 12386–12394.
[49] J. Kim, H. Choi, Y. Chang, C. Jeon, Feasibility of using ash-free coal in a solid-oxide-electrolyte direct carbon fuel cell, Int. J. Hydrogen Energy 37 (2012) 11401–11408.
[50] N. Wijaya, L. Zhang, A critical review of coal demineralization and its implication on understanding the speciation of organically bound metals and submicrometer mineral grains in coal, Energy Fuels 25 (2011) 1–16.
[51] A. Rady, S. Giddey, A. Kulkarni, S. Badwal, S. Bhattacharya, B. Ladewig, Direct carbon fuel cell operation on brown coal, Appl. Energy 120 (2014) 56–64.
[52] H. Shui, C. Lin, M. Zhang, Z. Wang, M. Zheng, Comparison of the associative structure of two different types of rich coals and their coking properties, Fuel 89 (2010) 1647–1653.
[53] A. Benk, M. Talu, A. Coban, Phenolic resin binder for the production of metallurgical quality briquettes from coke breeze: Part I, Fuel Process. Technol. 89 (2008) 28–37.

[54] L. Hao, P. Feng, W. Song, W. Lin, S. Yoon, I. Mochida, Modification performance of Hypercoal as an additive on co-carbonization of coal, J. Fuel Chem. Technol. 40 (2012) 1025–1031.
[55] T. Takanohashi, T. Shishido, I. Saito, Caking property of Hypercoal blending samples, in: Australian Institute of Energy National Conference, paper 4B4, (2006).
[56] T. Takanohashi, T. Shishido, I. Saito, Effects of HyperCoal addition on coke strength and thermoplasticity of coal blends, Energy Fuels 22 (2008) 1779–1783.
[57] H. Wachowska, M. Kozlowski, The influence of extraction with organic solvents on coking properties of raw and reduced coals, Fuel 75 (1996) 517–521.
[58] C. Chang, T. Whang, D. Huang, D. Wang, S. Tsai, M. Hung, Thermoplasticity and strength improvement of coking coal by addition of coal extracts, Fuel 117 (2014) 364–371.
[59] Wilson R.P. (2007) Clean coal diesel demonstration project. DE-FC21-94MC31260, Cambridge, MA, TIAX, 175 pp (March 2007).
[60] K. Nicol, The direct injection carbon engine, IEA Clean Coal Centre, IEACCC Ref: CCC/243, December 2014, ISBN 978-92-9029-565-5.
[61] L. Wibberley, DICE—the only option left for coal, Energy Generat. (2014) 23–27, January–April.
[62] F.S. Tong, Carbon Materials Production Question-Answering, Metallurgical Industry Press, Beijing, 1991.

Part II

Low-rank coal combustion, gasification, and pyrolysis

High-efficiency pulverized coal power generation using low-rank coals

5

Michalis Agraniotis, Christian Bergins, Malgorzata Stein-Cichoszewska and Emmanouil Kakaras
Mitsubishi Hitachi Power Systems Europe, Duisburg, Germany

5.1 Introduction

Low-rank coals such as lignites are without doubt one of the most abundant fossil fuel sources worldwide [1,2]. Due to their usually high moisture content and low heating value, the combustion of low-rank coals leads to higher specific CO_2 emissions compared to the combustion of other fuels like natural gas. Nevertheless, by using state of the art technology to reach the highest possible plant efficiency, specific CO_2 emissions can be considerably reduced, even when utilizing low-rank coals of extremely poor quality. Moreover by implementing state of the art Air Quality Control Systems (AQCS) flue gas emissions including CO, NOx, SO_2, and dust are reduced, so that the strictest European environmental limitations can be kept [3,4]. In this sense, using state of the art technologies for the utilization of low-rank coals may contribute to an affordable, cost competitive, and environmentally friendly electricity production for every type of low-rank coal to be utilized.

In the present chapter, a review of state of the art technology in pulverized fuel fired power plants is presented. Pulverized fuel firing is the mostly applicable firing technology in large scale, utility power plants. Hence, grate firing or fluidized-bed firing systems are not part of this review. Thermal design aspects, which are significant for achieving state of the art efficiency rates, are analyzed in Section 5.2, Thermal design aspects. In Section 5.3, Firing arrangements for lignite power plants, the firing systems that are suitable for a broad range of fuel qualities are presented and compared. Lignite predrying technologies and novel power plant concepts toward increasing flexibility in lignite power plants are presented in Section 5.4, Lignite predrying and Power Plant concepts. Technologies for pollution abatement and air quality control are discussed in Section 5.5, State of the art AQCS. Finally, a technology outlook is discussed in Section 5.6, Conclusions—technology outlook.

Low-Rank Coals for Power Generation, Fuel and Chemical Production
DOI: http://dx.doi.org/10.1016/B978-0-08-100895-9.00005-X

5.2 Thermal design aspects

5.2.1 State of the art ultra supercritical (USC) thermal cycles for lignite power plants

Although modern lignite power plants have a similar design and are comprised of the same basic components as with the plants commissioned decades ago, considerable improvements in all main components including boiler, turbine, and AQCS have been attained in the recent years. Through these development efforts, a considerable increase of the net plant efficiency and of their environmental performance has been achieved. As a highlight it can be stated that the today's state of the art net plant efficiency for a modern lignite boiler (42%–44%) [4], compared with the relevant efficiency of a similar plant in the 1980s (33%), indicates a relative improvement of efficiency for more than 30%.

A schematic diagram with all basic systems present in a typical lignite fired power plant is presented in Fig. 5.1. Following systems are depicted:

- air preheating system,
- water/steam cycle,
- cooling water cycle,
- lignite pretreatment including flue gas recirculation, milling, and drying system, and
- ash handling system.

In addition, the electrostatic precipitator (ESP) and the flue gas desulfurization (FGD) are also depicted as main parts of the AQCS.

5.2.1.1 Water/steam cycle

Regarding the water–steam cycle, today's state of the art live steam parameters in hard coal and lignite power plants are presented in Fig. 5.2. The increase of steam parameters has been made possible through the development of a new generation of austenitic steels. This in turn leads to state of the art net plant efficiency of about 43% in lignite power plants and of 46% in hard coal power plants [5].

5.2.1.2 Air/flue gas cycle

Apart from increasing steam temperatures, optimizing heat utilization of flue gas after economizer is also in focus of development of state of the art lignite boilers. The reason is that due to the high flue gas volume flow, the waste heat of the flue gases cannot be fully utilized for combustion air preheating in the air preheater. Two additional measures can be usually followed, in order to improve boiler efficiency:

1. Utilizing available flue gas heat from a LUvo Bypass after ECOnomizer (LUBECO), in order to preheat feed water at low-temperature and high-temperature levels.
2. Utilizing available flue gas heat before the FGD unit, in order to preheat the combustion air, by means of a dedicated heat recovery heat exchanger.

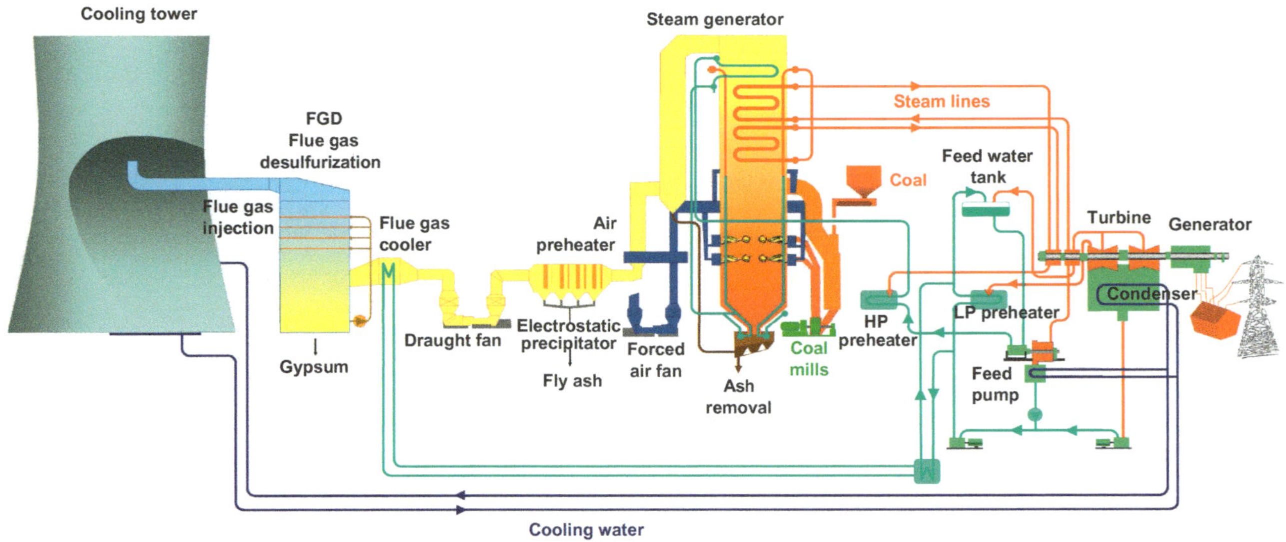

Figure 5.1 State of the art steam cycle with USC steam temperatures.

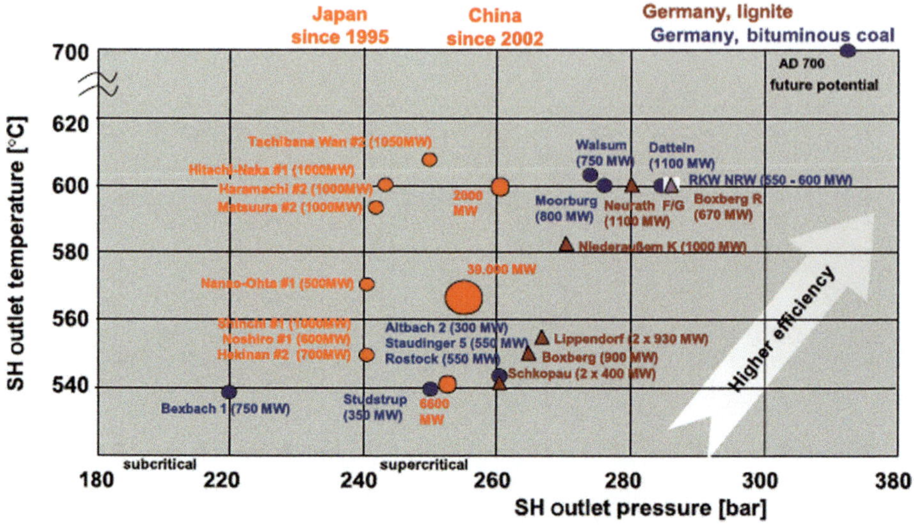

Figure 5.2 Evolution of steam temperatures and efficiencies' evolution.

The concept presented in (2) has been followed in unit R of Boxberg power plant [6], while a combination of both concepts has been adopted in the 1000 MW unit K of Niederaussem power plant [7] and also in a new unit currently under construction, where extremely low-quality lignite is utilized [8,9] (Fig. 5.3).

5.2.2 Boiler design

A typical pulverized coal boiler consists of (1) the furnace, which is the lower boiler part, where the burners and the firing system are installed and where the combustion takes place and (2) the convective section, where the convective heat exchanger surfaces are installed. The boiler membrane walls have usually the function of the boiler evaporator. The heat exchangers sections are placed in the convective part. In a typical lignite boiler design, the heat exchangers are placed in the below order following the flue gas path [10].

- Superheater 2 (as a hanging super heater),
- Superheater 4,
- Reheater 2,
- Superheater 3,
- Reheater 1,
- Economizer 2,
- Economizer 1.

The boiler sketch of hard coal and one lignite boiler from two different 1100 MWe power plants is presented in Fig. 5.4, in order to visualize the difference in the design and size of a typical hard coal and lignite boiler [11,12].

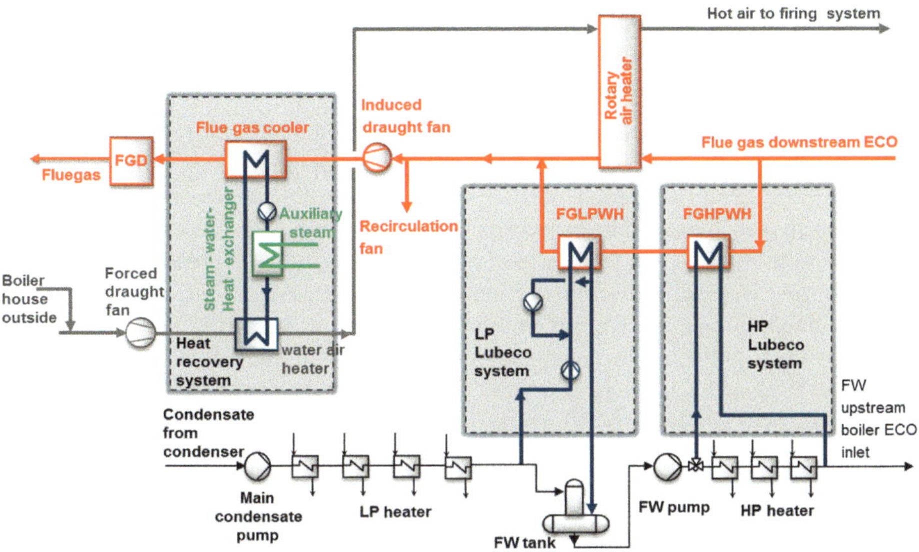

Figure 5.3 Combined system of high and low pressure of Luvo Bypass Economizer (Lubeco) and waste heat recovery heat exchanger.
Source: New build projects of coal-fired Power-Plants in Poland & Greece—Technology and latest experiences in execution, W. Schreier, VGB conference, September 2015 [9].

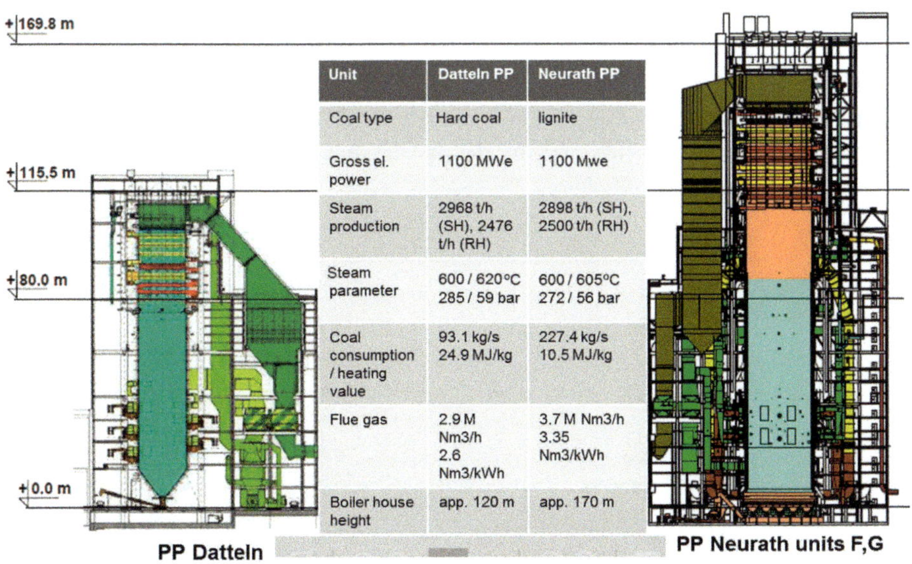

Unit	Datteln PP	Neurath PP
Coal type	Hard coal	lignite
Gross el. power	1100 MWe	1100 Mwe
Steam production	2968 t/h (SH), 2476 t/h (RH)	2898 t/h (SH), 2500 t/h (RH)
Steam parameter	600 / 620°C 285 / 59 bar	600 / 605°C 272 / 56 bar
Coal consumption / heating value	93.1 kg/s 24.9 MJ/kg	227.4 kg/s 10.5 MJ/kg
Flue gas	2.9 M Nm3/h 2.6 Nm3/kWh	3.7 M Nm3/h 3.35 Nm3/kWh
Boiler house height	app. 120 m	app. 170 m

Figure 5.4 Comparison of main technical parameters for a 1100 MWe hard coal and lignite boiler.
Source: Effiziente Kraftwerkstechnik in NRW zur umweltfreundlichen Stromerzeugung aus Kohle, K. Busekrus, 4th TÜV NORD Kraftwerksymposium, 2008 [11].

5.3 Firing arrangements for lignite power plants

5.3.1 Lignite characteristics influencing power plant design

Solid fuels can be characterized by their degree of carbonization. In Fig. 5.5 different solid fuels are presented in a Van Krevelen diagram. Based on this figure, the atomic H/C and O/C ratios are high for fuels with low degree of carbonization such as peat and lignite. The respective ratios decrease with increasing degree of carbonization. In Fig. 5.6 a classification of solid fuels is presented based on their heating value and their volatile matter content. Based on our experience with different fuels in recent years for the design of boilers and firing systems, several areas are identified that are characteristic for the main solid fuel types. The net calorific value varies between 3.5 and 33 MJ/kg and the range of volatile content

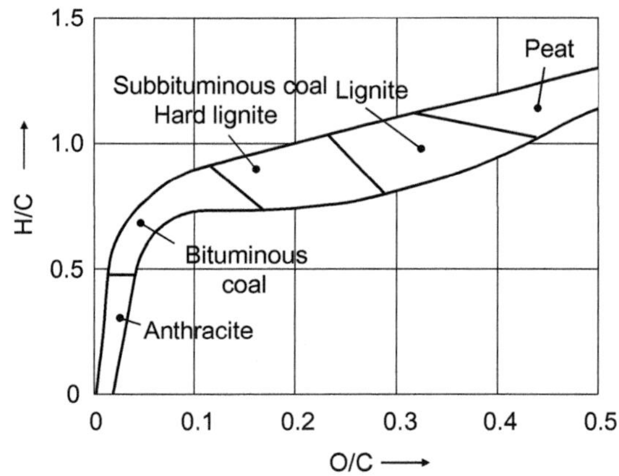

Figure 5.5 Solid fuels characterization by H/C and O/C atomic ratios in Van Krevelen diagram.

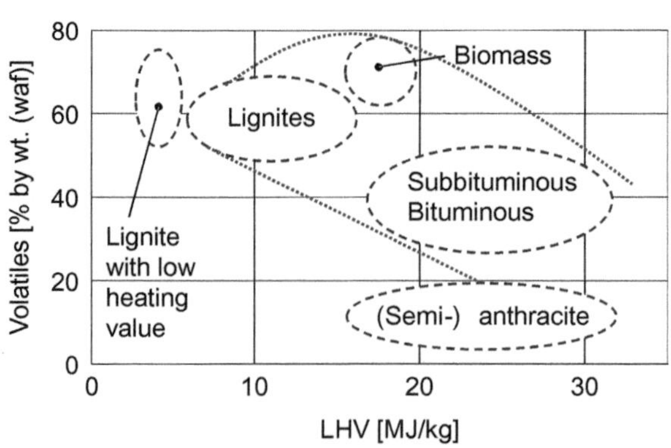

Figure 5.6 Characterization of solid fuels by heating value and volatile content.

is 4%–70% by wt. (dry ash free). Based on our experience, biomass fuels with a volatile content of more than 70% by weight can be also utilized. Moreover, other coal characteristics, such as grindability, devolatilization behavior, char reactivity, and ash composition, have to be considered for the design of adequate firing systems. The dotted line in the figure marks a range of fuels, which can be utilized with modern firing systems without any additional processes such as external fuel preparation or fuel drying.

However, there are fuels that must be treated in a special manner. Low-grade lignite with very high content of noncombustible substances (ash, water) and very low heating value (down to 4 MJ/kg) requires special processing, see Section 5.3.3, Typical raw lignite firing configuration.

The utilization of low-grade coals, as lignite has been of high importance in several European countries utilizing their indigenous fuel sources. In Europe the largest lignite reserves are available in Germany, Poland, Greece, Czech Republic, Romania, and Bulgaria [3]. A comprehensive overview of MHPSE's state of the art firing technologies for the combustion of low-grade fuels has been provided in a previous paper [13], while an IEA report can be also found in the following literature [14].

5.3.2 Lignite milling systems

Pulverizing lignite is particularly challenging in view of its high moisture content and the large throughputs involved. For this reason lignite milling and drying is performed with fan beater mills, which use hot flue gas extracted at the furnace exit level. MHPSE has developed two types of state of the art fan beater mills, the DGS mills (Fig. 5.7) and the NV mills (Fig. 5.8).

The DGS technology comes to various sizes up to a 240 t/h and involves the raw coal and flue gases being initially put through the precrushing beater section of the beater wheel. This ensures excellent air and coal dust distribution into the beater wheel. Crushing in the DGS mill is usually so intensive that adequate grinding fineness can be obtained in one working operation—without the need for an additional classifier. This both raises the pressure balance and cuts back on energy requirements. The DGS mill has stood the test of time particularly with difficult-to-work lignites, with superior durability and performance.

The NV mills (Fig. 5.8) ranging from 10 to 140 t/h throughputs are for crushing high moisture content lignites for direct injection into the furnace. Together with the flue gases, the raw lignite is brought for the crushing into the beater wheel. It acts as a fan impeller fitted with beater plates. Baffle-type classifiers ensure that the required fuel fineness for combustion is fulfilled.

5.3.3 Typical raw lignite firing configuration

Depending on the individual water and ash content, different firing systems for lignite have to be applied, in order to secure stable operation. In Fig. 5.9 the application range of the firing systems is presented according to the lignite quality

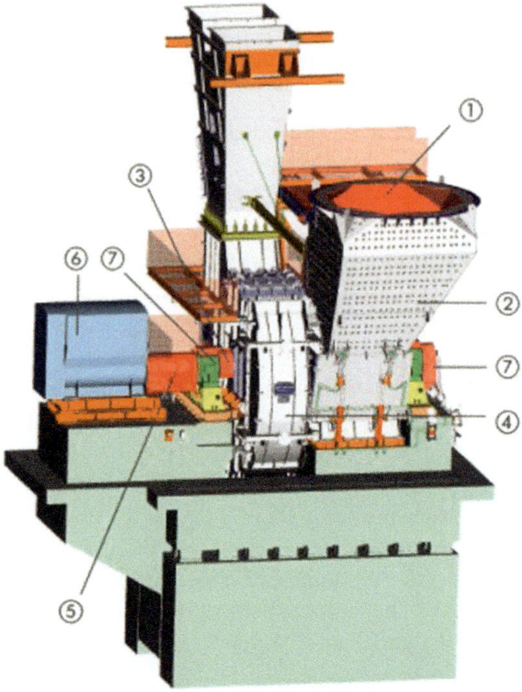

Figure 5.7 DGS fan beater mill: 1, flue gas gate valve; 2, inlet housing; 3, beater shaft; 4, beater wheel; 5, coupling; 6, motor; 7, bearing.

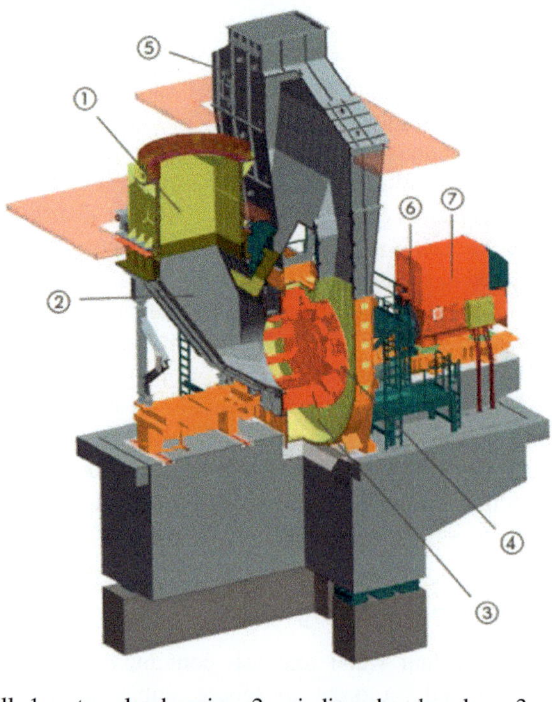

Figure 5.8 NV mill: 1, gate valve housing; 2, grinding chamber door; 3, grinding chamber; 4, beater wheel; 5, baffle-type classifier; 6, double bearing; 7, motor.

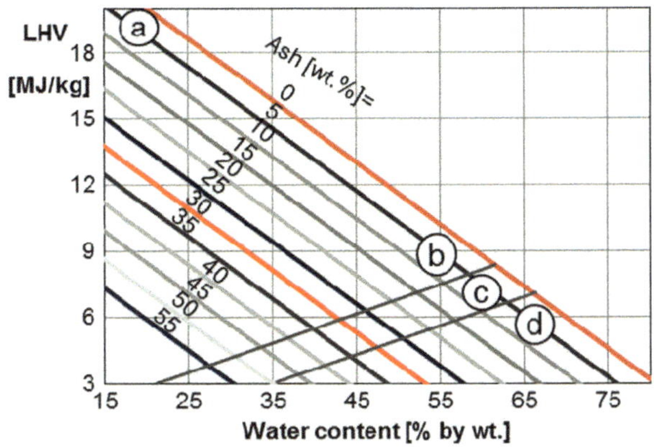

Figure 5.9 Coal quality ranges and respective firing systems applied.

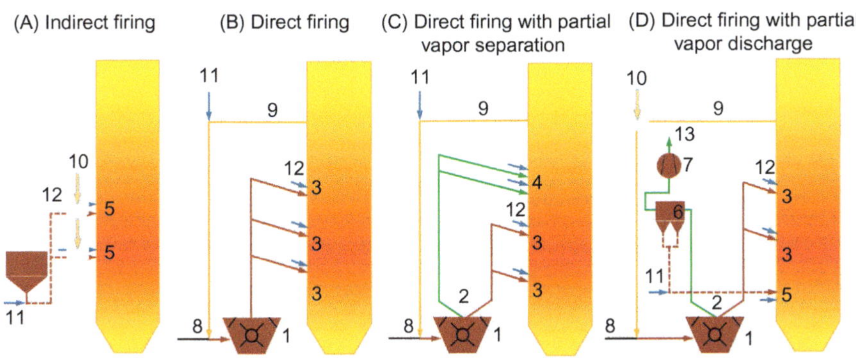

Figure 5.10 Direct and indirect firing systems for lignite: (A) indirect firing, (B) direct firing without vapor separation, (C) direct firing with partial vapor separation and vapors into the furnace, (D) direct firing with partial vapor separation and vapors discharged (1, beater fan mill; 2, vapor separation; 3, RS burners; 4, vapor burners; 5, burner for indirect firing/ DST-Brenner; 6, vapor filter; 7, vapor fan; 8, raw lignite; 9, flue gas recirculation; 10, cold flue gas; 11, primary air; 12, secondary air; 13, to FGD/Stack).

range. The corresponding firing schemes are then presented in Fig. 5.10. Area (a) of Fig. 5.9 corresponds to predried lignite utilization. Predried lignite is used predominantly in industrial scale boilers and for special purposes in utility boilers. For this reason, the firing and utilization technology is described in a separate section.

The expected lignite qualities utilized in large-scale utility boilers correspond to the areas (b), (c), and (d) of the technology review starts from technical solutions for firing of predried lignite, which has the highest heating value (area (a) in Fig. 5.9) and continuous with firing concepts for the lower quality lignites down to lignites with a heating value of less than 1000 kcal/kg (area (d) in Fig. 5.9).

An overview and comparison of the different lignite firing systems available are depicted in Fig. 5.10. Predried lignite is fired by using an indirect firing system composed of (1) a dedicated silo for storage of dry lignite dust, (2) a dedicated dosing and pneumatic feeding system, and (3) dedicated dry lignite burners (Fig. 5.10A). Predried lignite is usually produced in a dry lignite production facility located near an open cast mine and transported by truck or train to the relevant plant. Although this type of firing has not been applied up to now in large-scale power plants, several references are found in the smaller scale, including industrial boilers and combined heat and power installations. In Fig. 5.10B a scheme of *direct firing without vapor separation* is presented. This scheme is typical for lignite plants utilizing lignite with a sufficiently high heating value and low ballast. Furthermore in Fig. 5.10C a lignite *firing system with partial vapor separation* and reinjection is depicted, which is usually suitable for low-quality lignites. In Fig. 5.10D a firing system with a partial vapor discharge is depicted, which is applied in few power plant cases, where lignites with extremely low heating value will be utilized. Specific references for each of the firing system cases are provided in the next chapter.

5.3.3.1 Direct firing without vapor separation

As a typical example of a lignite fired power with a direct firing system, the 670 MWe Boxberg unit R in Germany is provided (Fig. 5.11). MHPSE's RS burners have been installed in the lignite boiler.

5.3.3.2 Direct firing with partial vapor separation

Lignite with higher ballast (i.e., ash and water) content requires vapor separation, Fig. 5.10C, in order to improve ignition and flame stability. *Vapor separation* is recommendable for lignite qualities with ballast content >65% by weight, if the water content is higher than 50% by weight. Vapors can be introduced into the furnace through dedicated vapor burners. As a typical example of a boiler with vapor separation, the new 660 MWe Ptolemais unit V currently under construction in Greece is presented in Fig. 5.12.

5.3.3.3 Direct firing with partial vapor discharge

Combustion of low-grade lignite with high ballast content >80% by weight and very low heating value requires *vapor separation and discharge of vapors* (e.g., to FGD plant, Fig. 5.4D). Remaining coal particles are separated from discharged vapors in ESPs and then transported to dedicated vapor dust burners. As a typical example of a boiler with vapor separation and discharge the scheme of the boiler and firing system of Ebistan B unit in Turkey is provided in Fig. 5.13.

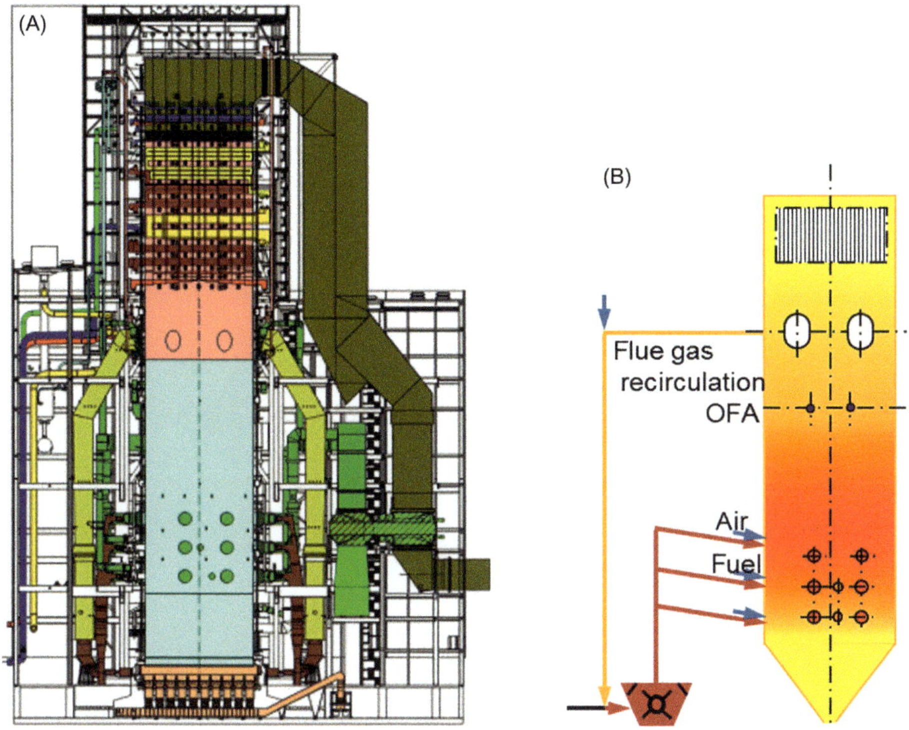

Figure 5.11 (A) Boiler scheme Boxberg unit R, (B) milling and firing scheme Boxberg unit R.

5.3.4 Lignite burners

Burners are key elements of any combustion system and their design is usually adjusted to the fuel properties to be utilized. In most of the lignite boilers worldwide, the conventional jet burner technology and the tangential firing arrangement are applied. Jet burners are composed of a number of injection ports vertically arranged in such a way that the coal stream jet entering the combustion chamber is mixed with air jets from above and below (Fig. 5.14A,B). Therefore in a typical jet burner configuration, the flame ignition happens 1–2 m inside the furnace. The secondary air velocity is usually higher than the velocity of lignite dust and carrier gas and this velocity difference enhances turbulence levels and mixing. Furthermore, the jet burners have a special arrangement, so that the ignited coal particles can enter a vortex flow field caused by the tangential arrangement (Fig. 5.14C), so that the overall residence time of the particles increased. One of the main characteristics of the jet burners is that ignition location cannot be controlled as it changes depending on coal quality and load. This in turn leads to limitations of operation at minimum loads and with lower coal qualities. The operation at boiler loads with less than 40% is usually not possible without supporting firing. In addition due to the intense mixing between coal and air, high-temperature peaks can be expected.

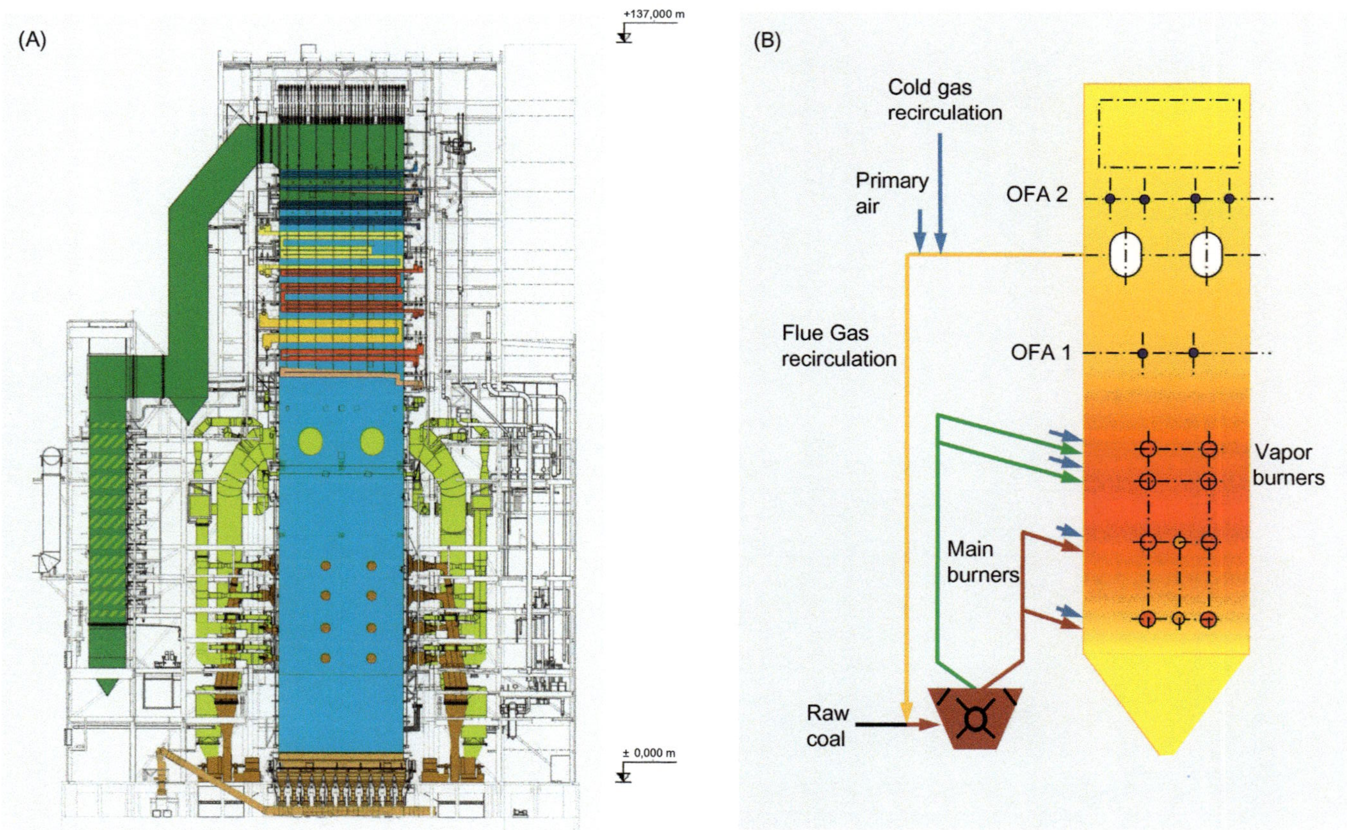

Figure 5.12 (A) Boiler scheme Ptolemais unit V, (B) milling and firing scheme Ptolemais unit V.

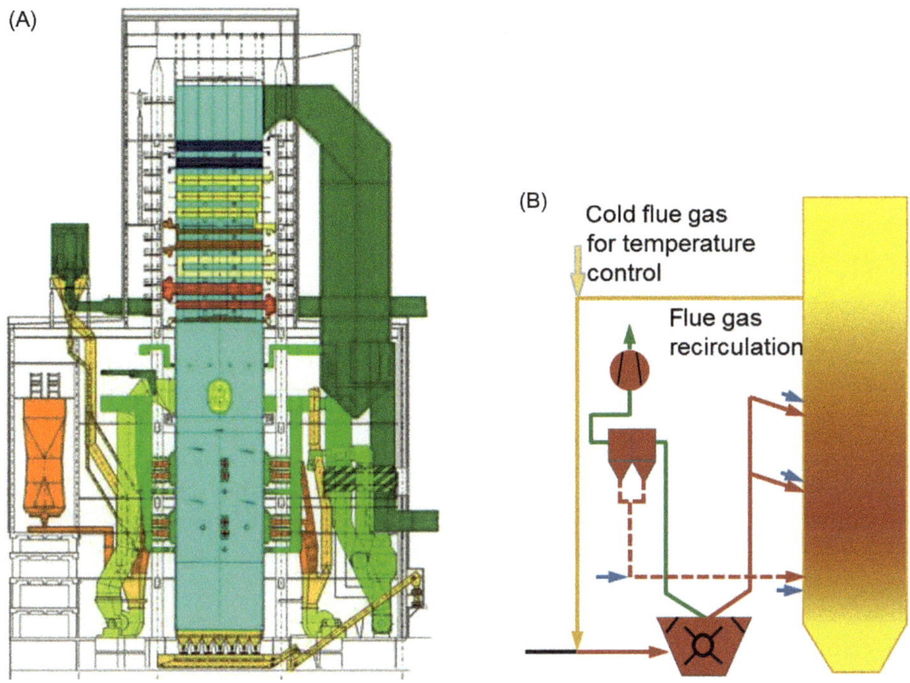

Figure 5.13 (A) Boiler scheme Elbistan unit B, (B) milling and firing scheme Elbistan unit B.

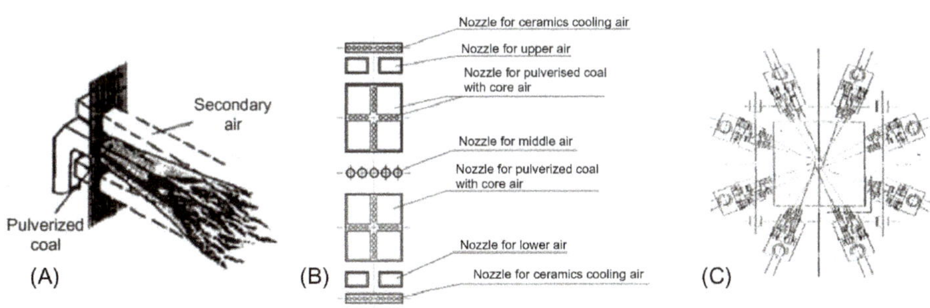

Figure 5.14 (A) Simplified sketch of an operating coal jet burner, (B) typical 2D burner sketch, (C) typical tangential firing arrangement for jet burners.
Source: Wroclaw University scripts on coal combustion. http://fluid.wme.pwr.wroc.pl/~spalanie/dydaktyka/combustion_MiBM/BCS/COAL_COMBUSTION.PDF [15].

This may lead to higher tendency for NOx formation and also to higher tendency on slagging in the near burner region, if the particular lignites utilized have a high slagging propensity. In order to overcome these issues, air staging or air and fuel staging in parallel is usually applied in tangential firing systems.

In order to overcome the drawbacks of a typical jet firing system concept, MHPSE has developed in the last years a new lignite burner based on the swirl burner concept. *RS burner* is the MHPSE's state of the art burner for low-rank coals. It is based on the same design principles like the *DS burner*, which is the MHPSE's state of the art burner for the combustion of hard coal, biomass, and other dry solid fuels [16]. Fig. 5.15A shows a sketch of the RS burner. Concentric setup and swirling of combustion air flow are essential features of this burner type. The functional principle of the RS Burner is determined by the definition of ignition, as well as its start and progress in a low-oxygen, primary reaction phase. The process-relevant preparation of the fuel particles is in the focus of the burner function as well as the pyrolysis and oxidation of the pyrolysis products. Due to the interaction between swirl device and fuel nozzle, as well as the resulting heat transfer in the nearby burner zone, the prerequisites required for the process are created. Early oxidation of the pyrolysis products with a defined oxygen volume is the prerequisite for stable low-NOx combustion in the core flame. The core flame is surrounded by the swirled secondary air flow, while the combustion process being supplied with the necessary oxygen by a delayed continuously supply from the peripheral burner flow. This is of particular importance for low-NO_x combustion. Thus, stoichiometry defined by the primary oxygen amount and oxygen demand of combustible volatiles released from the fuel has to be controlled, in order to support low-NO_x formation in the oxygen-lean devolatilization zone.

The patented fuel nozzle [17] comprises of three burner components: front part of the primary air tube, secondary air deflector cone, and flame stabilizer ring.

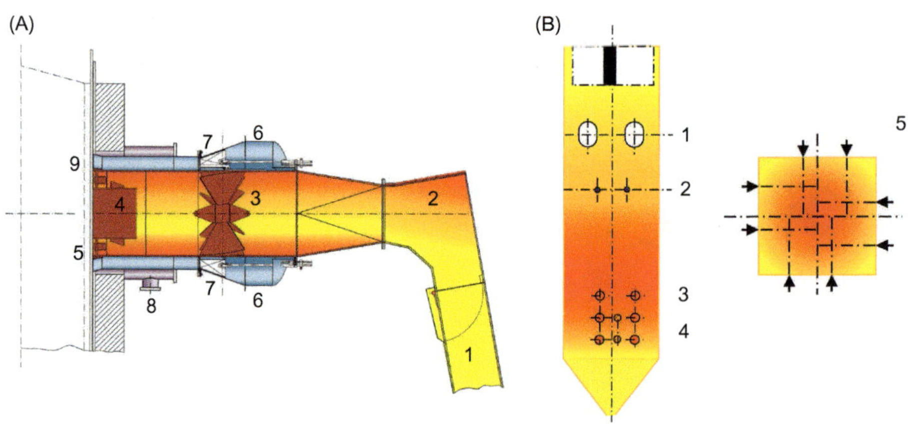

Figure 5.15 (A) RS burner for lignite (1, primary air inlet; 2, inlet elbow; 3, primary air swirler; 4, radiation protection tube; 5, heat resistant fuel nozzle; 6, secondary air; 7, secondary air swirler; 8, cooling air; 9, secondary air deflector, (B) furnace design and RS burner arrangement of lignite fired power plant (1, flue gas extraction; 2, over fire air; 3, lignite burners; 4, oil burners; 5, all wall firing arrangement top view).

Unlike the DS burner, the RS burner does not have a core air tube due to the larger cross section necessary for the transport of the vapor and lignite dust mixture. Oil burners are usually installed next to the main burners in lignite fired furnaces (Figure 5.15B).

5.3.5 Dry lignite utilization concepts

5.3.5.1 Hybrid firing for increased flexibility

Dry lignite boiler technology

As previously mentioned, conventional lignite power plants are usually the least flexible coal plants due to their restrictions in the mill-drying and firing system. More specific, the high water content of lignite imposes specific limits in the milling process and the operation of fan beater mills. The reduction of load is usually achieved through shutting down a number of mills, nevertheless an operation of four out of eight mills is usually considered by the manufacturers as the marginal operation, where a stable flame can be assured. Under these conditions, a minimum load for modern, state of the art lignite boilers usually corresponds to 35%–40%. In order to reduce further this minimum load, a stable flame is required. This can be produced by the combustion of a dry fuel through dedicated burners and the operation of two–three mills at part load in parallel. This concept is referred as "hybrid" firing and combines (1) the direct firing of raw lignite by operating a limited number of mills and (2) the indirect firing of predried lignite stored in an intermediate bunker through dedicated burners, like the "DST-Brenner" burner [18]. Fig. 5.15 shows a typical burner arrangement of RS burners in new built power plants. Burners are located on each wall with respect to the pulverize arrangement. Tangential firing as applied in the case of jet burners is no longer required, since intense mixing of air and fuel is secured through the swirling flow pattern. A typical burner arrangement for "hybrid" firing is presented in Fig. 5.16.

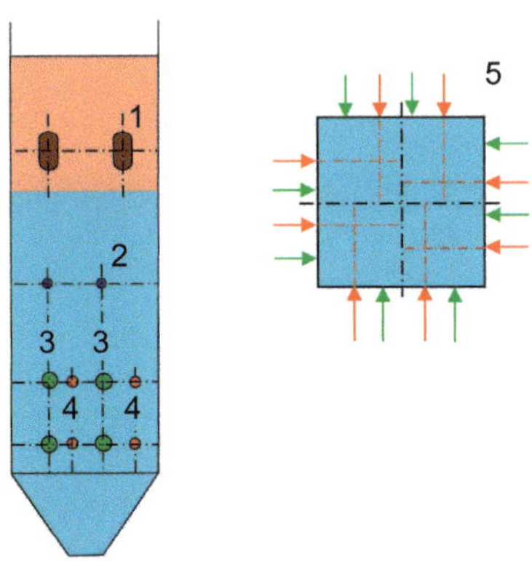

Figure 5.16 Burner arrangement in a hybrid concept [1, flue gas extraction; 2, over fire air; 3, raw lignite burners (RS burner); 4, predried lignite burners (DST-Brenner); 5, all wall firing arrangement top view].

Similar to the burner arrangement for raw coal firing, a dedicated dry lignite burner is installed near the raw coal burners, in order to support flame stability during low-load operation. The dedicated dry lignite burners have also an additional fuel lance installed for the utilization of natural gas or oil for ignition and start-up.

5.4 Lignite predrying and Power Plant concepts

5.4.1 State of the art utility scale lignite dryers

Lignite predrying has been considered for a long time as a key technology toward efficiency increase of future lignite power plants. Several highly sophisticated technologies for large-scale lignite drying in power plants have been under development and testing during the last decades [2]. From those activities, the steam heated fluidized-bed dryer (Wirbelschichttrocknung mit interner Abwärmenutzung, WTA) developed by RWE is still considered as one of the most energy efficient drying processes. A prototype WTA dryer with 110 t/h dry lignite capacity has been in operation at RWE's Niederaussem power plant. Nevertheless, for the further commercialization of these high-end predrying concept, further development and demonstrations steps are required, in order to achieve in these complex systems a reliable operation and high availability times equivalent with the ones of the power plants. In addition, the current low prices of CO_2 emissions and the low electricity market prices provide no incentives and no business case for new projects targeting to further efficiency increase in existing and new lignite power plants.

A new market opportunity for lignite predrying systems is currently arising, due to the flexibility requirements in lignite power plants and the fuel savings achieved by the utilization of dry lignite as auxiliary and start-up fuel instead of oil or natural gas. In this framework, the installation of a new lignite predrying system and the utilization of low-cost predried lignite shall be compared with auxiliary fuels' consumption for a longer time period, whenever low-load operation is requested. Hence, the savings in operating costs through the utilization of predried lignite as start-up fuel shall be compared with the annualized costs of the particular investment. Each particular power plant has to be specifically considered as a different case, since a slight change in the operation mode and the number of hot/warm or cold starts per year can have a significant impact on the business case and the bankability of the whole project.

In this new market framework, it is expected that a suitable, competitive lignite predrying system may not have the same technical characteristics like the high-end fluidized-bed drying technology referred previously. For example, conventional, reliable, and lower cost drying technologies can be also taken into consideration for the production of dry lignite as start-up fuel in future retrofits of existing power plants. In this context, most relevant solid fuel drying technologies, which available in the industrial scale, are presented below. The heating medium is used as classification parameter in this analysis. The first two concepts presented, the tubular dryer (1) and the fluidized-bed dryer (2) utilize steam as heating medium, while the two last systems, rotary (drum) dryer (3), and fan beater mill (4) utilize flue gas as a heating medium.

1. The tubular dryer

   The tubular dryer consists of a drum, which is slightly inclined and rotates during the drying process (Fig. 5.17). A high number of tubes is packed in the drum. The heating medium, steam flows in the drum outside the tubes and heats up in this way indirectly the raw lignite. The raw lignite is continuously inserted into the dryer tubes and transported to the exit by the rotary motion. The energy required for the moisture evaporation is provided by low-pressure steam, which by entering the dryer along the axis is expanded on the tube surfaces. The air, which goes through the tubes in the same direction as the brown coal, absorbs the evaporated moisture. A typical size of tubular dryer has an evaporation capacity of about 25 t/h, a drum length of 8 m, drum diameter of 5.6 m, and weights about 280 t. Detailed information for the thermodynamic modeling of a tubular dryer can be found in the literature [13].

2. The fluidized-bed dryer

   RWE's WTA dryer (Fluidized-bed drying with internal waste heat utilization) is one of the most developed drying technologies, which has been also demonstrated in the industrial scale. In the applied WTA process, lignite is first milled to a fine particle size by hammer mills and enters afterwards the fluidized-bed dryer. The dried fuel exits the stationary bed from the bottom of the dryer and is then cooled down to a temperature allowing an additional milling step down to the required particle size. The vapor stream exits the dryer from the top. The fine dust transported in the vapor is then separated from the main stream by filter components like ESPs or multicyclones. The heat needed for the drying of the fuel can be supplied by external steam, which is normally taken from

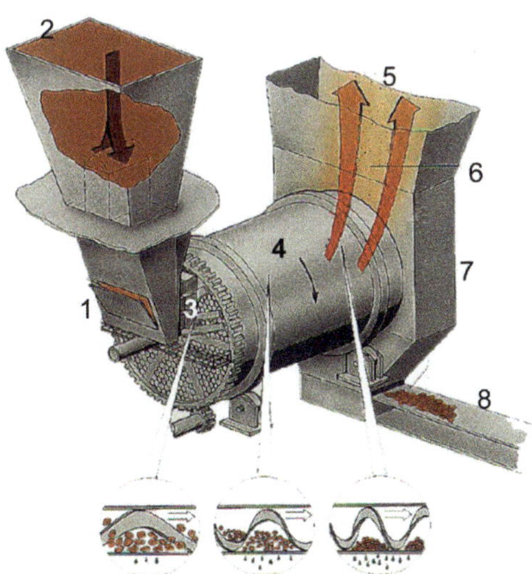

Figure 5.17 Scheme of a tubular dryer. 1, Filling chute; 2, raw coal bunker; 3, inlet air; 4, tubular dryer; 5, outlet to ESP; 6, particle-loaded vapor; 7, dryer outlet; 8, conveyor system for dried lignite.

Used with permission from RWE Power, WTA Technology, www.rwe.com.

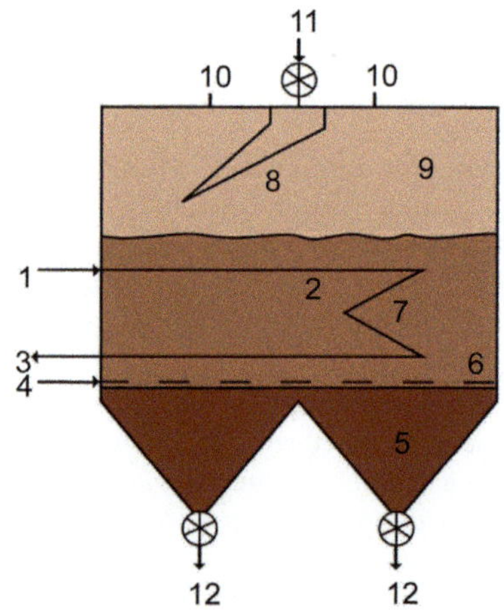

Figure 5.18 Scheme of a fluidized bed dryer. 1, Heating steam; 2, heat exchanger; 3, condensate; 4, fluidizing medium; 5, fixed bed; 6, fluidizing bottom; 7, fluidized bed; 8, revolving chute distributor; 9, freeboard; 10, vapor; 11, raw lignite; 12, dried lignite. Used with permission from RWE Power, WTA Technology, www.rwe.com.

the turbine or by a vapor compressor used to compress the produced vapors and utilize them as a heat source in the dryer, following in this way the principle of a heat pump. A schematic drawing of the RWE's WTA dryer is presented in Fig. 5.18. In addition, a further concept, which could lead to additional efficiency gains, is the pressurized fluidized-bed drying concept currently under development by Vattenfall [14]. A pilot scale prototype with a capacity of up to 10 t/h has been developed and tested up to now, while a future project for a first industrial scale prototype up to 50 t/h is currently under development. Furthermore, another fluidized-bed drying concept suited for subbituminous coals with water content up to 35% has been developed and demonstrated in Coal Creek power station in the United States. Additional information can be found in Ref. [15].

3. The rotary dryer or drum dryer

Rotary dryers can be of either direct or indirect type. In many industrial plants, rotary dryers are used for the drying of peat, coal muds as well as sewage sludge. Generally, in order to avoid self-ignition during operation, a cocurrent mode is applied in these dryers. Natural gas or coal is used to generate flue gas to be used as the drying medium. The layout of a typical rotary dryer is shown schematically in Fig. 5.19. The main component of this dryer is the steel drum lined with a refractory lining and setup on rollers by means of bandages located on the drum. A toothed gear is typically used to rotate the drum. The drum is set-up obliquely with a slope of 2°−5° to the horizontal. Inside the drum, there are lifters installed at the inner surface of the drum. During operation, as the drum rotates these lifters lift the coal granules so that a homogeneous mixing of coal with the hot flue gas is achieved.

4. Conventional mill-drying systems (such as MHPSE's DGS and the NV beater fan mills)

Conventional mill-drying systems are included in the usual scope of supply of MHPSE DGS mills (Fig. 5.20) with raw coal capacity up to a 240 t/h and NV mills raw coal capacity ranging from 10 to 140 t/h can be used as the drying technology in the overall

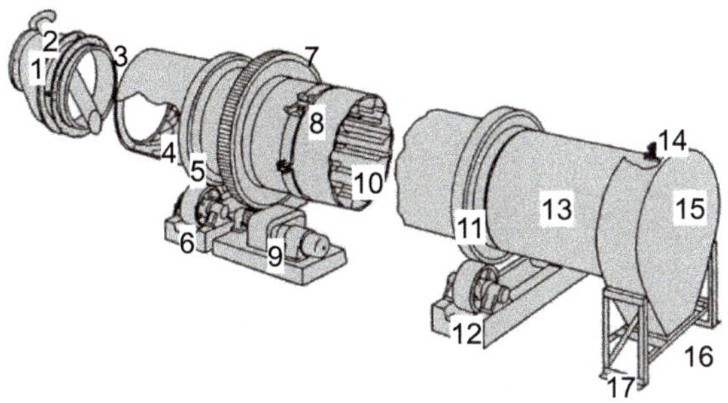

Figure 5.19 Scheme of a rotary dryer. 1, Inlet head (counter flow only); 2, feed chute; 3, friction seal; 4, spiral flights; 5, first riding ring; 6, roll assembly; 7, girt gear; 8, knocker; 9, drive assembly; 10, lifting flights; 11, second riding ring; 13, trunnion roll assembly; 14, shell; 15, breeching seals; 16, breeching; 17, discharge.
Used with permission from Arvos Raymond Bartlett Snow LLC.

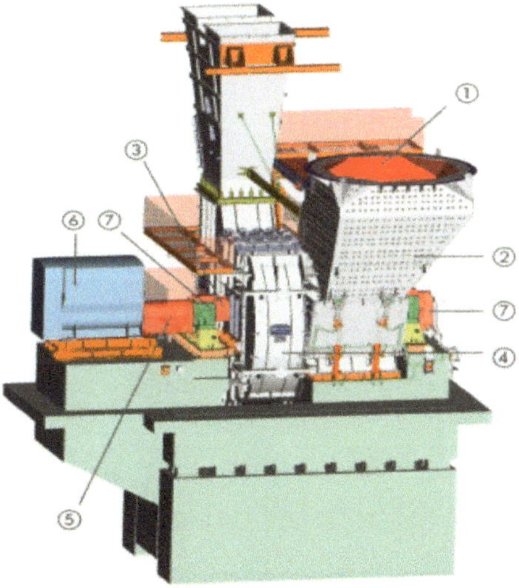

Figure 5.20 DGS fan beater mill. 1, Flue gas gate valve; 2, inlet housing; 3, beater shaft; 4, beater wheel; 5, coupling; 6, motor; 7, bearing.

predrying concept. DGS technology involves the raw coal and flue gases being initially put through the precrushing beater section of the beater wheel. This ensures excellent air and coal dust distribution into the beater wheel. Crushing in the DGS mill is usually so intensive that adequate grinding fineness can be obtained in one working operation—without the need for an additional classifier. NV mills (Fig. 5.21) are suitable for crushing high moisture content lignite and achieving in parallel fine particle size distribution through the installed classifier.

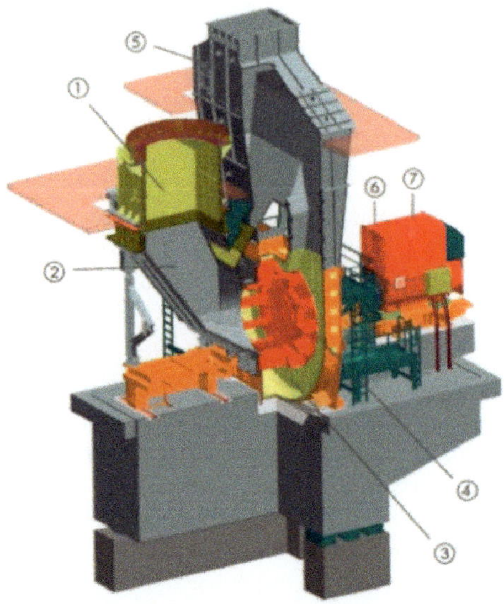

Figure 5.21 NV mill. 1, Gate valve housing; 2, grinding chamber door; 3, grinding chamber; 4, beater wheel; 5, baffle-type classifier; 6, double bearing; 7, motor.

Table 5.1 Comparison of dryers' main technical characteristics

	Tubular dryer	Fluidized bed dryer	Rotary dryer	Fan beater mills
Heating medium	Steam	Steam or flue gas	Flue gas	Flue gas
Direct/indirect	Indirect	Direct or indirect	Direct	Direct
Nominal steam evaporating capacity per dryer	Up to 25 t/h	Up to 110 t/h	Up to 50 t/h	Up to 110 t/h
Industrial scale references	>50	4[a]	>50[b]	>100

[a]Including industrial scale prototypes in Germany [12], Australia, and United States [15].
[b]Only for other solid materials, like sewage sludge and peat.

A comparison of the dryers' main technical characteristics is presented in Table 5.1 and a qualitative evaluation of the technologies' specific characteristics based on a comparative basis is presented in Table 5.2. The key message from this type of comparison is that every technology has its own characteristics and has to be evaluated in combination with the technical and nontechnical restrictions and boundary conditions applied in each case.

Table 5.2 **Qualitative evaluation of predrying technologies**

	Tubular dryer	Fluidized bed dryer	Rotary dryer	Fan beater mills
Requirements on coal pretreatment	++	+	+	0
Auxiliary components for pretreatment of heating medium (flue gas or steam)	0	0	++	++
Requirements on regular maintenance due to rotating parts	++	0	+	+
Further optimization requirements (technology maturity)	0	++	+	0

Note: 0, very small/modest; +, medium; ++, high.

5.4.2 Dryer integration concepts

"Hybrid" firing refers to a combination of existing technologies, whose application can provide definitive advantages in relation to flexibility and minimum load operation. These concepts only need smaller amounts of dry lignite compared to 100% indirect firing concepts. In this way the boiler can be operated at full load even, at times, when the drying system is out of operation. Most benefits regarding flexibility can be achieved by using dedicated burners for the *indirect firing of predried fuels*. In markets with increasing requirements on flexible operation of the hard coal and lignite power plants, the retrofit of existing boiler and the installation of an indirect firing system on parallel with the conventional direct firing will allow a reduction of boiler's minimum load lower than 30%. In this way, a kind of "idle" operation can be achieved, where the plant stays in the grid on a very low--load providing primary and secondary control services and is able to ramp up again, whenever required by the system operator.

In contrast to the above-mentioned case in typical lignite power plants, only direct firing systems without any additional split of the vapor mass flow and lignite dust separation are installed. Thus, the implementation of a "hybrid" system in a typical lignite power plant will require among other the installation of an external lignite predrying system, the thermal and process integration into the plant cycle and additional infrastructure for the intermediate storage and handling of dry lignite. More specific, following additional systems for lignite handling will be required apart from the drying system: (1) lignite precrushing system, (2) dry lignite handling and storage system, (3) dry lignite feeding system from the dryer site up to the boiler house, (4) intermediate storage at the boiler house, (5) lignite feeding and dosing to the dedicated dry lignite burners. Furthermore, additional effort is required for the handling of the produced vapor coming out from the dryer including the separation of fine particle dust from the vapor. The process scheme of the steam heated fluidized-bed drying and its integration in a hybrid firing system is shown in Fig. 5.22. The lignite is dried in a steam heated fluidized bed.

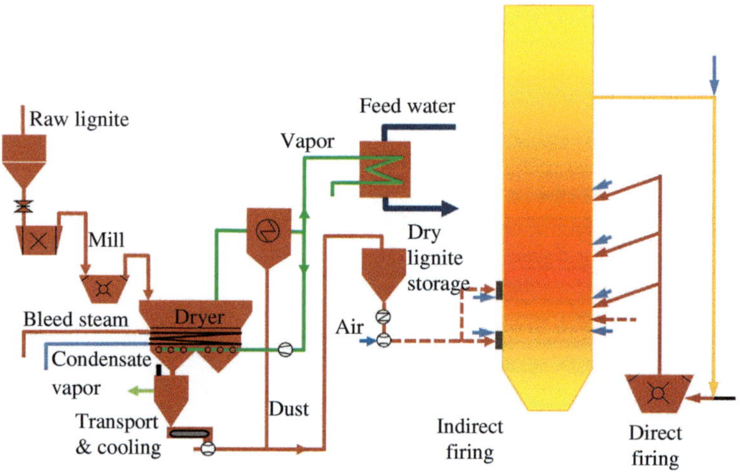

Figure 5.22 External predrying system based on the fluidized bed drying technology.

The vapor resulting from coal water evaporation is cleaned and partially used for fluidization and for heat recovery in the low-pressure preheaters of the power plant. A system similar to that shown in is in operation for more than 10 years in the Niederaußem K power station, where a prototype fluidized-bed dryer has been operated additionally.

An alternative concept to the fluidized-bed drying is the installation of an additional dryer utilizing hot flue gas, such as a rotary dryer or a fan beater mill. In conventional lignite fired power plants, and particular in the older ones, there is always some excess heat on the cold flue gas side, which cannot be used for air preheating, due to the mismatch of the mass flows of air and flue gas. Only in new lignite boilers, this excess heat is used for feed-water preheating in some cases. Therefore, the use of the available heat for the predrying of a small lignite flow should be considered for retrofits of older plants. In addition, for reasons of safety and reliability the drying medium shall have (1) lowest oxygen concentration and (2) lowest vapor content to facilitate the heat and mass transfer process. Therefore, one of the preferred solutions is the utilization of a slip stream of cleaned and dried flue gas as heating medium in the process, after being reheated with excess heat from the cold flue gas side. The process flow for this design is presented in Fig. 5.23. A partial flow of cleaned flue gas from downstream of FGD is used for lignite drying. Therefore, it is cooled down in a flue gas cooler and enters the mill after being reheated via a gas/gas heater. Specially designed beater mills are used for grinding and drying of a partial raw lignite flow, which afterwards is separated from the vapors, cooled down and filled into a silo for the use in indirectly fired support burners. There are also other dryer systems available, like drum dryers which can be used for similar applications. The optimum setup has to be determined case by case depending on individual project boundary conditions and possible arrangements (see Fig. 5.24).

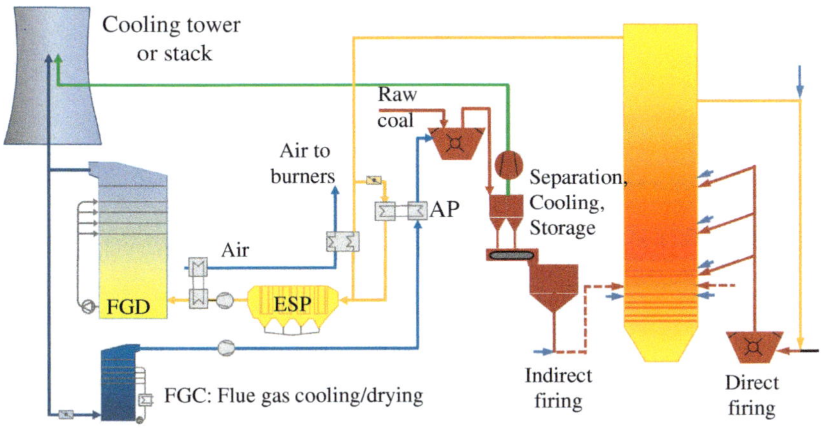

Figure 5.23 External predrying system based on drying with hot flue gas (mill or rotary/drum dryer).

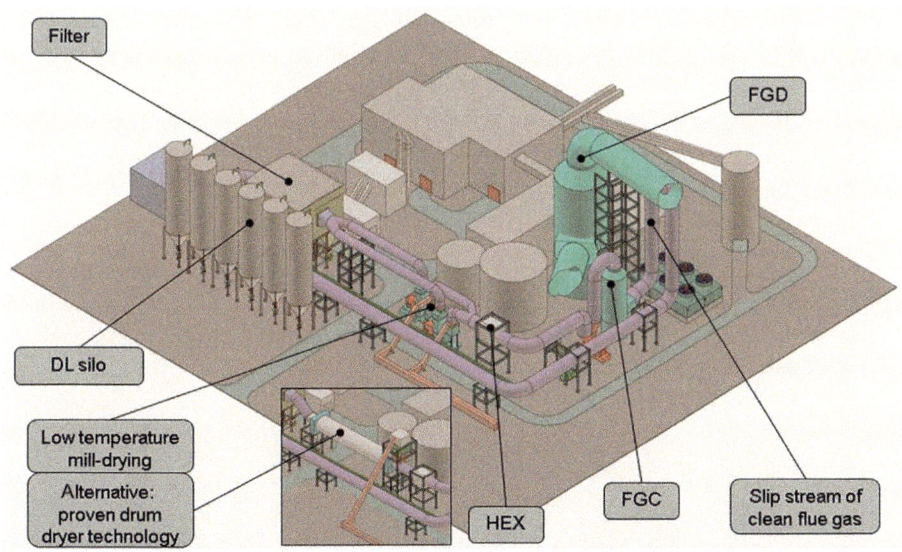

Figure 5.24 3D Layout plan on the integration of an external predrying system (fan beater mill) in an existing power plant.

5.5 State of the art AQCS

In this chapter, the state of the art AQCS are described with the focus on the DeNOx and FGD systems. A short reference on the new Best Available Techniques (BAT) Reference Document (BREF) on Large Combustion Plants (LCP) is also made. Stricter emission limits are imposed and new emissions classes are defined by that regulation affecting new and existing coal and lignite power plants in Europe.

5.5.1 DeNOx systems

State of the art lignite power plants have been up to now designed only with primary NOx reduction measures. The installation of low-NO_x burners and air staging has been usually sufficient to achieve the emission limit of 200 mg/Nm^3. Only in a few retrofits, where the primary measures could not be effectively applied secondary measures have been installed. These include Selective Catalytic Reduction (SCR) or Selective NonCatalytic Reduction (SNCR) systems. The new European BREF document imposes stricter NOx emission limits in new and existing lignite plants making secondary measures inevitable. For this reason, a brief description of the operating principle of $DeNO_x$ systems is provided below.

Fig. 5.25 explains the chemistry and Fig. 5.26 explains the principle of the $DeNO_x$ process. In this process, Ammonia (NH_3) is usually used as reducing reagent for NO_x. NH_3 reacts with NO_x activated by the catalyst to form harmless nitrogen (N_2) and steam (H_2O) as shown in the figure.

In the boiler system, the flue gas temperature level after economizer is usually most appropriate for the $DeNO_x$ installation (Fig. 5.27). The required temperature by the catalyst is in the range of 320–420°C, a range which is found at this location in the entire operation range of the boiler. Thus most of the $DeNO_x$ catalysts for boiler systems are installed between economizer outlet and air preheater inlet. A controlled amount of NH_3 is injected into the flue gas prior to the $DeNO_x$ reactor taking the following factors into consideration:

- $DeNO_x$ inlet NO_x concentration.
- Flue gas volume from boiler load signal.
- Required reduction of NO_x.

In the beginning of the development, apart from the catalyst components and composition, catalyst formulation was the other important factor for the commercialization of the $DeNO_x$ process. Today plate-type and honeycomb-type catalyst units are dominating the market.

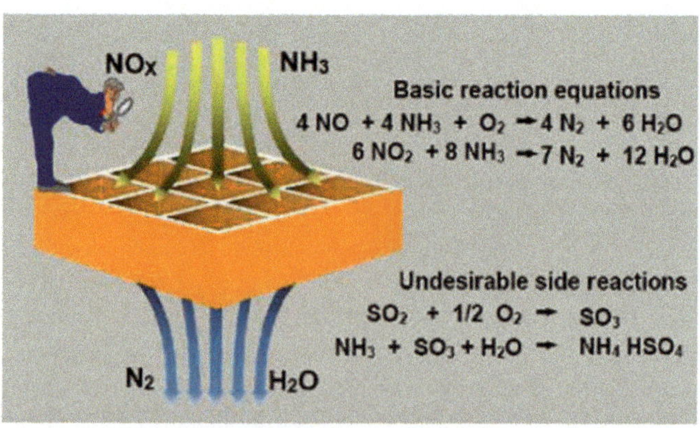

Figure 5.25 Chemistry of the DeNOx process.

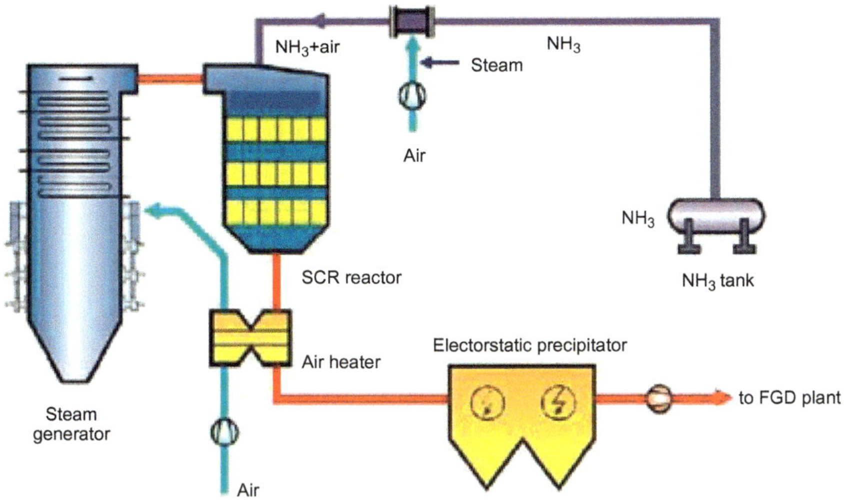

Figure 5.26 Principle of SCR system.

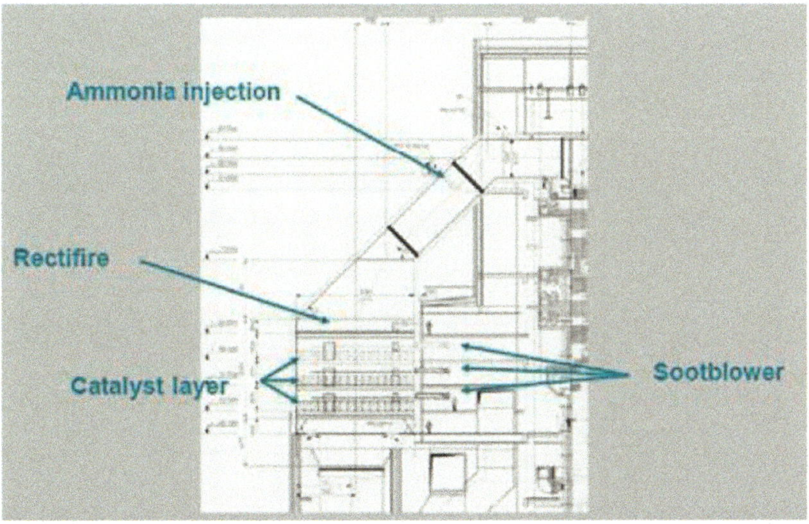

Figure 5.27 Arrangement of SCR reactor.

MHPS has developed the plate-type catalyst, which has many unique and profitable characteristics (Fig. 5.28), but also honeycomb modules are applied depending on customer's preferences. The DeNOx reactor is technically designed so that standard units or blocks of both types can be installed.

Figure 5.28 Catalyst units and their arrangement in the DeNOx system.

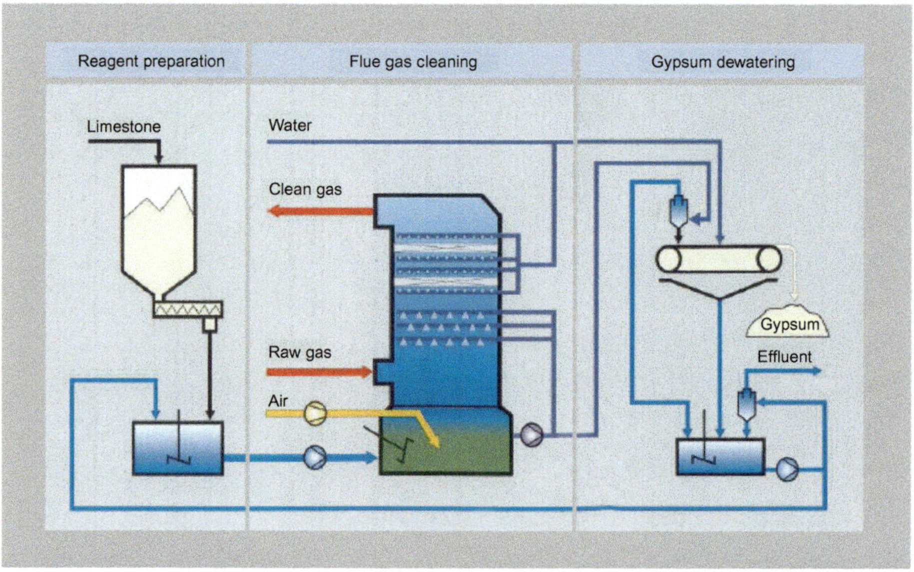

Figure 5.29 Wet scrubbing system.

5.5.2 Flue gas desulfurization

The principle of the wet scrubbing FGD system is given in Fig. 5.29. Limestone as reagent is mixed together with water to a suspension, which is pumped into the sump of the reactor. This fluid is pumped to different spray levels where it is injected into the flue gas, which flows through the reactor from bottom to top.

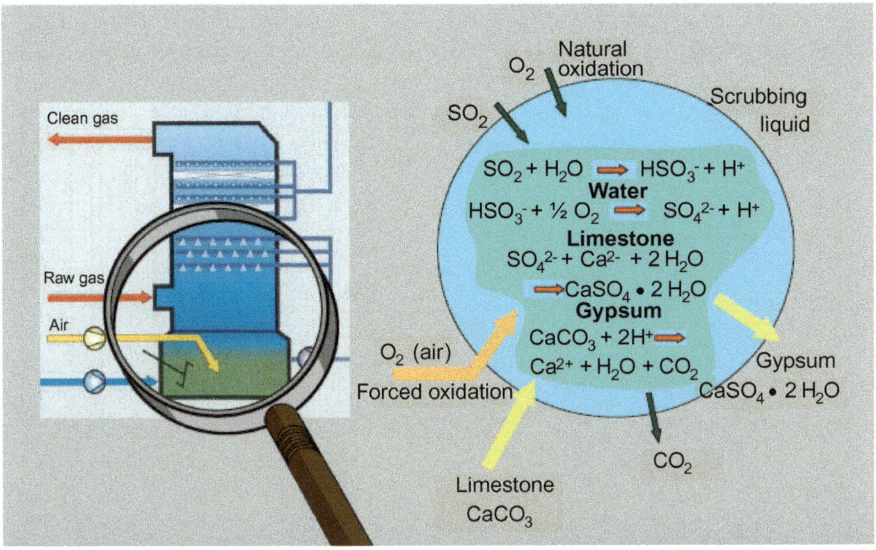

Figure 5.30 Wet limestone/gypsum, process-reaction mechanism.

Air is added into the sump to transform the $CaSO_3$ into $CaSO_4$. Due to the reaction time of the process, the fluid is recirculated several times. Depending on the size of the sump, approximately 5% of its content is outwarded per hour.

Corresponding to the chemical reactions described in Fig. 5.30, gypsum is formed as the main outcome out of the process besides the cleaned flue gas. The gypsum is dewatered so that it can be used for products for the house building industry.

Low-cost carbon steel with corrosion protection is used for ducts, absorber shell, and tanks. Concrete is applied for the underground pit. The selection of liner material varies depending on site conditions. Preferable is rubber lining or flake resin lining. Lining with power plant on concrete base material is also applicable. Rubber lining is used for slurry piping. Alternatively plastic piping such as FRP can also be used for this.

Depending on temperature and chloride concentration, stainless steel or nickel alloy material is preferred for absorber shell, internals and some parts of flue gas duct. With these materials initial investment cost will be higher, but there is no maintenance cost as usually required for flake resin or rubber lining.

5.5.3 New EU legislative framework (BREFs)

The new BREFs for LCP [19,20], which are currently in draft version and are expected to be officially adopted in 2017, have a considerable impact in the operation of new and existing coal and lignite power plants in Europe. After the official adoption of the BREF, EU Member States have a transitional period of 4 years to ensure that (1) all permit conditions have been updated to reflect

Table 5.3 Comparison of former IED emission limits and recent BAT-associated emission levels (AELs) (yearly average values for coal and lignite plants >300 MWth)

Emissions	Fuel	Former limit IED ELVs (existing/new plants) (mg/Nm^3)	New BAT-AEL (mg/Nm^3) (existing plants)		New BAT-AEL (mg/Nm^3) (new plants)	
			Lower end	Higher end	Lower end	Higher end
NOx	Lignite	200	<85	175	10	75
	Hard coal	200	65	150	50	85
SO_2	Lignite/ hard coal	200/150	10	130	10	75
Dust		20/10	2	8	2	5
HCl			1	5	1	3
HF			0.1	3	0.1	2
Hg	Lignite		<1 μg	7 μg	1	4
	Hard coal		<1 μg	4 μg	1	2

the new BAT Conclusions and (2) all installations comply with the new permit conditions. Stricter limits are foreseen by the new BREF for the conventional flue gas pollutants (SO_2, NO_x, dust), while new emission classes are adopted for CO, Hg, HCl, and HF. Hard coal and lignite plants that will be put in operation after December 20 2018 are considered as "new plants." A comparison of the previous limits foreseen in the Industrial Emissions Directive and the new limits adopted in the BREF document are given in Table 5.3. The change of the emission limits of large-scale thermal plants following the revision of the BREFs has a considerable effect particularly in the "new plants" where a considerable increase of investment is necessary compared to the new plants following the IED. For the effective implementation of emission reduction measures, a combination of actions is required in all AQCS components. The European Power Plant Supplier's Association has published specific documents about flue gas cleaning retrofits and about mercury removal to be used as guideline for assessment and design recommendations [21,22]. Furthermore, as an example of the required modifications in new lignite plants PGE's Turow 11 Power Plant currently under erection can be used. More information about the retrofit of the AQCS design, in order to meet the new BREF requirements, carried out by MHPSE, the supplier of the powertrain (boiler, turbine, generator, AQCS), is provided in the literature [23]. Overall, it shall be noted that the stricter emission limits imposed by the new BREF documents, particularly for new plants are challenging but manageable requiring comprehensive engineering and high efforts from power plant supplier's side and good cooperation with plant owner and future operator.

5.6 Conclusions—technology outlook

In this chapter, an overview of the pulverized lignite power plant technology is provided. Low-rank coals including lignites are considered as one of the most abundant fossil fuel sources worldwide. Nevertheless, due to their high water and ash content their efficient utilization by keeping high environmental standards has been always a challenging task. Today's state of the art technology can guarantee net plant efficiency rates between 42% and 44% in modern lignite power plants. Furthermore, current and future lignite power plants face new challenges toward improving their flexibility in operation imposed by the increasing share of renewable energy sources in the electricity grids. Further reduction of minimum load is a common technical measure, toward improving flexibility and economic competitiveness of new and existing plants. The firing system design is a key parameter on this aspect. Indirect firing systems, which utilize stored predried lignite, can therefore play a key role in this. Several lignite predrying technologies and dryer integration options are addressed in this context. Last but not least, the new EU legislation on the adoption of the new BREF on LCP sets new challenges in AQCS of existing and new lignite plants. Considerable upgrade in the AQCS system is required for new plants currently in the design phase, while also in existing power plants retrofits will be in some cases required. Power plant manufacturers in Europe have the experience and know how to successfully carry out this kind of retrofit and upgrade projects and achieve state of the art efficiency and environmental performance in new power plants firing low-rank coals such as lignite.

Abbreviations

AEL associated emission levels
AQCS Air Quality Control Systems
BAT Best Available Techniques
BREF Best Available Techniques Reference Documents
ESP electrostatic precipitator
FGD flue gas desulfurization
IED Industrial Emissions Directive
SCR Selective Catalytic Reduction
USC ultra supercritical

References

[1] Global perspective on the use of low quality coals, IEA Clean Coal Centre, ISBN 978-92-9029-500-6, January 2011.
[2] Utilisation of low rank coals, IEA Clean Coal Centre, ISBN 978-92-9029-502-0, April 2011.
[3] Coal Industry across Europe, 6th edition, 2017, Euracoal (https://euracoal.eu/library/publications/).

[4] Reference document for large combustion plants, draft, June 2016.
[5] Bituminous coal fired USC power plants for the European Market, J. Klebes, PowerGen conference 2007.
[6] Dampfturbinenanlage Boxberg R, Erfahrungen bei der Projektabwicklung und Ergebnisse, VGB Powertech, October 2014.
[7] Betriebserfahrungen mit dem Block Niederaußem K, VGB Powertech, November 2006.
[8] Ptolemais Unit V Greece, Emmanouil Kakaras, Academy of Athens, October 2015.
[9] New build projects of coal-fired Power-Plants in Poland & Greece—Technology and latest experiences in execution, W. Schreier, VGB conference, September 2015.
[10] Habermann M, Götte C, Tigges K D, Scheffknecht G (2004) Supercritical lignite-fired boiler for RWE's new 1,100 MW unit, in: Proceedings of the 12th Annual Power-Gen Europe Conference and Exhibition: Power-Gen Europe 2004, 25–27 May 2004, Barcelona, Spain, PennWell Corporation, Essex, UK, May 2004, 18 pp.
[11] Effiziente Kraftwerkstechnik in NRW zur umweltfreundlichen Stromerzeugung aus Kohle, K. Busekrus, 4th TÜV NORD Kraftwerksymposium, 2008.
[12] K. Busekrus, K.D. Tieges, F. Klauke, State of the art coal fired Steam Generators for low emissions CO_2, SO_2 and NOx, in: Coal Gen Europe 2008.
[13] C. Bergins, A. Leisse, S. Rehfeldt, How to utilize low grade coals below 1000 kcal/kg?, in: POWER-GEN Europe 2014.
[14] Update on lignite firing, Qian Zhu, IEA Clean Coal Centre, ISBN 978-92-9029-521-1, June 2012, Best Available Techniques (BAT).
[15] Wroclaw University scripts on coal combustion: http://fluid.wme.pwr.wroc.pl/~spalanie/dydaktyka/combustion_MiBM/BCS/COAL_COMBUSTION.PDF.
[16] A. Leisse, H. Gräwe, Pulverized coal burner for low NOx emissions, European Patent 1 741 977, 2006.
[17] H. Gräwe, A. Leisse, Burner for brown coal, European Patent 0 670 454, 1999.
[18] S. Rehfeldt, A. Leisse, A. Saponaro, Ignition of solid pulverized fuel by heated surfaces, in: Clearwater Clean Coal Conference 2015, 1–5 June 2014.
[19] Best Available Techniques (BATs) Reference Document (BREFs) for Large Combustion Plants, Draft, June 2016.
[20] Implications of the new LCP-BREF/BAT conclusions, EPPSA, VGB Congress, Power Plants 2016.
[21] Flue Gas Cleaning Retrofit—guideline for assessment and design recommendations, EPPSA, report, December 2015.
[22] Mercury Removal—a guideline for assessment and design recommendations, EPPSA Report, June 2015.
[23] LCP BREF for the Power Sector: Suppliers Point of View and resulting solutions LCP, in: Bernd Vollmehr, PowerGen Europe 2016, June 216, Milan.

Co-combustion of low-rank coals with biomass

6

Jarosław Zuwała and Janusz Lasek
Institute for Chemical Processing of Coal, Zamkowa, Zabrze, Poland

6.1 Introduction

Striving for a minimization of nonrenewable energy resources consumption and greenhouse gases (GHG) emissions is crucial from the point of view of sustainable mankind development. World CO_2 emissions statistics (2008) illustrate 41% of the total CO_2 emission coming from electricity and heat generation industries. Share of energy industry in total anthropogenic GHG emissions volume reaches 83%, of which 94% is CO_2 [1]. Thus, reduction of fossil fuels consumption and GHG emission in energy supply sector is becoming a very important issue. One of the approaches aiming at this is partial substitution of fossil fuels with solid biomass, commonly regarded as CO_2-neutral fuel. However, it must be borne in mind that when considering the full life cycle of biomass, emissions due to its cultivation, pre-treatment, and transport must not be neglected. During these stages, mostly fossil fuels are used which implies the nonnegligible contributions to primary resources consumption as well as GHG emissions. Useful products, including final energy carriers (electricity and heat), are generated as a result of power, technological, and transport processes implementation, creating a network of interdependences. Consequently, they should be burdened not only with the direct consumption of chemical energy in the processes of fuel combustion in utility boilers, but also with indirect energy consumption typical for processes of fuel acquisition, their transport and manufacturing of the machines and equipment used in all of the aforementioned stages (upstream processes). The same is valid for the emissions of GHG.

As long as biomass is used, all the stages from its acquisition until processing to the final physical form prior to use should be considered. A different biomass origin requires a diversified approach (e.g., forest residues or short rotation coppices applied). For residual biomass, the elimination or reduction of GHG emissions related to the natural biodegradation (CH_4 emissions) can be achieved using this residue biomass resource for energy generation.

Sustainability requirements for solid biomass for energy production were proposed in European Commission Report [2]. GHG performance for various biomass sorts and energy conversion technologies was identified by the European Commission as a base for EU MS recommendations concerning their national support schemes. So far none of the existing support schemes was found to be based on the assessment of real environmental benefits/burdens in the full life cycle of

Low-Rank Coals for Power Generation, Fuel and Chemical Production
DOI: http://dx.doi.org/10.1016/B978-0-08-100895-9.00006-1

biomass combustion or co-combustion. Mostly, support schemes were based on the level of investment costs of renewable energy units.

In this paper, the assessment of nonrenewable energy resources consumption and GHG emissions resulting from biomass co-firing in a co-generation plant with a backpressure turbine is developed based on life-cycle approach. Two different sorts of biomass, forest residual biomass (woodchips) and short rotation willow coppice, were considered as supplementary fuel.

The computational models were built and the simulations were carried out for three energy systems differentiated by combusted fuels. GHG emission and nonrenewable resources depletion were selected as impact categories based on SETAC guidelines [3].

6.2 State-of-the-art of co-firing of low-rank coal with biomass

Lignite is an important energy source. The combustion of lignite in power plants is widespread in many countries. It is assumed that low-rank coals (LRCs) constitute about 45% of the total global coal reserves [4]. Among many counters where lignite is extracted as energy crop, the largest lignite producers (data from 2013) are Germany (~183 Mt), Russia (~73 Mt), United States (~70 Mt), Poland (~66 Mt), ahead of Turkey (~63 Mt), Australia (~63 Mt), Greece (~54 Mt), India (~45 Mt), the Czech Republic (~40 Mt), and Serbia (~40 Mt) [5]. Co-combustion of LRCs with biomass in terms of biomass production capacity has potential ability to enrich the requirements. For example, it was estimated that the total above-ground *Miscanthus giganteus* yield was 20–35 t/ha/year in Europe [6]. The capacity of Polish producers of biomass reached 1 million dm^3/year [7]. In Latvia 75,530 ha of abandoned agricultural land suitable for short rotation woody crops that were dedicated to co-generation plants [8] was estimated. From around 2004 the development of energy crop areas in Germany is observed [9]. In Estonia, the volume of harvested spruce stumps (Norway spruce, *Picea abies*) reached up to 130 m^3/ha and potential energy content amounted to 290 MW h/ha [10]. On the contrary, energy crops areas and agricultural processing should be balanced. The rule of sustainable development must be taken into account due to such challenges like *food vs fuel* and coastal eutrophication problems [11,12]. Other important issue is the status of potential support for co-combusted producers of energy as the economic mechanisms supporting the development of renewable energy resources. For example, in Poland the system so-called green certificates created a support mechanism and it relied on requiring the energy companies to obtain the proof of biomass origin or to pay a compensatory fee otherwise. The green certificates constituted property rights that can be sold to generate profits, thus co-combustion was profitable for energy producers. After 2011 the market of green certificates noticed unstable dropping down, as presented in Fig. 6.1 [13]. However, the co-combustion of fossil fuels with biomass is still perspective solution to increase the application of renewable fuels.

Co-combustion of fossil fuels (including LRC) is widely applied over the world. Al-Mansour and Zuwała [15] reported that in 2010 a total number of co-firing

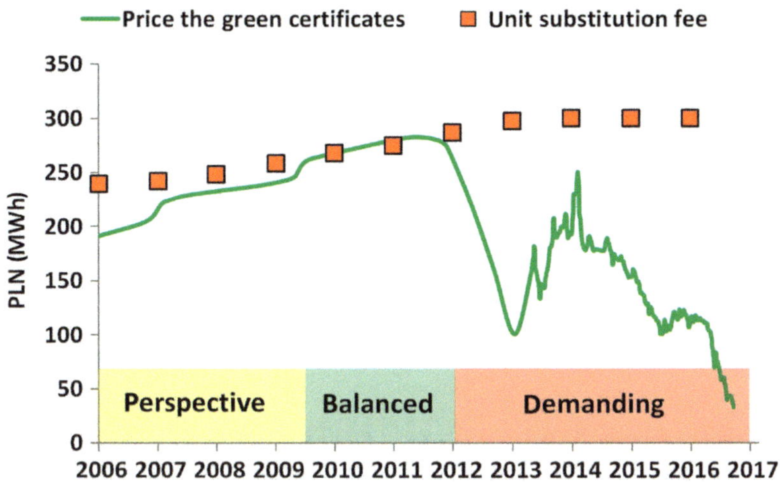

Figure 6.1 The market value of green certificates listed on Polish Power Exchange and the unit compensatory fee in the period 2005–16 (assumption: 1 PLN = 0.24 EUR).
Data from M. Dzikuć, A. Piwowar, Ecological and economic aspects of electric energy production using the biomass co-firing method: the case of Poland, Renew. Sustain. Energy Rev. 55 (2016) 856–862, https://www.tge.pl/pl/437/nowe-indeksy-dla-zielonych-certyfikatow (accessed 05.10.16), http://gramwzielone.pl/trendy/14848/ure-podal-wartosc-oplaty-zastepczej-dla-oze-na-2015-r (accessed 05.10.16) and M. Ćwil, Ustawa o OZE i projekty aktów wykonawczych, POLAGRA, Poznań, 2016 (21.01.2016) [14].

boilers (of the capacity in the range of 1–700 MW_e) exceed 220 commercial units. Finland was recognized as a country where the highest number units are operated (i.e., 78 units capacity in the range of 20–310 MW). Location of US co-firing units and energy crops capacity were presented by Dumortier [16]. Among four main types or furnaces, i.e., pulverized fuel (PF), circulating fluidized bed (CFB), bubbling fluidized bed (BFB), and grate (GRATE), the PF boilers are the most common ones in co-firing technology. There are three basic co-firing options for biomass materials in coal-fired boilers [15,17]:

- Direct co-firing
- Indirect co-firing
- Parallel co-firing

Fig. 6.2 shows scheme of three co-firing options. All of these ways are commercially applicable. Direct co-firing is the most commonly applied approach. It is possible to burn biomass and coal using the same or separate burners. However, it was estimated that co-combustion of biomass with coal at biomass fraction of approximately up to 3% (energy based) did not need significant investment costs. Indirect co-firing includes the combination of a fossil fuel furnace with the gasification reactor of biomass. Next, the process gas from gasifier is combusted in coal boiler. The third option, i.e., parallel co-firing, concerns two separate boilers, i.e., coal and

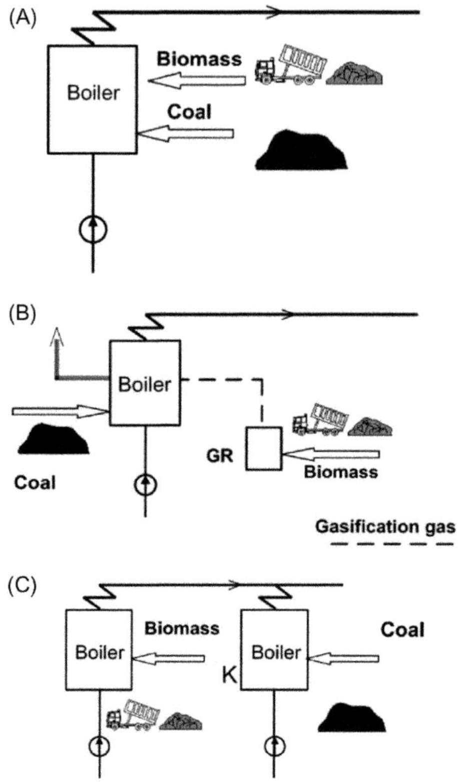

Figure 6.2 Biomass co-firing technologies. (A) Direct co-firing, (B) indirect co-firing, and (C) parallel co-firing. *GR*, gasification reactor [15].
Printed upon permission of Elsevier, License Number 3847580622965.

biomass fired, and combines these boilers into one system to produce of steam [15]. Another issue is the adaptation or retrofitting of existed fossil fuel boilers into co-combusting units. It is known that grate and CFB boilers can combust a relatively wide range of fuel types (e.g., in terms of particle size, composition, and moisture content). After minor modifications, these types of boilers can be usually adapted for efficient co-combustion of fossil fuels and biomass [18]. Tokarski et al. confirmed that co-combustion of hard coal with biomass in the ratio of up to 10% (energy basis) had no negative effect on the boilers efficiency [19]. It was concluded by the Karampinis et al. [17] co-firing of lignite with biomass having the potential to improve combustion conditions in the furnace.

Co-firing of LRC (mainly lignite) and biomass is valuable due to the possibility of increase of production efficiency. Zuwała [20] and Tokarski et al. [19] analyzed different boilers and impact of biomass fraction on boiler efficiency was presented. It was proved that in case of lignite co-combustion with biomass, the boiler efficiency increased when biomass fraction increased. The analysis was based on commercially operated boilers when sawdust, pellets of sunflower husk, pellets of sawdust, pellets of straw, and willow chips were co-combusted. The analyzed range of biomass fraction (energy basis) was 0%–25%. The empirical formula was estimated (see Eq. (6.12)) that proved the positive effect of lignite and biomass co-combustion on boiler thermal efficiency.

Co-firing of LRC is also considered by demonstration-scale projects. The example is the project entitled "Demonstration of Large Scale Biomass Co-Firing and Supply Chain Integration", DEBCO. DEBCO was a collaborative project in the framework of FP7 program (2008–12) that involved 17 partners from 8 different European countries. One of the project aim was to carry out industrial scale (demonstration) test of co-firing the Greek lignite with cardoon and olive kernel at biomass thermal input up to 10%. The tests were carried out in Kardia power plant (PC dry-bottom boiler, capacity of $300MW_{el}/856MW_{th}$, biomass milled and injected with the main fuel) [http://cordis.europa.eu/project/rcn/90325_en.html (accessed 05.10.16), http://www.debco.eu/attachments/093_Article%20Parliament%20Magazine%20June%202012.pdf (accessed 05.10.16)]. It was observed that the boiler efficiency was not influenced by co-firing biomass at a co-firing level of 10%. Moreover, no significant changes were observed in terms of slagging and flue gas temperatures at various positions in the boiler. However, the significant change of unburned carbon in bottom ashes was observed. The beneficial effect was obtained for ash properties. The co-firing caused more consistent ash chemical composition compared to 100% lignite firing. As a result, the quality of ash was increased and its use in concrete or cement would be considered after further processing [21]. Currently, examples of the power plants during the transformation from coal-fired into co-fired or biomass-fired power plant are Drax (United Kingdom), Langerlo (Belgium), and Lynemouth (United Kingdom). Drax is located in Selby, Yorkshire (United Kingdom) and is owned by Drax Power Limited (the operating subsidiary of Drax Group). It is capable to provide 4000 MW_e and the co-firing fuel is wood-derived biomass (e.g., willow) [http://www.power-technology.com/projects/drax/ (accessed 05.10.16)]. The Langerlo power plant is situated in Flanders region of Belgium. This year (2016) it was announced that Estonia's Graanul Invest bought the Langerlo power plant and invested about 250 million euros to switch the unit from coal-fired unit to pellet burning unit [http://www.baltic-course.com/eng/energy/?doc=121752 (accessed 05.10.16)]. It is suspected that full transformation of power plant will be finished in 2018 [http://www.business.conbio.info/pellet-producer-graanul-invest-buys-genk-langerlo-power-plant-in-belgium/ (accessed 05.10.16)]. Lynemouth (United Kingdom) is another power plant that is converted from coal-fired unit to biomass-fired unit. It is assumed that power plant consists of three 140 MW boilers and they will consume approximately 1.4 million metric tons of wood pellets annually for the next 10 years [http://biomassmagazine.com/articles/13314/lynemouth-conversion-progresses-doosan-babcock-awarded-contract (accessed 05.10.16)]. The combustion and emission systems will be delivered by Doosan Babcock company [http://www.doosanpowersystems.com/en/media/news_view.do?pressSeq=20160519181714009571; http://biomassmagazine.com/articles/13314/lynemouth-conversion-progresses-doosan-babcock-awarded-contract (accessed 05.10.16)].

It was found that the biomass materials have higher thermochemical reactivity than lignite and shorter devolatilization times that can ensure their complete conversion within operating lignite-fired boilers [22,23]. Table 6.1 shows benefits and barriers that were observed during co-combustion of lignite and biomass. Table 6.2 presents an overall view on the phenomena observed during co-combustion of LRCs and biomass. The results were ranked from laboratory-scale investigations up to industrial scale tests. It is known that the main challenge of biomass combustion is its negative effect on slagging, fouling, and corrosion of heat exchangers surfaces. Besides sodium and potassium, the presence of phosphorus can also greatly reduce the melting point of ash. The melting temperatures of the mineral compounds in ash are ordered as follows: $Al_2O_3 > MgO > CaO > P_2O_5 > Na_2O > K_2O$ [39]. Alkali metals, K and Na, in biomass play an important role in the formation of agglomeration during co-combustion of biomass in fluidized beds. It was reported by Varol and Atimlay [30] that high temperature of SiO_2 melting point (i.e. 1450°C) can be significantly reduced when SiO_2 reacted with alkali oxides (874°C) or alkali salts (764°C) in the ash. It was explained by the formation of eutectic with melting temperatures lower than the melting temperature of individual components of the eutectic (i.e. 891°C for K_2CO_3 and 851°C for Na_2CO_3). Moreover, Varol and Atimlay noticed interesting behavior during co-combustion of high-sulfur lignite coals and olive cake

Table 6.1 Main benefits and barriers/challenges connected with co-combustion of low-rank coal with biomass

Benefits	Barriers/challenges
Reduction of the emissions of SO_2, NO_x, and CO_2 per unit energy generated [24–27]	The alkali metals and halogens could cause operational problems with regard to slagging, fouling, or corrosion [28–30]
Reduction of CO_2 emission (adding approximately 20% of the biomass decreases the carbon dioxide emissions for 11%–25%) [31]	In case of high nitrogen content, co-combustion leads to high emissions of NO, NO_2, and N_2O [28]
Limitation of NO_x emission due to lower nitrogen content, local reburning effect and reduction of flame temperature [32,33]	Possible higher CO emission [25,27]
Compensation of a negative effect of low calorific value and ash composition of low-rank coal [28]	Reduction of combustion efficiency [34]
Higher thermochemical reactivity than lignite and shorter devolatilization times that can ensure biomass complete conversion within operating lignite-fired boilers [23]	Increase of unburnt losses in the fly and bottom ash in case of larger biomass particles (3 mm) combustion [33]
Increase of boiler efficiency [19,32]	
The reduction of the amount of limestone required for decrease of SO_2 emissions [35]	

Table 6.2 Experimental investigations of co-combustion of low-rank coal with biomass

Investigated fuels	Scale	Main results	Ref.
Lignite from Kardia Power Plant in Western Macedonia + pine needles, cardoon	TGA	Blending lignite with biomass fuels increased its thermochemical reactivity. The burnout temperature, T_b was lower and the burnout time t_b was shortened when pine needles were co-combusted with lignite	[36]
Low quality Turkish lignites + woody shells of sunflower seed	TGA	The blends of Turkish lignites with sunflower seed shell did not behave in very different way, they can be co-combusted in the existing combustion system	[28]
Turkish lignite from Soma-Denis region + woody biomass *Robinia pseudoacacia*	TGA	Carbonization of lignite and biomass eliminated segregation in weight loss properties as well as heat flows during co-combustion. Generally, the combustion performance was more uniform for carbonizated fuels. The carbonization process allows co-utilization with high substitution ratios of biomass	[37]
Bursa–Orhaneli lignite + olive cake and Denizli-Kale lignite + olive cake, with and without limestone addition	Laboratory-scale CFB combustor (inside diameter of 108 mm and a height of 6 m)	Sulfur in coal prevented the agglomeration in a fluidized bed combustor. Addition of woodchips increased CO emission but reduced char formation, NO and SO_2 emissions	[26,27,30]

(*Continued*)

Table 6.2 (Continued)

Investigated fuels	Scale	Main results	Ref.
Lignite + foot cake [solid residues from the olives (skim, meat, and pit), some oil retained (about 2%–4%) and water (typically 60%–70%)]	Bubbling Fluidized Bed combustor (inside diameter of 0.2 m and height of 3 m)	Increase of combustion efficiency (E_c) when temperature was increased, the share of foot cake has not any significant effect on E_c. Decrease of the SO_2 emissions due to SSR effect (calcium and potassium in ashes). Slight limitation of NO_x emission due to local reburning effect and reduction of flame temperature	[32]
Tunçbilek lignite + apricot stones or peach stones	BFBC (inside diameter of 102 mm and a height of 900 mm)	Decrease of SO_2 emission, increase of CO emission thus a secondary air addition is required for the system.	[25]
Greek lignite from Ptolemais reserve + natural waste wood, MDF residues and power poles	13.8 MW_{th} industrial steam boiler, moving grate, with 17 series of stairs (PINDOS SA)	A waste wood co-combustion is a promising option for industrial and district heating boilers. In all cases, emissions of CO, NO, and SO_2 were lower than the legislative limits. MDF residues resulting in a small improvement of emissions and combustion efficiency, under the same operating conditions.	[38]
Lignite + willow (*Salix viminalis*)	Industrial CFB boiler	Co-combustion caused decrease limestone feed for SO_2 capture	[35]

in a CFB combustor. It was found that sulfur in coal prevented agglomeration in the bed. Thus, it was advantageous to add high-S coals to the combustor to prevent the negative effect of biomass addition (i.e., agglomeration) [30].

Next, significant benefits of lower emission of such gaseous pollutants like NO_x and SO_2 can be obtained [27]. Varol et al. investigated the impact of excess air ratio (λ) and woodchips fraction (during co-combustion with lignite) on NO, CO, and SO_2 emissions. Addition of biomass (lignite and woodchips mixtures with woodchips as 10, 30, and 50 wt% of the mixture) caused the increase of CO emission above the limits (Turkish regulation) for coal combustion. However, if optimal λ is kept (i.e., 1.18, 1.32, and 1.41 for biomass fraction of 10%, 30%, and 50%, respectively) the CO emission was below the limits. The addition of biomass caused no change in NO emission and significant change in SO_2 emission. The results obtained by Varol et al. are presented in Fig. 6.3 [27]. It can be noticed from this

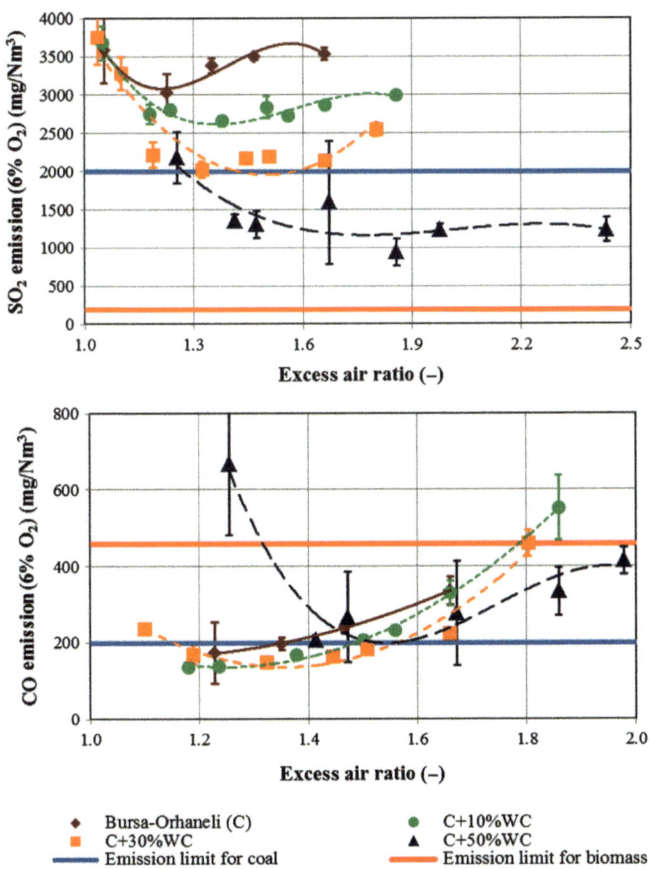

Figure 6.3 Effect of woodchips addition on CO and SO_2 emissions with respect to excess air ratio and comparison of emissions to the limits in Turkish regulation. *WC*, woodchips; dotted lines represent trend of emission [27].
Printed upon permission of Elsevier, License Number 3847580964500.

figure that SO_2 emission was reduced more than the corresponding limitation of sulfur content in the fuel. In other words combustion of lignite and biomass mixture at 50% of the biomass fraction caused slightly more than 50% reduction of SO_2 emission at $\lambda > 1.3$. Sulfur content (S_{total}) in the lignite was 1.76 wt%, whereas sulfur content in the biomass was below the detection limit. Lasek and Kazalski [35] underscored the role of sulfur self-retention (SSR) on reduction of SO_2 emission from industrial combustion chambers. SSR involves reactions between the mineral matter in coal ash and the SO_2 evolved during coal combustion, and consequently, the emission of SO_2 may be significantly reduced. The positive effect of SSR during co-combustion of LRCs was noticed by other researchers. The obtained results were briefly presented in Table 6.3. Efthimiadou and co-workers suggested that

Table 6.3 Limitation of SO_2 emission during co-combustion of low-rank coals and biomass due to SSR process

Type and capacity of combustor	Fuels	Obtained SO_2 limitation	Ref.
20 kW PF, entrained flow reactor	Kakanj brown coal ($S = 0.73\%$ ar) and a lignite blend of Dubrave–Šikulje ($S = 2.28\%$ ar) + spruce sawdust ($S = 0.33\%$ ar); Coal/biomass mixtures at 93:7 and 80:20 wt%	at 1140°C, there was 15% less SO_2 for the 93% − 7% Kakanj−spruce blend and 28% less for the 80% − 20% Kakanj−spruce blend than for Kakanj coal alone	[40]
BFBC 102 mm inside diameter and 900 mm height, fuel feeding rate of 10 g/min (for co-combustion)	Tunçbilek lignite ($S = 2.1\%$ db) + olive cake ($S = 0.08\%$ db), mixing ratios of 25%, 50%, and 75% of olive cake by weight	For lignite (separate) combustion results was not provided, for 25%, 50%, and 75% of olive the emission was ~2600 mg/m_n^3, ~1800 mg/m_n^3 and ~55 mg/m_n^3 at $\lambda \approx 1.5$	[41]
LAB-CFBC with inside diameter of 108 mm and a height of 6 m	Bursa–Orhaneli lignite ($S = 1.76\%$ db) and woodchips ($S = 0.00\%$ db) mixtures containing 10%, 30%, and 50% by wt. of woodchips	25% and 49% decreasing of SO_2 emission when fuel was switched from lignite to mixture of lignite and woodchips (10% and 30%, respectively) (emission decreased from 3676 ± 222 mg/m_n^3 to	[27]

(*Continued*)

Table 6.3 **(Continued)**

Type and capacity of combustor	Fuels	Obtained SO_2 limitation	Ref.
		2752 ± 128 mg/m_n^3) when λ was increased from 1.05 to 1.18 at woodchips 10% emission decreased from 3750 ± 355 mg/m_n^3 to 2023 ± 106 mg/m_n^3 when λ was increased from 1.04 to 1.32	
13.8 MW_{th} industrial steam boiler, moving grate, with 17 series of stairs (PINDOS SA)	Greek lignite (S = 1.1% db) from Ptolemais reserve + natural waste wood (S = 0.2% db), MDF residues (S = 0.3% db) and power poles (S = 0.0% db)	In all cases, emission of SO_2 was below 200 mg/m_n^3	[38]
BFBC (inside diameter of 0.2 m and height of 3 m)	Lignite (S = 0.8% db) + foot cake (S = 0.1% db) [solid residues from the olives (skim, meat, and pit), some oil retained (about 2%–4%) and water (typically 60%–70%)]	~16% decreasing of SO_2 emission (emission decreased from ~1020 mg/m_n^3 to ~860 mg/m_n^3 when oil cake fraction increases from 10 wt% to 25%)	[32]
BFBC (inside diameter of 102 mm and a height of 900 mm)	Tunçbilek lignite (S = 1.81% db) + Apricot Stones (S = 0.15% db) or Peach Stones (S = 0.14% db)	~33.6%, 57.2%, and 74.0% SO_2 emission reductions when 25 wt%, 50 wt%, and 75 wt% peach stone addition to the coal was used respectively. ~39.5%, 67.4%, and 83.5% SO_2 emission reductions when 25 wt%, 50 wt%, and 75 wt% apricot stone addition to the coal was used respectively.	[25]
Lignite + willow (Salix viminalis)	Industrial CFB boiler	Co-combustion caused decrease limestone feed for SO_2 capture.	[35]

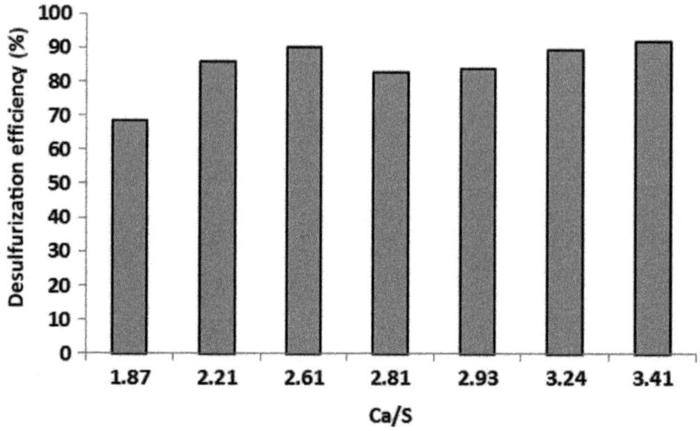

Figure 6.4 Natural desulfurization efficiency related to Ca/S ratio at load of 299–316 MW_e. Data from A. Efthimiadou, I.P. Nikolaides, A. Tourlidakis, A proposal for SO_2 abatement in existing power plants using rich in calcium lignite, Appl. Therm. Eng. 74 (2015) 119–127.

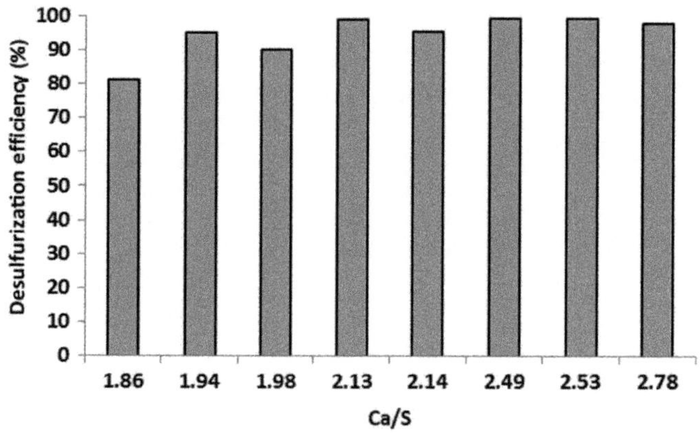

Figure 6.5 Natural desulfurization efficiency related to Ca/S ratio at load of 239–283 MW_e. Data from A. Efthimiadou, I.P. Nikolaides, A. Tourlidakis, A proposal for SO_2 abatement in existing power plants using rich in calcium lignite, Appl. Therm. Eng. 74 (2015) 119–127.

high-ash content lignite can be applied as natural desulfurization agent. The percentage of desulpurization experienced varies from 70% to almost 96% for lignite from Ptolemais–Amyndeon basin. The lignite can be mixed with other lignite or hard coal. The idea was applied in real scale boilers. Figs. 6.4 and 6.5 represent obtained desulfurization efficiency. The data was collected in real scale boilers in the period of 2009 and 2010. They also noted very important and practical rule that operational cost is rather small compared to investment cost of desulfurization system. When the Ca/S ratio is high (i.e., Ca/S $>$ 3) the cost of sorbent is dominant compared to the investment cost [42]. It proves the motivation of the research of SSR process to receive as high $deSO_2$ efficiency as possible at minimal cost of sorbents and cost of ash utilization.

6.3 Comparison of technological options—Co-firing in PF and CFB boilers—Trial test real data, energy, and environmental boiler performance

It is known that CFB boilers have higher potential for more flexible operational range in terms of fuel type and biomass fraction. Ericsson [18] compared different co-firing units in Poland. Due to higher number of PF boilers, efficient operational ranges (in terms of biomass fraction) should be taken into account. On the contrary, retrofitting of CFB boiler into co-combusted unit costs less than retrofitting of PF boiler. It was estimated that in case of Polish energy system, in 2007 the total cost of retrofitting of CFB 300 MWth boiler into co-combusted unit (20%) was in range of 1.1−3.1 €/GJ_{bio}. The total cost of PF 600 MWth boiler into co-combusted unit (10%) was in range of 1.9−3.4 €/GJ_{bio}. More particular information about parameters of this analysis can be found in the work of Ericsson [18]. CFB technology gives more possibilities of dry desulfurization by injection of limestone or other deSO_2 agent. It was described in the previous section that sulfur retention can be obtained by the reaction of SO_2 with some contaminants of ash (mainly Ca and Mg species). This process, called SSR, was confirmed at real, industrial conditions. Table 6.4 shows the results of industrial trial test. Lasek and Kazalski [35] and Kazalski et al. [43] noticed that limestone feed (to CFB boiler) was reduced by ~37% when 6.3 wt% of willow was added to main fuel (lignite). The phenomenon was confirmed at lower (157 MWth) and higher (261 MWth) capacity of the boiler. The parameters of produced stem were temperature of 565°C and pressure of 16.7 MPa.

Table 6.5 shows the main specifications of other Polish units where co-combustion of LRC with biomass is carried out or was carried out in near past.

6.4 Influence of low-rank coal and biomass co-firing on the depletion of nonrenewable energy resources and GHG emission—A life-cycle approach

Useful products, including final energy carriers (electricity and heat), are generated as a result of power, technological, and transport processes implementation, creating a network of interdependences. Consequently, they should be burdened not only with the direct consumption of chemical energy in the processes of fuel combustion in utility boilers but also with indirect energy consumption typical for processes of fuel acquisition, their transport, and manufacturing of the machines and equipment used in all of the aforementioned stages (upstream processes). The same is valid for the emissions of GHG.

As long as biomass is used, all the stages from its acquisition until processing to the final physical form prior to use should be considered. A different biomass origin requires a diversified approach (e.g., forest residues or short rotation coppices

Table 6.4 Comparison of sulfur retention during combustion of lignite and co-combustion with *Salix viminalis* in commercial CFB boiler [35]

Parameter	Lignite combustion	Co-firing of lignite and salix (6.3 wt%)
Theoretical SO_2 concentration ($mg/m^3{}_n$) at 6% O_2	1226	1164
Capacity of 157 MWth		
Measured SO_2 ($mg/m^3{}_n$) at 6%	395	333
Sulfur retention (%)	68	71
Limestone feed (kg/s)	1.1	0.7
Ca/S[a]	3.7–3.9	2.9–3.1
Limestone savings (%)	–	36
Fuel feed (t/h)	118–122	113–117
Biomass feed (t/h)	0	7
Bed temperature (°C)	840	845
Capacity of 261 MWth		
Measured SO_2 ($mg/m^3{}_n$) at 6%	395	363
Sulfur retention (%)	68	69
Limestone feed (kg/s)	2.1	1.3
Ca/S[a]	4.2–4.4	2.9–3.1
Lime savings (%)	–	38
Fuel feed (t/h)	199–201	197–201
Biomass feed (t/h)	0	12
Bed temperature (°C)	870	875

[a]Ca/S calculated including the total amount of CaO introduced with lignite, biomass, and lime.

Table 6.5 Specification of Polish co-firing power plants (PP) and CHP plant where low-rank coal is used as a main fuel

Plant/operator	Electric power (Mwe)	Thermal power (MWt)	Primary fuel	Boiler type
PGE SA—Elektrownia Turów (PP)	1699	–	Lignite	CFB
PGE SA—Elektrownia Bełchatów PGE SA	4440	375.6	Lignite	PC
ZE PAK SA—Elektrownia Pątnów (PP)	1200	–	Lignite	PC
ZE PAK SA—Elektrownia Adamów (PP)	600	–	Lignite	PC
ZE PAK SA—Elektrownia Konin (PP)				BFB
PGE SA—Elektrociepłownia Zgierz (CHP)	32.4	160	Hard coal/ lignite	BFB

Data from A. Piwowar, M. Dzikuć, Outline of the economic and technical problems associated with the co-combustion of biomass in Poland, Renew. Sustain. Energy Rev. 54 (2016) 415–420 [44] and own elaboration.

applied). For residual biomass, the elimination or reduction of GHG emissions related to its natural biodegradation (CH_4 emissions) can be achieved by co-firing this biomass for energy generation.

In this chapter, the assessment of nonrenewable energy resources consumption and GHG emissions resulting from biomass co-firing in a co-generation plant with a backpressure turbine is developed based on life-cycle approach. Two different sorts of biomass, forest residual biomass (woodchips) and short rotation willow coppice, were considered as supplementary fuel.

The computational models were built and the simulations were carried out for three energy systems differentiated by combusted fuels. GHG emission and nonrenewable resources depletion were selected as impact categories based on SETAC guidelines [45].

Analysis covering the full life cycle should start from raw material acquisition through their production and use until the final disposal (waste deposition or their further application in other processes). The following definition of LCA, taken from ISO standards [45,46], can be given: "A systematic set of procedures for compiling and examining the inputs and outputs of materials and energy and the associated environmental impacts directly attributable to the functioning of a product or service system throughout its life cycle."

Generally, three groups of environmental burdens may be distinguished, namely:

- impact on the quality of the ecosystem (e.g., pollution of the atmosphere, soil and water, and waste disposal),
- the impact on human health (e.g., occupational diseases and occupational safety),
- depletion of natural resources and the associated land degradation, e.g., changing the structure of groundwater.

LCA analysis has been often recognized as an appropriate method for the evaluation of the sustainability of biomass-based energy systems compared to their fossil alternatives [47–52]. Most of the authors considered only electricity production (no need of allocation procedure then) as a useful product of the regarded utility plant. Real operational data from biomass combustion tests were underlined as an important component of the analysis. However only in [50] co-firing boiler efficiency drops were included in the analyses, only as discrete values instead of being a variable function of the biomass share in the co-fired blend.

No single LCA methodology was found to be a referenced one although Renewable Energy Directive [53] lays down the methodology for biofuels and bioliquids for transport. Methodological assumptions like type of feedstock and system boundaries conversion technologies have a strong effect on the results of the analysis. Also, computational models can be the reason for uncertainty of the results obtained from the analysis. This is the effect of the limited reliability of the model, as well as of the incompleteness of being mostly the result of inconsistency between its components [54]. Any mathematical model reproduces the reality only with limited approximation. Life-cycle analysis is always a compromise between scientific accuracy and practical application. Here, Bellman's statement concerning mathematical models can be evoked: a mathematical model is a compromise between the trap of simplicity and the swamp of details. Therefore, it is recommended to remember that in LCA studies there is always a group of processes, which has a strong influence on the results of the analysis. To distinguish the most important ones from the less important parameters, the sensitivity analysis should be applied.

The factors causing the uncertainty of input data can be divided into exterior and interior ones. As examples of exterior factors, the links of the regarded system to the other systems (e.g., other branches of the economy) can be given. The duration curve of ambient temperature determining the heat demand can be given as another example. The availability of fuels and raw materials, which can enforce the necessity of substitution in co-generation systems, is another external factor. Concerning the internal factors, mainly energy efficiency of plant's machinery or equipment availability can be named. The role of input data uncertainty is becoming more important when the equipment is getting older. This is a separate problem in the LCA study, similar to the problem of optimization under uncertainty of input data [55]. Also, the authors of [56] underlined the assets of the co-production of electricity and heat regarding environmental and energy benefits. Life-cycle management tools have been found to be a proper tool for combined heat and power (CHP) plant-based energy systems. The detailed methodology of life-cycle assessment (LCA) of biomass and hard coal co-firing in CHP plant is described elsewhere [57].

The target of the analysis performed is to evaluate the impacts of biomass and coal co-firing in a CHP plant with a backpressure turbine on the environment in terms of nonrenewable energy resources depletion and GHG depletion. Lignite combustion was taken as a baseline scenario (system 1) and two other variant systems were considered:

- System 2: Co-firing of LRC (lignite) with willow biomass.
- System 3: Co-firing of LRC (lignite) with residual biomass (forest woodchips).

In both of the systems, share of co-fired biomass ranges from 0% to 20% on energy basis which reflects the co-firing practice and enables the sensitivity analysis. Cradle-to-grave pathways of willow chips, residual biomass, and lignite were constructed to form a tree of process blocks to be covered by life-cycle analysis (Fig. 6.6).

Considerations in this chapter do not regard transmission and distribution systems which can be justified for large co-generation systems. Anyway, complex life-cycle analysis of decentralized CHP plants based systems (not analyzed here) with extensive district heating networks should require these additional process blocks.

The aim of the analysis is to compare the environmental footprint of combusted lignite and biomass & lignite co-firing CHP plant. Geographical context for the assessment done in this study can be any country using biomass and fossil fuels as operational data regarding biomass co-firing in utility boilers can be considered as universal. Only some country-specific parameters, e.g., lignite upstream burdens, national energy mix, and fuel transport options, reflect Polish conditions.

The analysis was carried out with the example of a backpressure turbine CHP plant presented in Fig. 6.7, adapted from [57]. Backpressure turbine based CHP plant analyzed that there is a 55 MW_{el} unit typical for Polish co-generation units

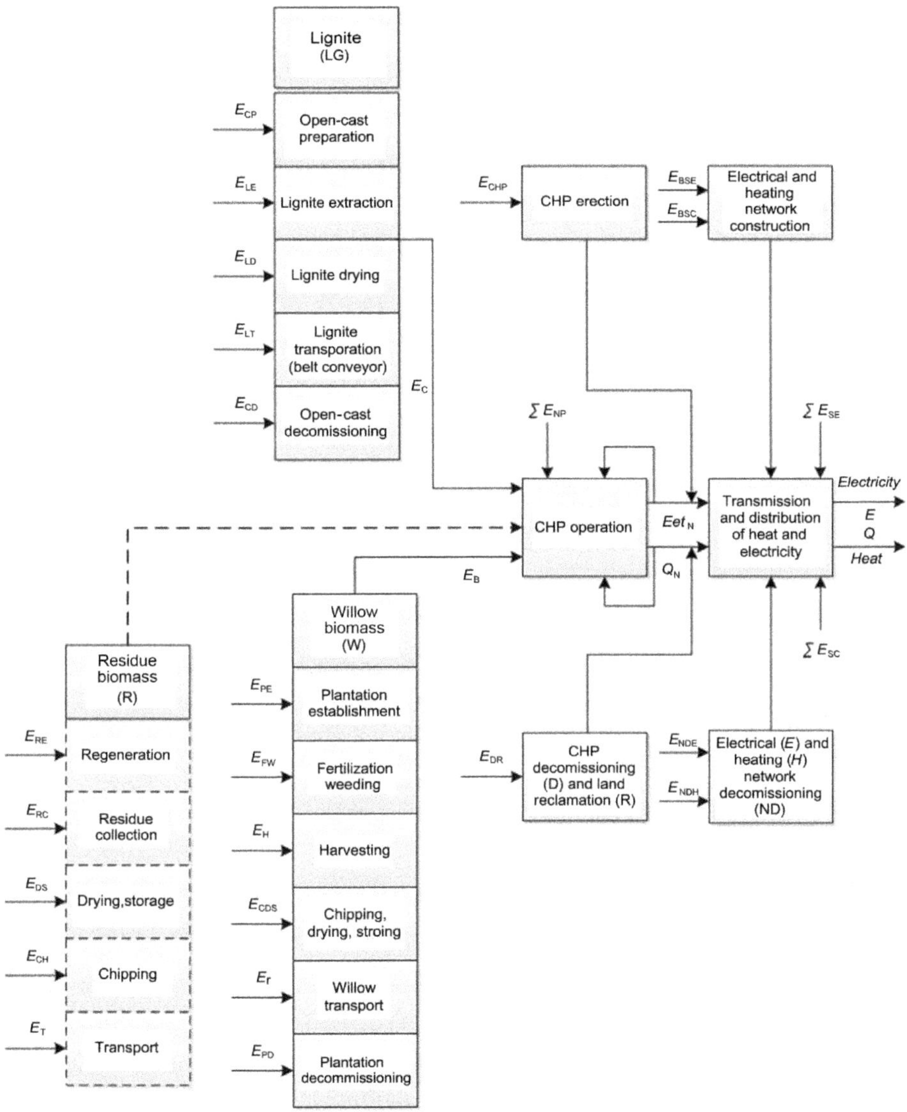

Figure 6.6 Conceptual drawing of the compared systems of coal and biomass co-firing in a CHP plant.
Scheme adapted from J. Zuwała, Life cycle approach for energy and environmental analysis of biomass and coal co-firing in CHP plant with backpressure turbine, J. Clean. Prod. 35 (2012) 164–175 printed upon permission of Elsevier, License Number 3847581198205.

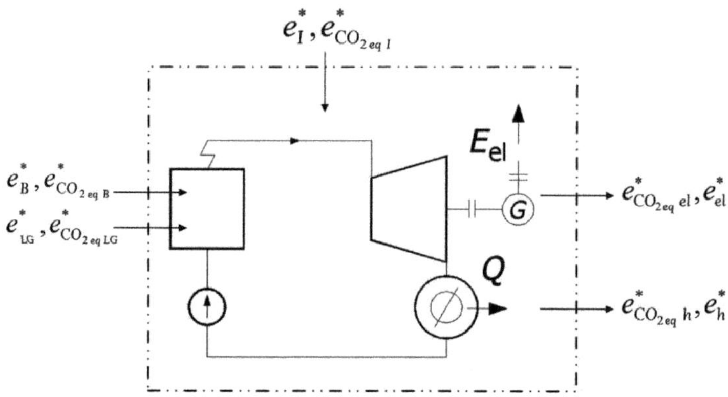

Figure 6.7 CHP plant with backpressure turbine—calculation layout for life-cycle analysis of nonrenewable resources depletion and GHG emissions.
Scheme adapted from J. Zuwała, Life cycle approach for energy and environmental analysis of biomass and coal co-firing in CHP plant with backpressure turbine, J. Clean. Prod. 35 (2012) 164–175 printed upon permission of Elsevier, License Number 3847581198205.

producing electricity and heat. District heating exchangers are fed by extraction and backpressure steam. The production of electricity is fully depending on heat demand, which in turn allows setting district heat as a main product and electricity as a by-product of such CHP plant-based co-generation system.

After setting the described boundaries of the system, the following input/output flows of the CHP plant can be defined:

- fuel "at the gate of CHP plant" with cumulative upstream burdens (nonrenewable energy resources consumption and GHG emission respectively) expressed by coefficients e^*_B, $e^*_{CO_{2eq}\ B}$ (for biomass) and e^*_{LG}, $e^*_{CO_{2eq}\ LG}$ (for lignite),
- electricity and heat "at the gate of CHP plant" with cumulative burdens (nonrenewable energy resources consumption and GHG emission, respectively) expressed by coefficients e^*_h, $e^*_{CO_{2eq}\ h}$ (for heat) and e^*_{el}, $e^*_{CO_{2eq}\ el}$ (for electricity),
- construction and decommissioning of CHP plant and biomass co-firing installation expressed by coefficients (e^*_I, $e^*_{CO_{2eq}\ I}$).

Functional unit chosen to compare the systems was 1 TJ of heat; additionally during the analysis, the influence of biomass addition on unit production of electricity, 1 MWh, was also discussed.

The LCA methodology applied for the research is based on the following approach: first, the cumulative coefficients of the nonrenewable energy consumption and GHG emission are evaluated for all the fuels and afterward the total burdens in the whole life cycle are estimated.

Obtained information can be used to prioritize biomass sorts used to improve the environmental performance of fossil fuel based energy generation. Intended users of the study results can be the policy makers within the field of renewable energy based support systems framework, especially based on biomass combustion and co-firing.

An important, well-known methodological problem concerning life-cycle analyses is an allocation of environmental burdens [58,59] of multifunctional processes. During allocation, environmental burdens are assigned to each useful product of such systems.

For the purpose of this study, the allocation method in which the harmful effects are partitioned between district heat and electricity in a way that reflects the avoided fuel input in a CHP plant [55,60] compared to the separate production of these energy carriers was applied. This method, commonly known as "avoided-process approach," is based on the assumption that, by co-generation of electricity and heat, electricity production replaces an "avoided" electricity produced during electricity-only production in utility plants and this in turn made it possible to define a primary function of a CHP-based energy system: production of district heat with production of "waste" electricity.

The analysis is done in a comparative way, where lignite alone combustion (system 1) was taken as the baseline case. Concerning the "lignite pathway," the following process blocks can be listed: open-cast preparation, lignite extraction, lignite drying, lignite transportation (belt conveyor), and open-cast decommissioning.

The total cumulative nonrenewable energy consumption (or cumulative GHG emission) E_{LG} burdening lignite supplied to the CHP plant can be calculated from the following equation:

$$E_{LG} = E_{CP} + E_{LE} + E_{LD} + E_{LT} + E_{CD} \tag{6.1}$$

The total energy consumption (or cumulative GHG emission) burdening willow and residue biomass E_B supplied to the CHP plant can be calculated from the following equations:

$$\begin{aligned} E_W &= E_{PE} + E_{FW} + E_H + E_{CDS} + E_T + E_{PD} \\ E_R &= E_{RE} + E_{RC} + E_{DS} + E_{CH} + E_T \end{aligned} \tag{6.2}$$

The software package used to track the material and energy flows between and within the individual process blocks of the system (Fig. 6.6) was Tools for Environmental Analysis and Management (TEAM), by Ecobalance, Inc., with manual modeling of the processes covering biomass procurement and supply. Proper data were collected during the research described in detail in [61], based on the literature and the real operational data and entered into TEAM model.

Process block "willow" comprises the following unit processes: plantation establishment, fertilization and weeding, harvesting, chipping, storing and drying, transportation and plantation decommissioning. In order to allocate the energy inputs for regeneration and harvesting of forest to the products (sawlogs, roundwood, and residues), an economic value approach (price per oven dry tone) was applied. An allocation key, respectively, 4:2:1 based on data of [62] was applied. Also other authors [59] back up such approach, mentioning value-based method as one of the methods for multiple products systems allocation issue (forest products processing system is just an example of such a system producing timber sawlogs and roundwood as main

Table 6.6 Fuel data elaborated based on the life-cycle inventory of the regarded fuels

Parameter	Value
Cumulative nonrenewable energy consumption for willow biomass (J/J)	0.0653
Cumulative nonrenewable energy consumption for residue biomass (J/J)	0.0306
Cumulative nonrenewable energy consumption for lignite (J/J)	1.083
Cumulative GHG emission for willow biomass ($kgCO_{2eq}/J$)	0.010
Cumulative GHG emission for residue biomass ($kgCO_{2eq}/J$)	0.002
Cumulative GHG emission for lignite ($kgCO_{2eq}/J$)	0.120

products and residual product like woodchips). Rosen [58] mentions also the advantages of value-based allocation method for owners of multiple purpose systems selling their products separately, which surely takes place in case of wood processing plants. Also, Guinee in [63] gives the possibility of economic allocation based on value.

Energy inputs for transport and milling of sawlogs were allocated to sawn timber; energy inputs to chipping of chunks were fully attributed to the chips derived, like those suggested in [62].

Including all the steps included in [45,46] the life-cycle inventory was made resulting in data presented in Table 6.6. Cumulative nonrenewable energy consumption was expressed in J/J of chemical energy of regarded fuel. Cumulative emission of GHG (CO_2, N_2O, and CH_4) for hard coal and biomass was also identified through all the process blocks presented in Fig. 6.6. Based on real operational data from biomass co-firing plants, concerning the individual emissions [64], the cumulative emission indices of GHG for both willow and residual biomass have been elaborated. Lignite coming from open-cast mines is burdened with GHG cumulative emission $e^*_{CO_{2eq}\,LG}$. To express cumulative emission coefficients in CO_{2eq} IPPC individual global warming potential for the individual GHG in 100-year time horizon [65] were applied.

Data in Table 6.6 have been elaborated in [61] where calculation procedures are presented in detail.

As the boundaries of the system are limited only to fuel pathways and the CHP itself (Fig. 6.7), the cumulative energy consumption (or cumulative GHG emission) relating to net electricity and heat production during biomass co-firing can be calculated from the following equation:

$$E_{LCA} = E_{LG} + E_B + \sum E_{N\text{-}F} + E_{CHP\ C} + E_{CHP\ D} \tag{6.3}$$

Basic parameters for CHP plant presented in Fig. 6.7 were identified and listed in [57].

Coefficients of gross chemical energy consumption burdening the production of electricity and heat for lignite combustion only are given by the following equations:

Table 6.7 Results of the calculations for lignite-fired CHP mode

Parameter		Value
CHP plant gross energy efficiency (energy utilization factor),	$\eta_{E\ CHP\ G}$	0.8963
Partial efficiency (gross) of electricity generation,	$\eta_{E\ el\ CHP\ G0}$	0.4431
Partial efficiency (gross) of heat generation,	$\eta_{E\ h\ CHP\ G0}$	1.5463
Cumulative energy consumption factor for heat generation (J/J),	$e^*_{h\ CHP0}$	0.8302
Cumulative energy consumption factor for electricity generation (J/J),	$e^*_{el\ CHP0}$	3.1949

$$e^*_{h\ CHP0} = \frac{1}{\eta_{E\ h\ CHP\ G}(1 - \varepsilon_h) \cdot \eta_{ht} \cdot \eta^*_{es\ LG}} \tag{6.3}$$

$$e^*_{el\ CHP0} = \frac{1}{\eta_{E\ el\ CHP\ G}(1 - \varepsilon_{el}) \cdot \eta_{tt} \cdot \eta^*_{es\ LG}} \tag{6.4}$$

Results of the calculations are given in Table 6.7.

Along with the increasing biomass share in the blend b, boiler energy efficiency changes and so do the auxiliaries' indicators. Based on lignite and biomass co-firing data, the following empiric correlations were evaluated [61]:

- boiler efficiency curve for biomass co-firing with lignite [19,20]:

$$\frac{\eta_{EB}}{\eta_{EB0}} = 1 + 0.0388 \cdot b \tag{6.5}$$

- auxiliaries (electricity consumption for lignite preparation, forced and induced draft fans drives) curve for biomass co-firing with lignite [20]:

$$\frac{\varepsilon_{el}}{\varepsilon_{el0}} = 1 - 1.0836 \cdot b + 2.6726 \cdot b^2 \tag{6.6}$$

Lignite combustion in a typical coal utility plant (Polish conditions) results in 6.5% of auxiliary electricity consumption. For a CHP plant, only a part of the auxiliaries is dependent on the biomass share. It has been assumed that out of total 16.5% CHP electric auxiliaries only about a half (8%) is dependent on the biomass share and the remaining 8.5% results from the operation of other equipment performance (e.g., district water pumping system), not dependent on the biomass share. Based on this assumption, the following can be evaluated:

$$\varepsilon_{el} = 0.165 - 0.0867 \cdot b + 0.2138 \cdot b^2 \tag{6.7}$$

The dependence of the cumulative energy consumption factors for heat and electricity generation on the biomass share in the blend can be expressed by the following relations [61]:

$$\frac{e^*_{\text{h CHP}}}{e^*_{\text{h CHP0}}} = \frac{\eta_{\text{EB0}}}{\eta_{\text{EB}}}\left[1 - b\left(1 - \frac{e^*_{\text{B}}}{e^*_{\text{LG}}}\right)\right] \tag{6.8}$$

$$\frac{e^*_{\text{el CHP}}}{e^*_{\text{el CHP0}}} = \frac{\eta'_{\text{EB0}}}{\eta'_{\text{EB}}}\cdot\frac{1-\varepsilon'_{\text{el0}}}{1-\varepsilon'_{\text{el}}}\left[1 - b\left(1 - \frac{e^*_{\text{B}}}{e^*_{\text{LG}}}\right)\right] \tag{6.9}$$

Figs. 6.8 and 6.9 present the dependences describing the influence of biomass co-firing on the factors of cumulative energy consumption burdening heat and electricity production for willow and residue biomass.

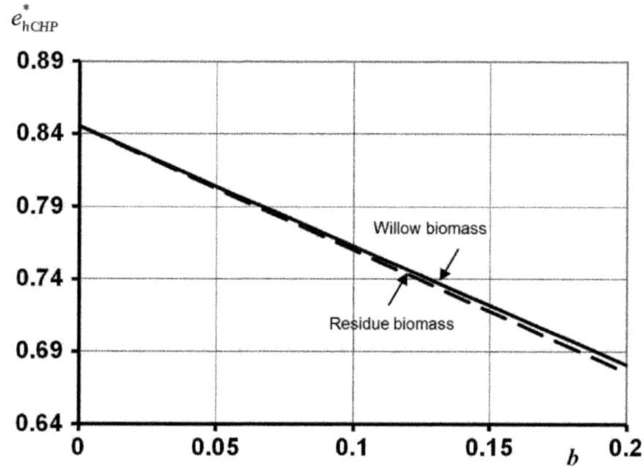

Figure 6.8 Coefficient of cumulative energy consumption burdening the production of heat in lignite-fired CHP plant, TJ/TJ.

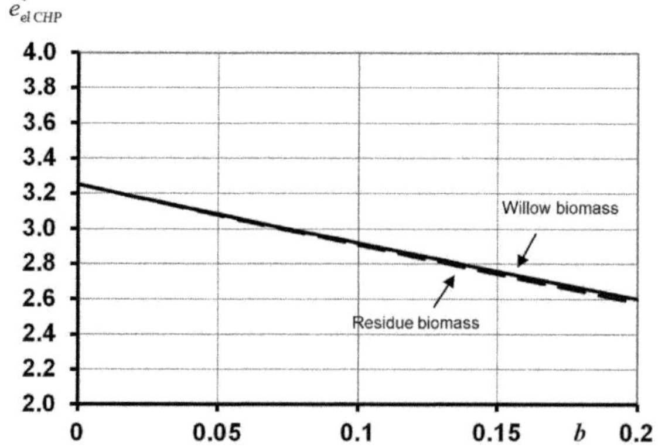

Figure 6.9 Coefficient of cumulative energy consumption burdening the production of electricity in lignite-fired CHP plant, TJ/MWh.

As it can be observed, biomass addition shows positive effect resulting in the decrease of nonrenewable primary energy resources consumption, both for the functional unit of heat and of the electricity.

In turn, Figs. 6.10 and 6.11 present the dependences describing the influence of biomass co-firing rate on cumulative GHG emission factors burdening heat and electricity production in CHP with a backpressure turbine, where the dependence of the cumulative GHG emission factors can be expressed in the analogical way, by the following relations [61]:

$$\frac{e^*_{CO_{2eq}\ el}}{e^*_{CO_{2eq}\ el0}} = \frac{\eta'_{EB0}}{\eta'_{EB}} \cdot \frac{1-\varepsilon_{el}}{1-\varepsilon_{el0}} \cdot \frac{1-\varepsilon'_{el0}}{1-\varepsilon'_{el}} \cdot \left[1 - b \cdot \left(1 - \frac{e^*_{CO_{2eq}\ B}}{e^*_{CO_{2eq}\ LG}}\right)\right] \tag{6.10}$$

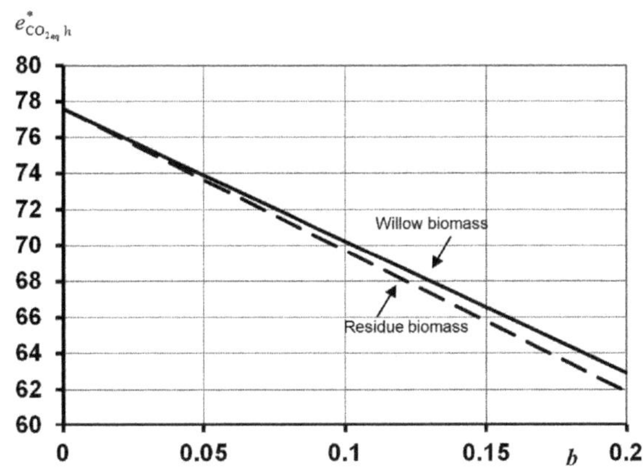

Figure 6.10 Coefficient of cumulative GHG emission burdening the production of heat in CHP plant with low-rank coal and biomass co-firing.

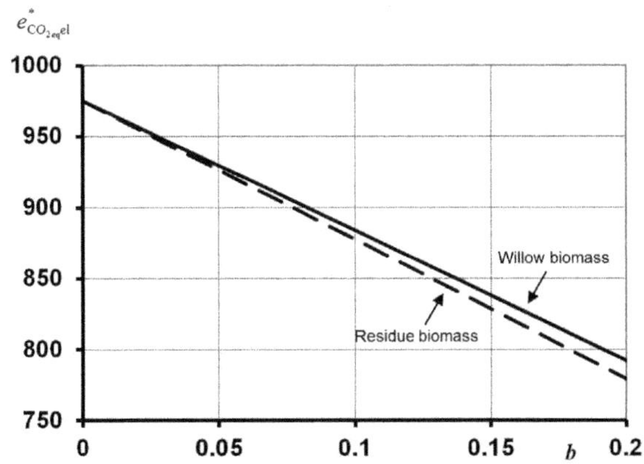

Figure 6.11 Coefficient of cumulative GHG emission burdening the production of electricity in CHP plant with low-rank coal and biomass co-firing.

$$\frac{e^*_{CO_{2eq}\ h}}{e^*_{CO_{2eq}\ h0}} = \frac{\eta'_{EB0}}{\eta'_{EB}} \cdot \frac{1-\varepsilon_{el}}{1-\varepsilon_{el0}} \cdot \frac{1-\varepsilon'_{el0}}{1-\varepsilon'_{el}} \left[1 - b \cdot \left(1 - \frac{e^*_{CO_{2eq}\ B}}{e^*_{CO_{2eq}\ LG}}\right)\right] \tag{6.11}$$

It can be observed that either for district heat or for electricity, the addition of biomass leads to the decrease of cumulative GHG emission. More positive effects can be noted for residual biomass due to the effect of the substitution of the landfilling with combustion (consequent substitution of released CH_4 with zero-emissive CO_2). Similar results were achieved in [50], however, the steam coal and biomass combustion was analyzed there.

Inventory analysis for the construction and decommissioning of CHP plant and biomass co-firing installation included the procurement of a CHP plant itself (lignite combustion baseline—system 1) and also the procurement of biomass co-firing installation for co-firing modes (systems 2 and 3).

Applying the approach described in [57] the cumulative emission of CO_2 during the construction phase can be calculated as $e_{ut} = 103{,}830$ kgCO_2/MW of a utility power plant. As mentioned in the previous paragraph, this is valid only for electric utility plants. To obtain values corresponding to the CHP plant, the same assumption as for cumulative nonrenewable energy consumption must be made. Finally, CO_2 emission burdening the CHP construction phase was estimated on the level of $E_{CHP\ C} = 730{,}000$ kgCO_2.

CO_2 emission burdening the construction phase of a biomass installation operating currently in one of Polish utility plants was estimated on the level of 397,000 kgCO_2 [59,66]. It can be seen that addition of biomass co-firing installation has a minor influence (approx. 5%) on the total CO_2 emission burdening the investment phase of the undertaken analysis. GHG other than CO_2 were not included in the analyses as well as all GHG emissions burdening the decommissioning phases of CHP and co-firing installations together with O&M bound emissions.

As residual biomass used for co-firing in System 3 has not been grown especially for this purpose, nonrenewable energy consumption and GHG emission during that would have occurred during the usual routes of its disposal which are avoided when it is combusted. Assuming that biomass is carbon-neutral, the whole CO_2 cycle can be regarded as a "closed loop" throughout all the stages, including landfilling. However, landfilled biomass is decomposed in anaerobic conditions with generation of CO_2 and CH_4, where methane has higher global warming potential compared to CO_2.

Therefore, the avoided-fate emissions concerning the landfilling have been credited in the life-cycle inventory [50,57]. Usual route of forestry waste disposal in Poland is landfilling. Landfilled biomass is decomposed in anaerobic conditions with generation of CO_2 and CH_4 in the following proportion: 40–70 CH_4 and 30–60 CO_2. During the landfilling process, only nonlignin fractions are subject to anaerobic decomposition; according to [50] it was assumed that only 50% of cellulose fraction is decomposed. Methane generated during the decomposition process is released to the atmosphere or used for electricity and heat generation by combustion in IC engines directly on site. Based on Central Statistical Office, CSO data [67] in Poland, currently 304 landfills are operating while 82 of them are equipped with gas combustion facilities (with or without energy recovery). All of this brings to the conclusion that the reduction of methane emission from landfilling by combustion into CO_2 reaches 35% in Polish conditions (compared to United States data of 40.5% [50]). After the estimation of the avoided emission of CO_2 resulting from the avoided fate of wood residues is estimated, it should be subtracted from the total life-cycle emission of residual biomass.

With 30 years of CHP plant operation with the nominal output, the net life-cycle production of heat and electricity as well as cumulative nonrenewable energy consumption and GHG emission can be estimated.

Life-cycle net production of heat and electricity can be calculated from the following equations, based on [48]:

$$Q_{\mathrm{LCA}} = 0.838\tau_{\mathrm{hs}} + 0.3(8760 - \tau_{\mathrm{hs}})Q_{\mathrm{N}}(1 - \varepsilon_{\mathrm{h}})\tau_{\mathrm{LCA}} \tag{6.12}$$

$$E_{\mathrm{el\ LCA}} = \sigma Q_{\mathrm{LCA}}(1 - \varepsilon_{\mathrm{el}}) \tag{6.13}$$

Cumulative nonrenewable energy consumption burdening the production of heat and electricity can be then calculated as:

$$E^{*}_{\mathrm{ch\ LCA\ h}} = Q_{\mathrm{LCA}}\eta_{\mathrm{ht}}e^{*}_{\mathrm{h\ CHP}} \tag{6.14}$$

$$E^{*}_{\mathrm{ch\ LCA\ el}} = E_{\mathrm{el\ LCA}}\eta_{\mathrm{tt}}e^{*}_{\mathrm{el\ CHP}} \tag{6.15}$$

Life-cycle parameters obtained for regarded CHP plant were summed up in Table 6.8. Cumulative GHG emission burdening the production of heat and electricity is calculated from the following equations:

$$E^{*}_{\mathrm{CO_2eq\ LCA\ h}} = Q_{\mathrm{LCA}}\eta_{\mathrm{ht}}e^{*}_{\mathrm{CO_2eq\ h}} \tag{6.16}$$

$$E^{*}_{\mathrm{CO_2eq\ LCA\ el}} = E_{\mathrm{el\ LCA}}\eta_{\mathrm{tt}}e^{*}_{\mathrm{CO_2eq\ el}} \tag{6.17}$$

All these values obtained for regarded CHP plant are given in Table 6.8.

The increase of the share of the chemical energy of biomass by 10% leads to the relative decrease of primary energy resources depletion amounting to about also 10% both for willow biomass and residual biomass which is a little more favorable value as compared to co-firing of such biomasses with hard coal [57].

Construction, O&M, and decommissioning component amounts to 0.9% of the total burdens for base case of LRC only combustion and to 1% for the co-firing case (regardless of the biomass sort). The share of operational (fuel bound) LCA component is dominant. Comparatively, in [68] 94% of coal energy in the total system burdens has been indicated, which authenticates the result obtained.

Table 6.8 Life-cycle parameters of CHP operation with lignite combustion vs lignite–biomass co-firing

Parameter/biomass energy share (%)	0	10	20
Heat production (net) (TJ)	50,234	50,234	50,234
Electricity generation (net) (TWh)	4.912	4.950	4.963
Total cumulative nonrenewable resources consumption (willow biomass co-firing) (TJ)	88,659	80,097	71,601
Total cumulative nonrenewable resources consumption (residue biomass co-firing) (TJ)	88,659	79,817	71,043
Cumulative emission for co-firing of willow biomass (mln tons CO_2)	9.73	8.820	8.750
Cumulative emission for co-firing of residue biomass, (mln tons CO_2)	9.73	7.900	7.770
Construction phase burdens (mln tons CO_2)	0.0073	0.077	0.077
Total cumulative emission for willow biomass co-firing, mln tons CO_2	9.7373	8.8970	8.827
Emission resulting from the avoided path of residue biomass (change of fate) (mln tons CO_2)	0	−1.571	−3.049
Total cumulative emission for residue biomass co-firing (mln tons CO_2)	9.7373	6.406	4.798

The increase of biomass share (in terms of chemical energy) by 10% leads to a decrease of GHG emission amounting to about 9% for willow biomass and almost two times more: 18% for residue biomass regarding heat. Concerning the full life-cycle burdens, it is clear that the operational component is deciding about the total cumulative emission.

The share of investment component (construction phase) is about 0.1% share in the total volume, which confirms the conclusions drawn by the authors of [68] where 1% impact of construction and decommissioning phases on the total life-cycle GHG burdens was found. Power plant operation was found to be responsible for 96% of life-cycle GHG emission (the remaining 4% belong to lignite mining and transportation).

6.5 Conclusions

State-of-the-art of co-firing of LRC with biomass was presented in this chapter. The application of co-combustion is still attractive in large LRC boilers due to advantages like reduction of the emissions of SO_2, NO_x, and CO_2 per unit energy, the increase of a boiler efficiency, and others. On the contrary, emission and operational challenges like higher CO emission, higher probability of slagging, fouling or corrosion in the boiler must be taken into account. Experiences from trial tests showed tangible benefits related to lower SO_2 emission and decrease of feeding rate of desulphurization agent (e.g., limestone).

Biomass co-firing in CHP plant equipped with backpressure turbine was considered in this chapter. The methodology of life-cycle analysis for energy processes has been presented and the influence of biomass addition on the depletion of nonrenewable energy resources as well as GHG emission was examined.

Brief discussion concerning allocation dilemma in terms of multifunctional process (production of electricity and heat in co-generation) was now done. Avoided-process approach applied in the consideration was based on the assumption that, by co-generation of electricity and heat, electricity production replaces electricity produced in "displaced" or "avoided" process of electricity-only production in a utility plant. For the simulations, the developed real operational empirical characteristics of boiler energy efficiency drop as well as the change of auxiliaries was applied.

It was proved that the partial substitution of lignite with biomass (in the considered range of 0%–20% on the thermal basis) leads to the decrease of the total life-cycle nonrenewable energy resources depletion and cumulative GHG emissions for generation and supply of 1 MJ of heat (functional unit) and 1 MWh of electricity generated during multifunctional process of CHP generation.

Along with the increasing biomass share in the blend, CO_{2eq} emission per unit of generated electricity and heat drops down. More positive effect was observed in the case of residue biomass mainly due to the elimination of GHG emissions resulting from the avoided landfilling processes being an alternative way of its use for energy purposes.

Taking into account the conclusions drawn by Ponton [69] that rather residual biomass (on the example of straw) should be co-fired as serious concerns can be raised for the net balance of GHG (compared to energy crops, mostly due to the fertilizers application), it can be said that the obtained results allow to back up this thesis. Residual biomass (on the example of woodchips) is a better fuel for co-firing in terms of life-cycle GHG net balance. Although the analysis presented in the paper was carried out for wood residues, an important conclusion can be that residue-like sorts of biomass for energy generation should be more prioritized in the renewable energy support systems than it is actually.

It has been found that the material requirements for the construction of biomass co-firing installation are at a very low level compared to the whole construction, O&M, and decommissioning energy consumption for CHP plant. The share of operational component dominates (99% of the total system burdens in the category of nonrenewable energy depletion). The component of construction, O&M, and decommissioning of biomass co-firing installations amounts less than 1%, which is a comparable level as estimated in [50].

LCA of biomass co-firing has been the subject of the studies of many authors. Different approaches were applied to assess the environmental impact of biomass and coal co-firing. Some of the authors applied the exergy method to quantify the exergy destruction in the co-firing processes [70]; some used directly LCA tools to assess GHG emissions (global warming), photochemical oxidation, and acidification impacts during biomass co-firing [71]. Positive effects have been found in any case when biomass was added to the base fuels combusted in the utility plants. It was confirmed that local biomass should be prioritized in terms of LCA due to the lower impact of long distance transportation-bound harmful emissions. Some publications are devoted to the analysis of coal and biomass transport stage alone, however, they are dealing mostly with the truck, rail [72], or sea transport [73]. Biomass torrefaction leading to the increase of biomass energy density is a way of lowering the impact of GHG emissions per the mass unit [74].

In case of lignite LRCs, the transport impacts are significantly lower as lignite is transported from the open-cast mine to the utility plant by the network of belt conveyors driven by electric motors.

On the other hand, it should be underlined that we have found no works presenting the results of the application of LCA-based methodologies for the assessment of the depletion of the nonrenewable resources and GHG emission related to biomass and low grade coal (lignites) co-firing, based on the real experimental data from biomass and coal co-firing trials for comparison in the literature.

Nomenclature

b biomass energy share in lignite–biomass fuel blend

e^*_{B} operational unit coefficient of cumulative energy consumption burdening biomass supplied to CHP plant, J/J

e^*_{LG} operational unit coefficient of cumulative energy consumption burdening lignite supplied to CHP plant, J/J

$e^*_{CO_2eq\,I}$ unit coefficient of cumulative emission of GHG burdening CHP plant and co-firing installation construction phases, J/MW

$e^*_{CO_2eq\,B}$ unit coefficient of cumulative emission of GHG burdening biomass supply to CHP plant, tCO_{2eq}/J

$e^*_{CO_2eq\,LG}$ unit coefficient of cumulative emission of GHG burdening lignite supply to CHP plant, tCO_{2eq}/J

$e^*_{CO_2eq\,el}$ unit coefficient of cumulative emission of GHG burdening electricity production in CHP plant, tCO_{2eq}/MWh

$e^*_{CO_2eq\,h}$ unit coefficient of cumulative emission of GHG burdening heat production in CHP plant, tCO_{2eq}/J

e^*_{el} operational unit coefficient of cumulative energy consumption burdening the production of electricity in CHP plant, J/MWh

e^*_{h} operational unit coefficient of cumulative energy consumption burdening the production of heat in CHP plant, J/J

e^*_{LG} operational unit coefficient of cumulative energy consumption burdening lignite supplied to CHP plant, J/J

e^*_{I} unit coefficient of cumulative energy consumption for CHP plant construction, J/MW

$E^*_{CO_2eq\,LCA\,h}$ cumulative GHG emission burdening the production of heat, tCO_{2eq}

$E^*_{CO_2eq\,LCA\,el}$ cumulative GHG emission burdening the production of electricity, tCO_{2eq}

$E_{el\,LCA}$ CHP life-cycle net electricity production, MWh

$E^*_{\mathrm{ch\,LCA\,h}}$ cumulative nonrenewable energy consumption burdening the production of heat, MJ

$E^*_{\mathrm{ch\,LCA\,el}}$ cumulative nonrenewable energy consumption burdening the production of electricity, MJ

E_{LCA} cumulative nonrenewable energy consumption (GHG emission) relating to the net electricity and heat production in a CHP plant, MJ (tCO_{2eq})

$\sum E_{N\text{-}F}$ cumulative nonrenewable energy consumption (GHG emission) burdening nonfuel raw materials supplied to the CHP plant, MJ (tCO_{2eq})

E_{W} total cumulative nonrenewable energy consumption (GHG emission) burdening willow biomass supplied to CHP plant, MJ (tCO_{2eq})

E_{R} total cumulative nonrenewable energy consumption (GHG emission) burdening residue biomass supplied to CHP plant, MJ (tCO_{2eq})

E_{LG} total cumulative nonrenewable energy consumption (GHG emission) burdening lignite supplied to CHP plant, MJ (tCO_{2eq})

E_{B} total cumulative nonrenewable energy consumption (GHG emission) burdening biomass supplied to CHP plant, MJ (tCO_{2eq})

E_{CP} total cumulative nonrenewable energy consumption (GHG emission) burdening lignite open-cast preparation, MJ (tCO_{2eq})

E_{LE} total cumulative nonrenewable energy consumption (GHG emission) burdening lignite extraction, MJ (tCO_{2eq})

E_{CDS} total cumulative nonrenewable energy consumption (GHG emission) burdening willow chipping, drying and storage, MJ (tCO_{2eq})

E_{LD} $E_{\mathrm{CHP\,C}}$ total cumulative nonrenewable energy consumption (GHG emission) burdening the CHP construction, MJ (tCO_{2eq})

$E_{\mathrm{CHP\,D}}$ total cumulative nonrenewable energy consumption (GHG emission) burdening CHP decommissioning, MJ (tCO_{2eq})

E_{FW} total cumulative nonrenewable energy consumption (GHG emission) burdening plantation fertilization and weeding, MJ (tCO_{2eq})

E_{H} total cumulative nonrenewable energy consumption (GHG emission) burdening willow harvesting, MJ (tCO_{2eq})

E_{CD} total cumulative nonrenewable energy consumption (GHG emission) burdening open-cast decommissioning, MJ (tCO_{2eq})

E_{PD} total cumulative nonrenewable energy consumption (GHG emission) burdening plantation decommissioning, MJ (tCO_{2eq})

E_{PE} total cumulative nonrenewable energy consumption (GHG emission) burdening plantation establishment, MJ (tCO_{2eq})

E_{RE} total cumulative nonrenewable energy consumption (GHG emission) burdening forest regeneration, MJ (tCO_{2eq})

E_{RC} total cumulative nonrenewable energy consumption (GHG emission) burdening residue biomass collection, MJ (tCO_{2eq})

E_{DS} total cumulative nonrenewable energy consumption (GHG emission) burdening residue biomass drying and storage, MJ (tCO_{2eq})

E_{CH} total cumulative nonrenewable energy consumption (GHG emission) burdening residue biomass chipping, MJ (tCO_{2eq})

E_{LT} total cumulative nonrenewable energy consumption (GHG emission) burdening lignite belt conveyor transportation, MJ (tCO_{2eq})

E_{T} total cumulative nonrenewable energy consumption (GHG emission) burdening biomass transportation, MJ (tCO_{2eq})

$E^*_{\mathrm{CO_2eq\,LCA\,h}}$ cumulative GHG emission burdening the production of heat, tCO_{2eq}

$E^{*}_{CO_2eq\ LCA\ el}$ cumulative GHG emission burdening the production of electricity, tCO_{2eq}
E_{el} CHP life-cycle net electricity production, GWh
Q_{LCA} life-cycle net production of heat, GJ
$\dot{Q}_N$ nominal thermal capacity, MW

Greek symbols

ε_{el} CHP electric auxiliary coefficient, %
ε_h CHP thermal auxiliary coefficient, %
η_{EB} CHP boiler energy efficiency, %
σ co-generation factor, %,
η_{tt} transformation and transmission efficiency of electricity, %
η_{ht} transmission efficiency of heat, %
$\eta_{E\ el\ CHP\ G}$ partial efficiency (gross) of electricity generation, %
$\eta_{E\ h\ CHP\ G}$ partial efficiency (gross) of heat generation, %
τ_{hs} duration of heating season (5400), h
τ_{LCA} duration of LCA analysis (30), years

Subscripts and superscripts

′ applying to substituted utility plant parameters
0 applying to lignite combustion in CHP plant
LG applying to lignite
B applying to biomass

References

[1] ISO14040, ISO14040:2006: Environmental Management—Life Cycle Assessment—Principles and Framework, *British Standards Institution, London*, 2006.
[2] European Commission 2010. Report from the Commission to the Council and the European Parliament on Sustainability Requirements for the Use of Solid and Gaseous Biomass Sources in Electricity, Heating and Cooling, Publications Office of the European Union.
[3] Van Der Hoeven, M. 2011. CO_2 emissions from fuel combustion—highlights, IEA Statistics < https://www.iea.org/media/statistics/CO2highlights.pdf> *(last accessed 20.07.2016)*
[4] N. Nikolopoulos, I. Violidakis, E. Karampinis, M. Agraniotis, C. Bergins, P. Grammelis, et al., Report on comparison among current industrial scale lignite drying technologies (A critical review of current technologies), Fuel 155 (2015) 86–114.
[5] M. Widera, Z. Kasztelewicz, M. Ptak, Lignite mining and electricity generation in Poland: The current state and future prospects, Energy Policy 92 (2016) 151–157.
[6] V.C. Pandey, O. Bajpai, N. Singh, Energy crops in sustainable phytoremediation, Renew. Sustain. Energy Rev. 54 (2016) 58–73.
[7] B. Igliński, A. Iglińska, W. Kujawski, R. Buczkowski, M. Cichosz, Bioenergy in Poland, Renew. Sustain. Energy Rev. 15 (2011) 2999–3007.
[8] E. Abolina, V.A. Luzadis, Abandoned agricultural land and its potential for short rotation woody crops in Latvia, Land Use Policy 49 (2015) 435–445.

[9] K. Bunzel, M. Kattwinkel, M. Schauf, D. Thrän, Energy crops and pesticide contamination: Lessons learnt from the development of energy crop cultivation in Germany, Biomass Bioenergy 70 (2014) 416–428.
[10] V. Uri, J. Aosaar, M. Varik, H. Becker, M. Kukumägi, K. Ligi, et al., Biomass resource and environmental effects of Norway spruce (Picea abies) stump harvesting: an Estonian case study, For. Ecol. Manag. 335 (2015) 207–215.
[11] J. Tomei, R. Helliwell, Food versus fuel? Going beyond biofuels, Land Use Policy (2016) (in print).
[12] J. Van Wijnen, W.P.M.F. Ivens, C. Kroeze, A.J. Löhr, Coastal eutrophication in Europe caused by production of energy crops, Sci. Tot. Environ. 511 (2015) 101–111.
[13] M. Dzikuć, A. Piwowar, Ecological and economic aspects of electric energy production using the biomass co-firing method: the case of Poland, Renew Sustain. Energy Rev. 55 (2016) 856–862.
[14] Ćwil, M. Ustawa o OZE i projekty aktów wykonawczych. POLAGRA, 2016 Poznań (21.01.2016).
[15] F. Al-Mansour, J. Zuwala, An evaluation of biomass co-firing in Europe, Biomass Bioenergy 34 (2010) 620–629.
[16] J. Dumortier, Co-firing in coal power plants and its impact on biomass feedstock availability, Energy Policy 60 (2013) 396–405.
[17] E. Karampinis, P. Grammelis, M. Agraniotis, I. Violidakis, E. Kakaras, Co-firing of biomass with coal in thermal power plants: technology schemes, impacts, and future perspectives, Wiley Interdiscip. Rev.: Energy Environ. 3 (2014) 384–399.
[18] K. Ericsson, Co-firing—a strategy for bioenergy in Poland? Energy 32 (2007) 1838–1847.
[19] S. Tokarski, K. Głód, J. Lasek, J. Zuwała, M. Kopczyński, Evaluation of models for estimation of heating values of solid fuels and their test in full-scale combustion chambers. Part II: Analysis of co-firing combustion chambers (in Polish: Ocena modeli do określania właściwości kalorycznych paliw stałych pod kątem ich zastosowania w bilansowaniu obiektów energetyki zawodowej–Cz. II bilansowanie obiektów współspalających biomasę), Rynek Energii 2 (2013) 65–68.
[20] J. Zuwała, Influence of biomass co-firing on the operational parameters of utility plants, Energetyka 2 (2010) 1–7 (in Polish).
[21] Gasperetti, S., Leoni, P., Smeeths, L., Fuller, A., Maier, J., Miller, E., et al. 2013. Advanced Biomass Co-Firing Guidebook <https://www.vgb.org/vgbmultimedia/322_DEBCO_D7_10_Guidebook-p-6752.pdf>.
[22] E. Kastanaki, D. Vamvuka, P. Grammelis, E. Kakaras, Thermogravimetric studies of the behavior of lignite–biomass blends during devolatilization, Fuel Process. Technol. 77–78 (2002) 159–166.
[23] S.G. Sahu, N. Chakraborty, P. Sarkar, Coal–biomass co-combustion: An overview, Renew. Sustain. Energy Rev. 39 (2014) 575–586.
[24] C.Y.H. Chao, P.C.W. Kwong, J.H. Wang, C.W. Cheung, G. Kendall, Co-firing coal with rice husk and bamboo and the impact on particulate matters and associated polycyclic aromatic hydrocarbon emissions, Bioresour. Technol. 99 (2008) 83–93.
[25] A.T. Atimtay, B. Kaynak, Co-combustion of peach and apricot stone with coal in a bubbling fluidized bed, Fuel Process. Technol. 89 (2008) 183–197.
[26] M. Varol, A.T. Atimtay, H. Olgun, Emission characteristics of co-combustion of a low calorie and high-sulfur-lignite coal and woodchips in a circulating fluidized bed combustor: Part 2. Effect of secondary air and its location, Fuel 130 (2014) 1–9.
[27] M. Varol, A.T. Atimtay, H. Olgun, H. Atakül, Emission characteristics of co-combustion of a low calorie and high sulfur–lignite coal and woodchips in a

circulating fluidized bed combustor: Part 1. Effect of excess air ratio, Fuel 117 (Part A) (2014) 792–800.
[28] H. Haykiri-Acma, S. Yaman, Effect of biomass on burnouts of Turkish lignites during co-firing, Energy Convers. Manag. 50 (2009) 2422–2427.
[29] H. Haykiri-Acma, S. Yaman, S. Kucukbayrak, M.H. Morcali, Does blending the ashes of chestnut shell and lignite create synergistic interaction on ash fusion temperatures? Fuel Process. Technol. 140 (2015) 165–171.
[30] M. Varol, A.T. Atimtay, Effect of biomass-sulfur interaction on ash composition and agglomeration for the co-combustion of high-sulfur lignite coals and olive cake in a circulating fluidized bed combustor, Bioresour. Technol. 198 (2015) 325–331.
[31] I. Andrić, N. Jamali-Zghal, M. Santarelli, B. Lacarrière, O. Le Corre, Environmental performance assessment of retrofitting existing coal fired power plants to co-firing with biomass: carbon footprint and emergy approach, J. Clean. Prod. 103 (2015) 13–27.
[32] L. Armesto, A. Bahillo, A. Cabanillas, K. Veijonen, J. Otero, A. Plumed, et al., Co-combustion of coal and olive oil industry residues in fluidised bed☆, Fuel 82 (2003) 993–1000.
[33] N. Nikolopoulos, M. Agraniotis, I. Violidakis, E. Karampinis, A. Nikolopoulos, P. Grammelis, et al., Parametric investigation of a renewable alternative for utilities adopting the co-firing lignite/biomass concept, Fuel 113 (2013) 873–897.
[34] B. Kaynak, H. Topal, A.T. Atimtay, Peach and apricot stone combustion in a bubbling fluidized bed, Fuel Process. Technol. 86 (2005) 1175–1193.
[35] J.A. Lasek, K. Kazalski, Sulfur Self-retention during Cocombustion of Fossil Fuels with Biomass, Energy Fuels 28 (2014) 2780–2785.
[36] D. Vamvuka, S. Sfakiotakis, Combustion behaviour of biomass fuels and their blends with lignite, Thermochim. Acta 526 (2011) 192–199.
[37] H. Haykiri-Acma, S. Yaman, S. Kucukbayrak, Does carbonization avoid segregation of biomass and lignite during co-firing? Thermal analysis study, Fuel Process. Technol. 137 (2015) 312–319.
[38] G. Skodras, P. Grammelis, P. Samaras, P. Vourliotis, E. Kakaras, G.P. Sakellaropoulos, Emissions monitoring during coal waste wood co-combustion in an industrial steam boiler, Fuel 81 (2002) 547–554.
[39] S. Du, H. Yang, K. Qian, X. Wang, H. Chen, Fusion and transformation properties of the inorganic components in biomass ash, Fuel 117 (Part B) (2014) 1281–1287.
[40] A. Kazagic, I. Smajevic, Synergy effects of co-firing wooden biomass with Bosnian coal, Energy 34 (2009) 699–707.
[41] A.T. Atimtay, M. Varol, Investigation of co-combustion of coal and olive cake in a bubbling fluidized bed with secondary air injection, Fuel 88 (2009) 1000–1008.
[42] A. Efthimiadou, I.P. Nikolaides, A. Tourlidakis, A proposal for SO2 abatement in existing power plants using rich in calcium lignite, Appl. Therm. Eng. 74 (2015) 119–127.
[43] K. Kazalski, J. Zuwała, K. Głód, Badania nad efektem auto-odsiarczania przy współspalaniu węgla i biomasy, Karbo 4 (2010) 168–172 (in Polish).
[44] A. Piwowar, M. Dzikuć, Outline of the economic and technical problems associated with the co-combustion of biomass in Poland, Renew. Sustain. Energy Rev. 54 (2016) 415–420.
[45] ISO14040 (2006) ISO 14040: 2006: Environmental Management—Life Cycle Assessment—Principles and Framework, International Organisation for Standardisation (ISO), Geneve.
[46] ISO14044 (2006) ISO 14044:2006: Environmental Management—Life Cycle Assessment—Requirements and Guidelines, International Organisation for Standardisation (ISO), Geneve.

[47] M.C. Heller, G.A. Keoleian, M.K. Mann, T.A. Volk, Life cycle energy and environmental benefits of generating electricity from willow biomass, Renew. Energy 29 (2004) 1023–1042.
[48] M.C. Heller, G.A. Keoleian, T.A. Volk, Life cycle assessment of a willow bioenergy cropping system, Biomass Bioenergy 25 (2003) 147–165.
[49] Y. Huang, D. Mcilveen-Wright, S. Rezvani, Y.D. Wang, N. Hewitt, B.C. Williams, Biomass co-firing in a pressurized fluidized bed combustion (PFBC) combined cycle power plant: a techno-environmental assessment based on computational simulations, Fuel Process. Technol. 87 (2006) 927–934.
[50] M. Mann, P. Spath, A life cycle assessment of biomass cofiring in a coal-fired power plant, Clean Prod. Process. 3 (2001) 81–91.
[51] Restrepo, A., Miyake, R.G., Kleveston, F. & Bazzo, E. Life cycle assessment for co-firing potential analysis in a pulverized coal power plant, in: 24th International Conference on Efficiency, Cost, Optimisation, Simulation and Environmental Impact of Energy Systems ECOS 2011, July 4–7 2011, Novi Sad, Serbia, 2011, pp. 1152–1163.
[52] F. Sebastián, J. Royo, M. Gómez, Cofiring versus biomass-fired power plants: GHG (Greenhouse Gases) emissions savings comparison by means of LCA (Life Cycle Assessment) methodology, Energy 36 (2011) 2029–2037.
[53] Directive_2009/28/Ec 2009. 2009 Directive 2009/28/EC of the European Parliament and of the Council on the Promotion of the Use of Energy from Renewable Sources.
[54] Szargut, J. 1983. Thermodynamic and Economical Analysis in Industrial Energy Analysis, Warszawa, Poland, WNT (in Polish).
[55] J. Szargut, A. Ziębik, Fundamentals of Thermal Engineering, PWN, Warszawa, Poland, 2000 (in Polish).
[56] J. Korhonen, A material and energy flow model for co-production of heat and power, J. Clean. Prod. 10 (2002) 537–544.
[57] J. Zuwała, Life cycle approach for energy and environmental analysis of biomass and coal co-firing in CHP plant with backpressure turbine, J. Clean. Prod. 35 (2012) 164–175.
[58] M.A. Rosen, Allocating carbon dioxide emissions from cogeneration systems: descriptions of selected output-based methods, J. Clean. Prod. 16 (2008) 171–177.
[59] A. Azapagic, R. Cliftb, Allocation of environmental burdens in multiple-function systems, J. Clean. Prod. 7 (1999) 101–119.
[60] J. Szargut, A. Ziębik, W. Stanek, Depletion of the non-renewable natural exergy resources as a measure of the ecological cost, Energy Convers. Manag. 43 (2002) 1149–1163.
[61] J. Zuwała, Energy and Environmental Effects of Fossil Fuels and Biomass Co-firing in Cogeneration Technologies, Central Mining Institute, Katowice, Poland, 2013.
[62] M. Elsayed, R. Matthews, N. Mortimer, Carbon and Energy Balances for a Range of Biofuels Options, Resources Research Unit, Sheffield Hallam University, Sheffield, England, 2003.
[63] J.B. Guinée, Handbook on life cycle assessment operational guide to the ISO standards, Int. J. Life Cycle Assess. 7 (2002) 311–313.
[64] A. Ziębik, D. Szegda, B. Qvale, B. Elmegaard, Thermodynamic simulation analysis of a multifuel CHP plant basing on the technological diagram of Avedøre unit 2, Arch. Thermodyn. 31 (2010) 79–93.
[65] Boer, J., Stouffer, R., Dix, M., Noda, A., Senior, C., Raper, S., et al. 2001. IPCC Third Assessment Report—Climate Change 2001. Chpt.
[66] Zuwała, J. & Ziębik, A. Estimation of greenhouse gases (GHG) emissions in the course of biomass co-firing in CHP plant by means of LCA (life cycle assessment)

methodology, in: 24th International Conference on Efficiency, Cost, Optimisation, Simulation and Environmental Impact of Energy Systems ECOS, 4–7 July 2011, Novi Sad, Serbia, 2011, pp. 3705–3718.
[67] Cso 2008. Central Statistical Office, Environmental Protection 2008. Central Statistical Office, Warsaw 2008. ISSN 0867-3217 (in Polish). Available at: <www.stat.gov.pl> (last accessed 17.04.12).
[68] Spath, P.L., Mann, M.K. & Kerr, D.R. 1999. Life Cycle Assessment of Coal-fired Power Production, Technical Report No.: NREL/TP-570-25119. Contract No. DE-AC36-98-GO10337, National Renewable Energy Laboratory, Colorado, USA.
[69] J.W. Ponton, Biofuels: Thermodynamic sense and nonsense, J. Clean. Prod. 17 (2009) 896–899.
[70] C. Martín, M.A. Villamañán, C.R. Chamorro, J. Otero, A. Cabanillas, J.J. Segovia, Low-grade coal and biomass co-combustion on fluidized bed: exergy analysis, Energy 31 (2006) 330–344.
[71] J. Royo, F. Sebastián, D. García-Galindo, M. Gómez, M. Díaz, Large-scale analysis of GHG (greenhouse gas) reduction by means of biomass co-firing at country-scale: Application to the Spanish case, Energy 48 (2012) 255–267.
[72] M. Bałazińska, J. Zuwała, S. Tokarski, Carbon footprint related with transport of fuels for energy production, Rynek Energii (2013) 68–73.
[73] J. Zuwała, M. Bałazińska, Carbon footprint associated with transport biomass by sea, Karbo (2012) 54–61.
[74] G.-A. Tsalidis, Y. Joshi, G. Korevaar, W. De Jong, Life cycle assessment of direct co-firing of torrefied and/or pelletised woody biomass with coal in The Netherlands, J. Clean. Prod. 81 (2014) 168–177.

CFBC and BFBC of low-rank coals

7

Zhongyang Luo and Wang Tao
Zhejiang University, Hangzhou, PR China

7.1 Introduction

For combustion processes, the particulate flow helps in burning and transporting heat to different portions of the burning regime by this circular/circulating motion. Therefore a fluidized bed combustor is a combustion chamber that has bed of solid particles fluidized by air or behaving as though a fluid itself. This is possible when a right amount or evenly distributed air/fluid passes through the bed and/or solid fuel particles. The speed/velocity of the flow should be high enough so that the particles are either suspended in air or circulate to imitate gases and liquids. If the air velocity is further increased, the bed becomes highly turbulent and the rapid mixing of particles occurs which appear like formation of bubbles in a boiling liquid and the process of combustion as a result is known as fluidized bed combustion. This velocity is simply not a speed or flow measure but it requires careful consideration and calculations of different parameters that include size of the fuel particles and density of air fuel mixtures, to enhance the fluidization and improve the burning process inside the combustor, which is commonly known as FBC or fluidized bed combustion technology [1]. FBC boilers can burn fuels other than coal, and the lower temperatures of combustion (800°C) have other added benefits as well [2].

To consider the effect of velocity on the fluidization characteristics during the process and the bed, a steady increase is required to study the behavior. The first condition is minimum fluidization velocity [3], where the fluidization just starts to appear to take shape and at this condition, the weight distributed per cross-section of the bed will be proportional to the drop in pressure. This minimum velocity is sometimes referred to as critical velocity. With further increase in velocity up to a certain number, the second condition of fluidization is achieved, which resembles a boiling liquid state. This can be achieved when the velocity reaches about three to five times of our minimum fluidization velocity or simply the critical velocity as the bed of the furnace begins to expand due to the increased velocity as more air is passing through in the form of bubbles. At this condition, the combustion efficiency and the heat transfer rates of the boiler/combustor are very high. The higher the velocity, the greater is the circulation and the more contact of the evaporator tubes with the reduced size coal particles. These tubes are in direct contact with the coal/fuel particles.

Low-Rank Coals for Power Generation, Fuel and Chemical Production
DOI: http://dx.doi.org/10.1016/B978-0-08-100895-9.00007-3

7.1.1 Types of fluidized bed combustion (FBC)

The fluidization technology application in the boilers and combustors has now become one of the most efficient systems in terms of combustion of complex fuel scenarios and is being employed for the very purpose in most of the modern industries due to its added benefit over conventional combustors. The two major types of FBC are atmospheric systems or simple FBC and pressurized FBC (PFBC). The more familiar terms, bubbling (BFB) and circulating fluidized bed (CFB), are categorized as minor subtypes of FBC technology [4].

Even though the FBC technology has been preferred for other types of combustion technologies due to its overall benefits, there are some issues concerning its effective application. One of them is power required by the fan to push in air/fluid constantly requires a high value to efficiently sustain a proper required fluidization by supplying a constant velocity and at higher pressure. Though this normally does increase the operating and maintenance cost of the auxiliary systems of the plants it is compensated by the high combustion efficiency of the FBC.

7.1.2 Advantages of circulating fluidized bed combustion

1. Fuel flexibility

   Circulating fluidized bed boiler possesses better gas–solid mixing quality due to its special hydrodynamic characteristics. The fuel which is hard to burn can be heated rapidly in the furnace to above the ignition temperature and it determines that the circulating fluidized bed boiler does not require auxiliary fuel to burn any fuel, due to which CFB boilers can accommodate both high as well as low quality coals even with high ash content. Different types of waste and other types of fuels can be processed.
2. High combustion efficiency

   Combustion efficiency of CFB boiler can be greatly improved by including a high temperature cyclone separator with it. This in turn increases the combustion area and the unburnt particles that have been carried out with the flue gas are returned to the furnace with the help of the cyclone separator. This way, the particles return the heat carried out by them and eventually contribute to a more complete combustion, due to which the efficiency of the CFB boilers is higher than in the case of BFB and it is achievable up to 99.5%.
3. SO_2 removal

   SO_2 removal and conversion is done by adding limestone in the CFB furnace where it converts it to sulfates during the combustion. In any other scenario, a flue gas desulfurization is required to efficiently remove SO_2 to keep the level of emissions within acceptable levels even though the FGD is a heavy investment and requires high operational costs [5]. Therefore a CFB has added advantage due to the limestone process; the sulfates are easily collected with the ash.
4. NO_x reduction

   The CFB boiler produces less NO_x emissions than PC boilers as the combustion temperature is lower. There is no additional requirement of separate NO_x capturing devices which reduces capital investment.

5. Ash comprehensive utilization

   Circulating fluidized bed combustion process is low temperature combustion, with boiler ash carbon content being low, as a result of good burning conditions, and it is easy to achieve the comprehensive utilization of ash. In addition, limestone added into the furnace, and the ash composition has changed, which contains certain $CaSO_4$ and unreacted CaO. Ash can be used to make cement admixtures or other building materials, raw materials, which are conducive to the comprehensive utilization of ash.

6. Better load regulation performance

   Owing to the high cross-section of wind speed and heat control easily, the load regulation is fast. Load adjustment range can be as low as 10% −25%; Load adjusting speed can reach (5%−10%)/min.

One major limitation is the wear and tear of the CFB boiler due to the travel speed of the increased concentration of the fuel particle which is a reaction of higher velocities to achieve fluidization. The second would be the N_2O emissions, generating up to 220 ppm, even though it actively stops and removes other environmentally problematic emissions such as NO_x and SO_x.

Separation of the circulating fluidized bed boiler circulating system is more complex, the wind plate and system resistance increases, and the boiler's power consumption is big, which is about 7% of the unit capacity, with an increase in operation cost.

The design of the circulating fluidized bed temperature generally is 850−900°C, and the bed temperature in the actual operation is commonly above 950°C, is even higher, more than 1000°C. For easy coking coal, this problem is very outstanding.

7.1.3 History of fluidized bed combustion

In December 1921, the German Winkler invented the first fluidized bed using coarse particles bed material. In December 1938, the Massachusetts Institute of Technology, Lewis and Girland, invented a fast fluidized bed. Circulating fluidized bed is truly a practical technology with industrial utility value in the 1950s and 1960s. The study of fluidized bed concept started in the 1960s [6]. In the late 1960s, Lurgi developed and operated an aluminum hydroxide roasting reactor for a Lurgi/VAW circulating fluidized bed boiler. In 1979, Ahlstrom produced a 20 t/h circulating fluidized bed boiler. In 1982, Lurgi's first 50 t/h commercial circulating fluidized bed boiler was put into operation. This marks the birth of a circulating fluidized bed boiler as a coal combustion equipment.

FBC technology knows two major developments: The first generation pressurized fluidized bed combustor uses a bubbling-bed technology. The second generation pressurized fluidized bed combustor uses circulating fluidized-bed technology and a number of efficiency enhancement measures [7]. There is a link between the two, but there are also differences. Subsequently, the circulating fluidized bed technology developed rapidly, especially in foreign countries developed rapidly, and to the direction of large-scale development.

China's circulating fluidized bed combustion technology is relatively late, but progress quickly. As in the early 1960s, bubbling (fluidized) bed boiler began to develop, but the circulating fluidized bed boiler started late, in 1981 the State Planning Commission issued a "coal fluidized bed combustion technology research" topic, marking China's circulating fluidized bed boiler research and product development technology officially launched. Until November 1989, the first 35 t/h circulating fluidized bed boiler developed by the Chinese Academy of Sciences and Jinan boiler plant, and put into operation in Shandong Ming water power plant. Chinese Academy of Sciences, Tsinghua University, Zhejiang University, Xi'an Institute of Thermal Engineering, Xi'an Jiaotong University, Harbin Institute of Technology, Huazhong University of Science and Technology and other colleges and universities and the boiler plant develop a variety of technology at 35–220 t/h medium pressure, secondary pressure, and high pressure circulating fluidized bed boiler.

7.2 Hydrodynamics

7.2.1 Overview of flow regimes

In a gas–solid fluidized bed, it has long been understood that fixed bed flow transforms to fluidized bed flow at the minimum fluidization velocity, U_{mf}. For fine particle systems, bubbles appear at the minimum bubbling velocity, U_{mb}, while slugs start to form at the minimum slugging velocity, U_{ms}. The transition to the turbulent fluidization regime is generally assumed to occur when the superficial velocity, U, reaches U_c, the superficial velocity at which the standard deviation of pressure fluctuations passes through a maximum. The termination of batch operation of fluidized beds is marked by significant entrainment of bed particles beyond U_{se} where particles can no longer be maintained in the column unless entrained particles are captured and returned to the bed efficiently. Such a flow transition process is depicted in Fig. 7.1.

$Ar = \rho_G(\rho_P - \rho_G)gd_p^3/\mu_G$

As the superficial gas velocity, U, is increased stepwise beyond U_{mf} for a bed of solid particles, one encounters a number of different flow regimes. In the most general case they are, in order of increasing U:

- Bubble-free bed expansion
- Bubbling fluidization
- Slug flow
- Turbulent fluidization
- Fast fluidization and dense suspension upflow

These flow regimes are shown diagrammatically in Fig. 7.1, together with dilute pneumatic conveying, reached at even higher gas velocities. In the rest of this section, we consider the transitions between these flow regimes and the corresponding transitional superficial gas velocities [8].

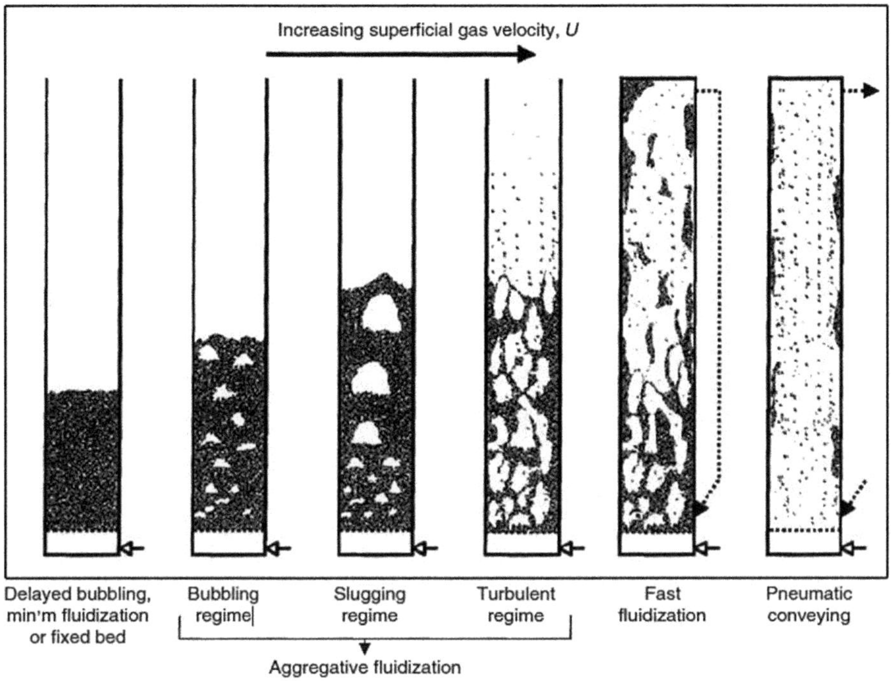

Figure 7.1 Flow chart showing regime transitions in gas–solid fluidized beds with little or no overflow of solids DB; *max*, maximum stable bubble size.

7.2.2 Axial flow structure

The axial distribution of particle concentration is influenced by many factors, such as particle circulation rate and operating gas velocity and particle, particle and gas import and export structure, diameter and shape of the riser, riser height, total storage capacity is solid particles, internals, and secondary air. The increase in the axial distribution of particle concentration inside from the bottom up may (but not necessarily at the same time) be in four different areas: dense phase section, transition section at the bottom of the bed or accelerated section, fully developed section near the top exports relatively dense phase section (only when using strong constraint exports).

Some researchers think about the axial distribution of particle concentration distribution model for S (the bottom is the dense phase section with high particle concentration, the upper is the dilute phase section, and the middle is the transition section), and subject to the effects of operating gas velocity and particle circulation rate.

However, in the wider operating conditions (the structure of the particles and the equipment within the scope of the experiment), the axial distribution of the tube is not always the S type distribution but under certain conditions it may be S type distribution, and is likely to be the exponential distribution, and also may be the

C type distribution. The bottom dense region may not always be present and a denser region may exist at the top for constricted exits. Operating variables, such as solids circulation rate, total solids inventory, particle size and density, solids inlet configuration, riser exit structure, secondary air injection, and the level of solids reintroduction into the riser, affect the axial voidage profile [9]. Li [10] proposed that the shape of the axial voidage profile depends on the relative magnitude of the solids circulation rate to the saturation solids carrying capacity of the gas. An exponential shape can be observed for circulation rates less than the saturation carrying capacity, while an S-shape or an almost straight line profile occurs for the circulation rates greater than the saturation carrying capacity. Profiles were classified into three types, a dilute phase transport regime (type I, exponential shape), the fast fluidization regime (type II, S-shape), and the dense transport regime (type III, straight line), all of which can be achieved in a circulating fluidized bed.

7.2.3 Gas and solids mixing in CFB

The downward motion of particles in the wall region described above causes backmixing of solids and may also lead to downflow (backmixing) of gas in the annulus. This results in non-uniform residence time distributions, which are undesirable for some chemical reactions. On the other hand, radial dispersion of gas and solids is usually beneficial to overall conversion.

Gas mixing has been studied by many researchers. The axial gas flow deviates significantly from plug flow. One-dimensional dispersed plug flow models have been employed by several researchers to correlate experimental data. Little information is available on the radial exchange of particles between the core and the annulus, and no data have been reported on the radial solids dispersion coefficient [5].

7.3 Combustion

Coal is primarily used as a solid fuel to produce electricity and heat through combustion. World coal consumption was about 7.25 billion tonnes in 2010 [11] (7.99 billion short tons) and is expected to increase 48% to 9.05 billion tonnes (9.98 billion short tons) by 2030 [12].

When coal is used for electricity generation, it is usually pulverized and then burned in a furnace with a boiler [13]. The furnace heat converts boiler water to steam, which is then used to spin turbines that turn generators and create electricity [14].

7.3.1 Stages of coal combustion

Low-rank coals are those that have undergone minimal metamorphic variation in the process of coal formation. Compared with the high-rank coals—bituminous and anthracite, they retain more moisture and volatile and contain less fixed carbon [15]. The low-rank coals are often characterized by:

High-moisture content (sometimes >50%)
Moderate to high ash (although some are low ash, <10%)
Low to moderate specific energy (12–25 MJ/kg, typically ~15 MJ/kg)
In many cases a very low ash fusion temperature (high Na/K content, high base/acid ratios)
A high propensity for spontaneous combustion
Liable to degrade to excessive fines during transport and handling.

For the coal fuels, it is a process of vigorous oxidation reaction between the inorganic and organic substances that compose the coal and the oxygen in the air, and release heat energy; in terms of energy conversion, it is a process by which the chemical energy in the coal is converted to heat energy. The coal burning is essentially a heterogeneous reaction of gas–solid phase. But due to the complexity of the nature, its combustion contains the drying, devolatilization, combustion of volatiles, and coke burning; therefore, it couples a number of processes of chemical reactions, heat transfer, mass transfer and fluid flow, etc. At the macro level, considering the balance estimate of relevant substances for coal burning, it can be calculated from the perspective of elements balance. Generally, the complete combustion of coal in the air can follow the reaction equation:

$$\begin{aligned} &C_{x1}H_{x2}O_{x3}N_{x4}S_{x5} + n_1H_2O + n_2(O_2 + 3.76N_2) \\ &= n_3CO_2 + n_4H_2O + n_5N_2 + n_6NO_x + n_7SO_2 + \text{Energy} \end{aligned} \tag{7.1}$$

The left second term of the formula represents the water entering with the fuel and air in the combustion process. The water is evaporated from liquid to enter the smoke in the combustion process. Although it is not directly involved in the combustion, it has a high latent heat of vaporization; therefore reducing the content is conducive to the combustion. The third term of the formula represents the air, represented by 21% oxygen and 79% nitrogen. The composition of the combustion products is complex. The equation only gives the macro-component and the major gas phase pollutants. The smoke in the actual combustion process usually contains the O_2 brought by excessive air, and the incomplete combustion of CO, HC and alkali metal chlorides, sulfates, carbonates, silicates, and other complex gray phase substances.

Despite the detailed knowledge about hydrocarbon oxidation chemistry developed over the years, the understanding of volatiles oxidation in solid fuel combustion remains limited. This can be attributed both to the complexity of the volatiles composition, involving a range of components from hydrogen to heavy organic compounds and tar, and to the complex physical and chemical interactions during oxidation. Volatiles conversion may occur on the internal or external particle surface, in the pore system, in the particle boundary layer, or in the bulk flow, where the oxidation rate may be either kinetically or mixing limited [16].

Different from caking bituminous coals, low-rank coal particles maintain shape and size during ignition, devolatilization, and initial combustion, rather than coalescing into plastic blocks. Low-rank coals allow uniform release of volatiles over the outer surface and enhance its reactivity due to the high porosity. In the process of devolatilization, low-rank coals produce low-tar, light oil, and some flammable gases including CH_4, CO, and H_2, and a large amount of inert gas which is mainly in the form of CO_2. The high moisture content influences the initial heating of coal particles, as a result, affecting the ignition. In addition, the ignition properties are considered to be more dependent on reactivity than volatiles, due to the lower yield of flammable volatiles which can stabilize the combustion of bituminous coal. Also, the ion-exchangeable cations have an important influence on the ignition properties.

7.3.2 Combustion process in BFB and in CFBC

Coal is fed to fluidized bed combustor and undergoes the following process: 1. drying and heating, 2. devolatilization and volatile combustion, 3. coke combustion, accompanied with particle expansion, primary crush, coke combustion, and secondary crush and abrasion. Fig. 7.2 shows the process of coal particles combustion in fluidized bed.

When at lower temperature like 1500°F or extended gas/solid contact, the sulfur content remained on ash is higher, and sometimes it is 100% for some low-rank coals. According to the alkaline component and the mass effect of sulfur content, the utilization efficiency for added sorbent or ash alkalinity varies from 25% to 90%. When the sulfur content reach 1%, it needs a 100% or more alkaline (alkali/sulfur = 2) to accomplish 90% sulfur control.

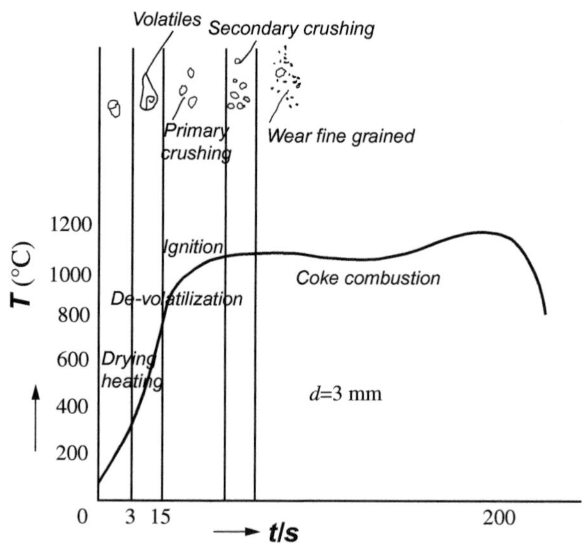

Figure 7.2 The process of coal particles combustion in fluidized bed.

7.3.3 *Efficiency of coal combustion*

The thermodynamic efficiency of the coal combustion process has been improved over time; some older coal-fired power stations have thermal efficiencies in the vicinity of 25% [16] whereas the newest supercritical and "ultra-supercritical" steam cycle turbines, operating at temperatures over 600°C and pressures over 27 MPa (over 3900 psi), can achieve thermal efficiencies around 43% (LHV basis) when using low-rank coals. Further thermal efficiency improvements are also achievable by improved pre-drying (especially relevant with high-moisture fuel such as lignite or biomass) and cooling technologies [17].

The integrated gasification combined cycle (IGCC) power plants are a potential method of using coal to generate electricity with higher efficiency. Instead of pulverizing coal and directly burning it, the coal is gasified to produce syngas and the syngas is burnt in a gas turbine to generate electricity. And it uses the hot exhaust gases to heat the steam in a heat recovery steam generator. The thermal efficiencies of IGCC vary from 39% to 42% (HHV basis) using low-rank coals. IGCC performs better in pollutions and is easier to capture carbon than conventional combustion.

In Denmark, a net electric efficiency of >47% has been obtained at the coal-fired Nordjyllandsværket CHP Plant and an overall plant efficiency of up to 91% with cogeneration of electricity and district heating [18]. The multifuel-fired Avedøreværket CHP Plant just outside Copenhagen can achieve a net electric efficiency as high as 49%. The overall plant efficiency with cogeneration of electricity and district heating can reach as much as 94% [19].

Coal-water slurry fuel (CWS) which was invented in the Soviet Union is another alternative approach to burn coal. There exist other ways like heat and power co-generation and MHD topping cycle.

The total known deposits recoverable by current technologies, including low-rank coals which is highly polluting and low-energy content, are sufficient for many years. Consumption is increasing and maximal production could be reached within decades. On the other hand, much may have to be left in the ground to avoid climate change [16,19].

7.4 Emissions

Based on the characteristics of CFB boiler combustion process and its characteristics, harmful substances in the flue gas can be roughly divided into four categories: dust, SO_x and NO_x, and other harmful substances (including acid gases, organic, toxic and harmful gases, heavy metals, etc.)

The number of emissions harmful substances and composition is concerned with the ways of mechanical structure, combustion system and combustion process, physical and chemical properties of material and purification technology.

7.4.1 SO_x formation mechanism

In the process of CFBC, the sulfur content in the fuel gas can form a variety of molecular structure of harmful gases containing sulfur components under different conditions. SO_x not only harms the natural environment, which causes acid rain, but can also be harmful to human respiration. In addition, SO_x causes the serious corrosion of the running equipment [20].

The sulfur content in the fuel gas is easy to transfer into sulfur oxide SO_x in the combustion process, SO_2 exists mainly in the form of elemental composition in the form of sulfur oxides (discharge), and SO_3 is often adsorbed on the exhaust of tiny dust particles or the device wall. Data indicated that SO_3 is difficult to generate under the condition that at 1127°C and excess air coefficient less than 1, beyond the temperature parameters, the production of SO_3 could be increased dramatically [20,21].

Total reaction equation of inorganic sulfur FeS_2 in the process of combustion:

$$FeS_2 \rightarrow FeS + S \tag{7.2}$$

$$FeS \rightarrow Fe + S \tag{7.3}$$

The temperature of the reaction conditions is generally above 500°C. The reaction must be carried out at high temperatures; if the temperature is too low, FeS will remain in the fuel. The speed of the decomposition has nothing to do with the local atmosphere.

Under the effect of hydrogen, reaction equations are as follows:

$$FeS_2 + H_2 \rightarrow FeS + H_2S \tag{7.4}$$

$$FeS + H_2 \rightarrow Fe + H_2S \tag{7.5}$$

When the temperature exceeds 400°C, Eq. (7.11) completed quickly, and the conversion rate is almost 100%. But the reaction of Eq. (7.6) rate is very slow even under the high temperature of 800°C, and this reaction process is often difficult to carry out in the actual process.

FeS_2 and CO reaction equation is as follows:

$$FeS_2 + CO \rightarrow FeS + COS \tag{7.6}$$

FeS_2 and O_2 reaction equation is as follows:

$$8FeS_2 + 22O_2 \rightarrow FeS + FeSO_4 + Fe_2(SO_4)_3 + 11SO_2 + 2Fe_2O_3 \tag{7.7}$$

Under the condition of above 500°C and oxygen atmosphere, the equation is

$$2FeS_2 + 5.5O_2 \rightarrow 4SO_2 + Fe_2O_3 \tag{7.8}$$

$$5\ FeS2 + 10.5O_2 \rightarrow FeSO_4 + Fe_2\ (SO_4)_3 + SO_2 + Fe_2O_3 \tag{7.9}$$

For above 650°C, $Fe_2(SO_4)_3$ will be broken down and the decomposed velocity will be faster under the atmosphere of H_2O, the equation as follows:

$$Fe_2(SO_4)_3 \rightarrow Fe_2O_3 + 3SO_3 \tag{7.10}$$

When up to 1000°C, FeS_2 can directly contact with C forming CS_2, the equation as follows:

$$2FeS_2 + C \rightarrow 2FeS + CS_2 \tag{7.11}$$

Excess air coefficient affects the reaction process, when the air surplus coefficient is less than 1; in general, the generation of H_2S is the key step. When the excess air coefficient is greater than 1, the formation of SO_2 is a key step.

At present the exact form of organic sulfur and the reaction mechanism is less clear. It is commonly believed that in the 100–1000°C range of organic sulfur acid molecules is carbon composition of C–SH and C–S–C and other compounds react with oxygen to generate CH–SH and CH–S–CH compounds [21].

7.4.2 Desulfuration

Within the broad scope of desulfurization, methods include fuel level forecast desulfurization, combustion desulfurization, and FGD. According to the working process and the different mechanical principles of the removal method, it can be divided into physical method, chemical method, and microorganism method. In a CFB boiler, the combustion process of the desulfurization method is the basic principle of the limestone and dolomite with fuel coal particles in a fluidized bed, combustion and desulfurization process at the same time. Among them, the formation of $CaSO_4$ to help reduce the SO_x generated content, so as to realize desulfurization. In the process of its removal, a desulfurization agent can be mixed with fuel in the bed for prolonged contact, so generally the desulfurization rate is higher. SO_x removal method in flue gas can be divided into two categories: dry method and wet method. Dry process desulfurization is suitable for large capacity exhaust treatment; this method has the advantages of no waste water treatment and corrosion, including the cupric oxide or iron oxide dry absorption method and highly active manganese oxide absorption method. The characteristic of the wet desulfurization include high desulfurization efficiency; the disadvantage is that wastewater discharge waste liquor, corrosion problems, low running cost, including aqueous solution of NaOH and Na_2CO_3 absorption method, sodium acetic acid aqueous solution absorption method, ammonia absorption and limestone absorption, and calcium hydroxide slurry. Based on the international comparison experiment and test results of CFB desulfurization, the effect factors of CFB desulfurization can be summarized as: Ca/S, bed temperature, particle diameter, oxygen concentration, combustion mode, wind speed, circulation ratio, SO_2 retention time, feeding method, load change, and so on.

7.4.3 The formation mechanism of NO_x

Nitrogen oxides include N_2O, NO, NO_2, and N_2O_5 molecular structure of gas, but in fact, CFBC mainly produces N_2O and NO. The two parts are collectively referred to as nitrogen oxides, with NO_x said [22]. The source of nitrogen oxide NO_x can be divided into two parts, which is due to the fuel itself contains nitrogen composition and in the combustion process in air when the nitrogen content of the generated nitrogen oxides in combustion process through chemical reaction. Due to the nitrogen composition of fuel contains nitrogen oxides in combustion generated when called fuel type oxide NO_x. Due to air combustion generated when nitrogen oxide type called thermal oxide NO_x. Fuel type nitrogen oxide NO_x formation directly related to the nitrogen content in the fuel and its chemical structure characteristics [23].

7.4.3.1 Thermal NO_x

Thermal NO_x originated from burning nitrogen in the air. If researchers use the Zeldovich NO_x combustion theory model, the product is called Zeldovich NO_x. Usually in more than 1800K of drier can generate the types of NO_x, the reaction equation has the following three:

$$N_2 + O \rightarrow NO + N \tag{7.12}$$

$$N + O_2 \rightarrow NO + O \tag{7.13}$$

$$N + OH \rightarrow NO + H \tag{7.14}$$

The NO_x formation rate is slower than the rate of combustion. According to the Zeldovich combustion theory, the speed can be generated to the NO will rapidly increases with the rise of temperature.

7.4.3.2 Prompt NO_x

Although many studies have reported only C—N fuel will produce dramatically prompt NO_x in the process of combustion, but there are many scholars pointed out that the C—N fuel also appears similar prompt NO_x in the process of coal combustion [24]. According to chemical dynamics analysis and calculation and the experimental test results confirmed, at the beginning of the coal volatile process, diffusion flame will have $(100–200) \times 10^{-6}$ type prompt NO_x formation in volatile combustion area and around the particles combustion diffusion area. That part of the nitrogen quantity is less, but the subsequent process of the formation of NO_x has very important influence. In addition, the low-NO_x combustion control is also nots allow to ignore. To analyze it, the mechanism can be thought as: the volatile composition contains a large number of C—N compound, burning with oxygen in the air, generates the intermediate such as NON and NH_3, after produce prompt NO_x [25].

7.4.3.3 Fuel NO_x

Generated in the process of combustion of NO_x by nitrogen in fuel composition forms through chemical reaction, called fuel NO_x. Nitrogen content of coal fuels in CFB between 0.4% and 3.0%, the nitrogen content of fuel about 0.1%–0.3%. The nitrogen element in the process of combustion is first pyrolysis overflow generate HCN and NHi, etc. Products, then the combustion reaction, further generate NO and N_2. Nitrogen composition in the fuel converted intothe NO_x, its conversion ratio and excess air coefficient, temperature, pressure, and fuel types related physical and chemical properties [26].

When burning, coal particles simultaneously pyrolysis and combustion reaction after heated, which contain nitrogen composition generally most volatile and decomposition into NH_3 and HCN first, then after oxygen composition or NO chemical reaction generated NO_x, transformation to the amount of nitrogen in the coke rarely, are negligible.

A large number of studies have shown that generated in the process of CFB particle combustion, NO_x transformation to generate most of the nitrogen in the fuel components. Generally, at the temperature below 1350°C condition fuel type NO_x conversion rate was over 99%.

7.4.4 Denitration

NO_x generation of harmful substances mainly including NO, NO_2, and N_2O_3 kinds in the process of CFBC. For its processing, it can be divided into two methods: furnace combustion process of purification treatment technology and flue gas purification treatment technology.

7.4.4.1 NO_x control and removal in the furnace

According to the previous discussion, controlling thermal NO_x generated methods reduce the combustion process of oxygen concentrations in the atmosphere, reduce atmosphere of nitrogen concentration, reduce the temperature of the combustion process, and reduce the residence time of NO_x in the generation areas.

Prompt NO_x generated when mainly in the excess air coefficient between 1.0 and 1.4, chemical reaction generated by NH_3 and N_2 produce intermediate HCN in the burning flame surface. Its characteristic is the relationship with the temperature is small, fast generation. This can put forward to suppress the fast type generated NO basic method is as follows: add water vapor inhibit NO production, replace diffuse combustion with premixed combustion, Reduce excess air coefficient of combustion method NO generation.

In order to control the generation of fuel NO_x, you can use to adjust working condition of the combustion chamber combustion method. The change of the combustion air input, such as temperature, position, scale regulation and control in furnace combustion. Has certain influence, especially under the low-NO_x index, the producing rate must attach importance to it.

The excess air coefficient is one of the important factors affecting the fuel type oxides generated. Bed temperature of the CFB has a certain influence on the formation type fuel NO, especially under the low-NO_x index, the producing rate must attach importance to it. In addition, coal and fuel particle properties will have an effect on fuel type oxides generated. The main reason is that the composition of volatile matter, heat transfer characteristics and the characteristics of decomposition play a decisive role on the follow-up of NO_x formation [27].

7.4.4.2 NO_x removal and purification in flue gas

In ammonia catalytic reduction (selective catalytic reduction, SCR) removal technology, the main principle is to choose suitable reductant that will become a harmless reduction of NO_x in flue gas nitrogen oxides of nitrogen removal methods. The key question is the choice of good reductant. The main reactive using reduction of ammonia removal is as follows:

$$4NH_3 + 4NO + O_2 \rightarrow 4N_2 + 6H_2O \tag{7.15}$$

$$8NH_3 + 6NO_2 \rightarrow 7N_2 + 12H_2O \tag{7.16}$$

Nitrogen removal, according to the above reaction generally require the use of catalyst. If you don't use the catalyst, the temperature of 900–1000°C removal processThe process needs to consume large amounts of ammonia. If the use of titanium oxide as catalyst carriers, nitrogen removal process can work in between 200–450°C, about 1 in NH_3/NO, removal efficiency of 80% or more. In order to prevent the catalyst poisoning, actually in the process of removing temperature should be between 200–400°C.

No catalyst reduction (SNCR) namely did not use a catalyst under the condition of directly for NO_x reduction reaction will be converted to nitrogen removal method of N_2. The method of the basic reaction equations is as follows:

$$6NO + 4NH_3 \rightarrow 5N_2 + 6H_2O \tag{7.17}$$

Using the method of removal, NO into N_2 ratio is low, $NH_3/NO = 2$, theory of the maximum removal rate was 90%, generally in the temperature of 900–950°C, $NH_3/NO = 2$, residence time in 0.4 s, but the actual operation process of the removal rate is only 50%, Secondly the removal reaction generally use limited to under 900°C temperature, removal reaction speed is slow[28]. And the temperature is above 1000°C the reaction process of NH3 conversion into NO response, hence the need to avoid. This method uses simple equipment, at low operation cost, and is a very attractive pretreatment removal method.

7.5 Design of combustion system

It is an important issue to work on the mechanism of heat transfer in the CFB. The accuracy of the coefficient of heat transfer is decisive for the design, manufacture and operation of the CFB. The layout, number, and structure of the heated surfaces

Table 7.1 The capacity and coefficient of heat transfer in the CFB

Position	Position of the heated surfaces	The capacity of heat transfer	The coefficient of heat transfer of heat transfer [W/(m^2 K)]	Possible problems
Lower part of the secondary air	Horizontal or vertical	q_{db}	$h_{db} = 300-500$	
The wall of the secondary air	Vertical	q_1	$h_1 = 150-250$	
Upper secondary air suspension heating surface	Vertical	q_2	$H_2 = 150-250$	

are decided by the coefficient of heat transfer. There are different types of heat transfer process, including the particle and gas flow, particle and particle, the gas–solid multi-phase flow and the heating surface, and the gas–solid multi-phase flow and the gas flow into the furnace. In fact, the complicated heat transfer process is the summary of the process above. Only the heat transfer between the gas–solid multi-phase and heated surfaces is taken into consideration [29]. The capacity and coefficient of heat transfer distribution of the CFB are listed in the Table 7.1.

The heat balance of the combustor is as follows:

$$(Q_g + Q_p)_{in} + \Delta H_{rx} = (Q_g + Q_p)_{out} + q_1 + q_2 + q_{ex} + q_{ls} \tag{7.18}$$

q_1: Heat capacity of the internal surface, q_2: Heat capacity of the hanging heating surface, q_{ex}: Heat capacity of the external surface, Q_g: The enthalpy of the gas, Q_p: The enthalpy of the particle, ΔH_{rx}: Chemical heat release, q_{ls}: Heat loss.

In the BFBC and CFBC, the fuels are combusted in the combustor and part of the heat is transferred to the heated surface at the end of boiler by the high-temperature flue gas. However, the flue gas cannot carry all of the heat generated by the fuels combustion. There ought to be some heated surfaces in the particle loops [30]. The heat distribution depends on the heated surface layout.

There are mainly several heated surface layouts which are listed below. (1) There are water walls fixed up in the furnace which were commonly used in the past and are compatible to the small-capacity and low-parameter boilers. (2) There are double-Ω water walls and superheated heating surface in the furnace, including the high-capacity boiler from the Ahlstrom. It would make up the lack of absorption of heat of the heated surfaces in the end of the boiler. (3) There are heated surfaces in the furnace and external heat exchanger in the particle loops from Lurgi/CE and Battelle/ Riley companies. All of the schemes above are proved to be feasible.

In the development of the high-capacity boiler, there are not only water walls in the particle loops, but also superheated surfaces or reheated surfaces, including the No. 2 and No. 3 schemes. Each has its own advantages. When the platen

Table 7.2 The concentration and flow rate of the typical solids of types of boilers

Position	Concentration of the solid material (kg/m^3)	The velocity of the flue gas (m/s)
The dense phase of the CFB	100–1000	4.5–7
The dilute phase of the CFB	5–50	4.5–7
Convection pass of the CFB	<4	12–16
The dense phase of the BFBC	200–1000	1–3.5
Convection pass of the pulverized coal furnace	<2	20–25
Convection pass of the gas furnace	0	>30

The abrasion of the metal parts of the CFB.

superheater is applied, the abrasion problem should be noted. When the external heat exchanger is adopted, the operation of system is more complicated because the cold material and hot material must be controlled separately.

The abrasion problem is related to the concentration, velocity and property of the solid particle and geometry of the runner [30,31]. The heated surface and the refractory material is scoured by the large amount of the particles. The concentration and flow rate of the solid particle are listed in the Table 7.2.

Air distributor equipment: There are mainly two cases of the abrasion of the air distributor equipment. One is the abrasion of wind caps and the most serious part is around the part of returning material [32]. The surface of the water walls: The abrasion of the surface of water walls is one of the most serious abrasion which is related to the materials. There are mainly four cases: the abrasion of the rectory belt in the lower part of the furnace, the abrasion of four corners of the furnace, the abrasion of general water walls and the abrasion of irregular regions.

The abrasion of secondary air nozzle: The pulsation of the material in the dense phase would bring the bed material into the secondary air nozzle.

The abrasion of the cyclone: Since the surfaces of the cyclone are covered with the refractory material, the metal part of the cyclone is not very serious. The damage of the center cylinder of cyclone is caused by the thermal deformation [33].

The abrasion of the external exchanger: The external exchanger operates in the bubbling fluidized condition. The velocity of the gas is little higher than the critical velocity; therefore the abrasion is less serious.

The abrasion of the convection pass: It is similar to that of the PC boiler. As literature shows that once the CFB is under good condition the abrasion of the convection pass is not very serious.

The cyclones used for CFB plant are developed from the regular cyclones. As is known to people that the cyclone is invented according to the centrifugal force of the dusty gas [34]. The dust in the gas can be separated by using cyclones [35]. There are many kinds of cyclones, including high-efficiency cyclones, high-volume cyclones and general cyclones. There are also can be divided into three kinds of the

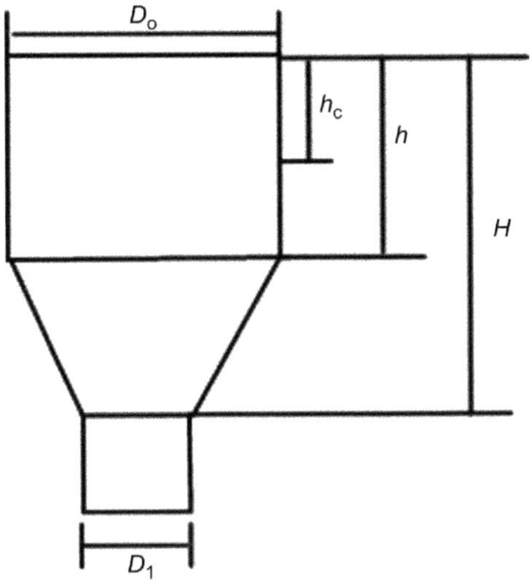

Figure 7.3 Typical cyclone of vertical axis import.

cyclones according to the air import situation, including cyclones of tangential flow and cyclones of axial flow.

The cyclones of vertical axis import are widely used in the CFB industrial which is shown in Fig. 7.3. The shape of the top and bottom of the cyclone is cylinder type and cone type. The gas–solid flow enters the cyclone through the rectangular duct and the gas flow enters the two concentric eddies of the cyclone, The external eddy is down while the internal eddy is up. Since the density of the solid is higher than that of gas, the solids leave the external eddy and move to the internal surface by the centrifugal force. The separated solids fall down along the surface and accumulated in the bottom of the cyclone. The gas goes up through the internal eddy and out of the cyclone.

Since a large amount of the high-temperature gas is handled by the cyclones, the smaller the diameter of the cyclone, the higher the efficiency of the cyclone. However, it is hard to lay the refractory material on the small-scale cyclones and it is common to adopt several cyclones for higher efficiency. The typical size of cyclones in industrial application is listed in Table 7.3.

The loopseal type return unit is used for transferring the separated high-temperature solids to the combustor and making sure the amount of gas to the cyclones is minimal. There are two parts of the return device, including dipleg and valve. The dipleg is mainly for overcoming the negative pressure gradient between the separate and furnace. The valve is for adjusting the particle flow. The design of the dipleg is according to the pressure performance and amount of the returning material. The diameter and height of the dipleg can be determined.

It is an important device for safe and steady operation of CFB which is shown in Fig. 7.4.

Table 7.3 **The typical size of cyclones in the industrial application**

Heat capacity (MW)	67	75	109	124	124	207	211	230	234	327	394	396	422
The number of cyclones	2	2	2	2	1	2	2	2	2	2	3	2	4
The diameter of cylinder D_0 (m)	3	4.1	3.9	4.1	7.2	7.0	6.7	6.8	6.7	7.0	5.9	7.3	7.1
The volume flow rate of the gas at 850°C (*10^6 m^3/h)	0.175	0.19	0.285	0.325	0.54	0.55	0.55	0.6	0.61	0.85	0.68	1.03	0.55
The velocity of the imported gas (m/s)	43	25	41	43	23	25	27	29	30	38	43	43	24

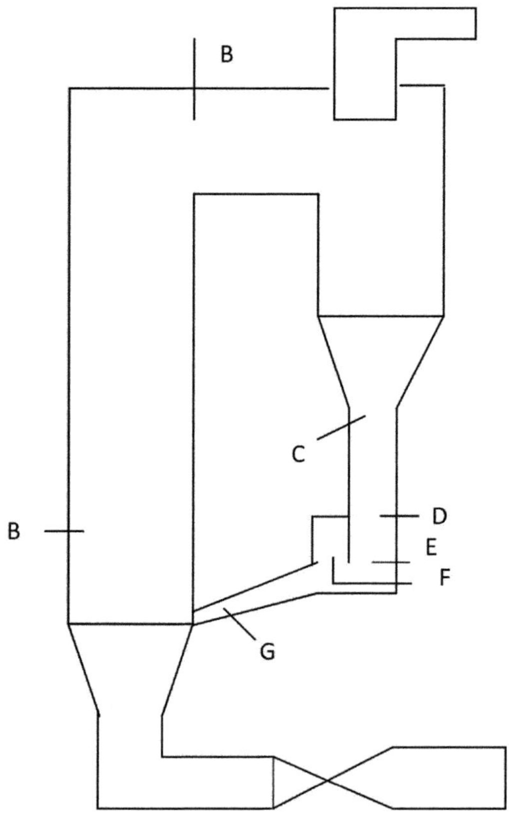

Figure 7.4 Typical diagram of the loop seal.

The pressure of dipleg (ΔP_{CE}) is equal to the sum of ΔP_{AB}, ΔP_{BC}, and ΔP_{EA}, while ΔP_{CE} is maximum when the flow is in the critical condition. The minimum value of the dipleg as followed:

$$H_{min} = (\Delta P_{AB} + \Delta P_{BC} + \Delta P_{EA})\max/(\rho_p(1 - \varepsilon_{mf}) * g) \quad (7.19)$$

When in the actual situation, H is defined as followed:

$$H = (1.5 - 2.0)H_{min} \quad (7.20)$$

References

[1] E.J. Anthony, Fluidized-bed combustion of alternative solid fuels—status, successes and problems of the technology, Prog. Energy Combust. Sci. 21 (3) (1995) 239–268.

[2] Y.M. Chen, Fundamentals of a centrifugal fluidized-bed, AICHE J. 33 (5) (1987) 722–728.

[3] V. Kain, K. Chandra, B.P. Sharma, Failure of carbon steel tubes in a fluidized bed combustor, Eng. Fail. Anal. 15 (1–2) (2008) 182–187.

[4] T.P. Chen, S.C. Saxena, Mathematical modelling of coal combustion in fluidized beds with sulphur emission control by limestone or dolomite, Fuel 56 (4) (1977) 401–413.

[5] P. Basu, Combustion and Gasification in Fluidized Beds, CRC Press, Boca Raton, 2006.

[6] M. Horio, Fluidization science, its development and future, Particuology 8 (6) (2010) 514–524.

[7] S. Oka, Fluidized Bed Combustion, CRC Press, Boca Raton, 2003.

[8] J. Koornneef, M. Junginger, A. Faaij, Development of fluidized bed combustion - An overview of trends, performance and cost, Prog. Energy Combust. Sci. 33 (1) (2007) 19–55.

[9] H.T. Bi, J.R. Grace, Flow regime diagrams for gas-solid fluidization and upward transport, Int. J. Multiph. Flow 21 (6) (1995) 1229–1236.

[10] Y. Li, M. Kwauk, Fluidization, The Dynamics of Fast Fluidization, Springer, USA, 1980, pp. 537–544.

[11] World Coal Consumption 2000–2011, EIA statistics.

[12] World Energy Projections Plus (2009), EIA.

[13] Total World Electricity Generation by Fuel (2006), Source: IEA 2008.

[14] Fossil Power Generation, Siemens AG, April 23, 2009.

[15] E.A. Sondreal, G.A. Wiltsee, Low-rank coal: its present and future role in the United States, Annu. Rev. Energy 9 (1984) 473–499.

[16] P. Glarborg, A. Jensen, J.E. Johnsson, Fuel nitrogen conversion in solid fuel fired systems, Prog. Energy Combust. Sci. 29 (2003) 89–113.

[17] M. Le Page, IPCC digested: just leave the fossil fuels underground, New Scientist, October 1, 2013.

[18] Nordjyllandsværket, Archived November 20, 2012 at the Wayback Machine (05.11.13).

[19] Avedøreværket, Ipaper.ipapercms.dk (05.11.13).

[20] M. Zetterdahl, et al., Impact of the 0.1% fuel sulfur content limit in SECA on particle and gaseous emissions from marine vessels, Atmos. Environ. 145 (2016) 338–345.

[21] K.F. Boersma, G.C.M. Vinken, J. Tournae, Ships going slow in reducing their NO_x emissions: changes in 2005–2012 ship exhaust inferred from satellite measurements over Europe, Environ. Res. Lett. 10 (2015) 074007.
[22] A. Sjodin, et al., Verification of expected trends in urban traffic NO_x emissions from long-term measurements of ambient NO_2 concentrations in urban air, Sci. Tot. Environ 189 (1996) 213–220.
[23] K.M. Han, et al., A comparison study between CMAQ-simulated and OMI-retrieved NO_2 columns over East Asia for evaluation of NO_x emission fluxes of INTEX-B, CAPSS, and REAS inventories, Atmos. Chem. Phys. 15 (4) (2015) 1913–1938.
[24] M.J. Melo, et al., Flow and combustion characteristics of a low-NO_x combustor model for gas turbines, J. Propuls. Power (2011) 1212–1217.
[25] M.D.M. Dutk, Emission characteristics of a novel low NOx burner fueled by hydrogen-rich mixtures with methane, J. Power Technol 95 (2) (2015) 105–111.
[26] D. Guan, et al., The socioeconomic drivers of China's primary PM2.5 emissions, Environ. Res. Lett. 9 (2014) 0240102.
[27] S. Biswas, et al., Chemical speciation of PM emissions from heavy-duty diesel vehicles equipped with diesel particulate filter (DPF) and selective catalytic reduction (SCR) retrofits, Atmos. Environ. 43 (11) (2009) 1917–1925.
[28] J. Krzywanski, W. Nowak, Modeling of bed-to-wall heat transfer coefficient in a large-scale CFBC by fuzzy logic approach, Int. J. Heat Mass Transf. 94 (2016) 327–334.
[29] J.R. Grace, Heat transfer in circulating fluidized beds, in: P. Basu (Ed.), Circulating Fluidized Bed Technology, Pergamon Press, Toronto, 1986.
[30] X.L. Yin, C.Z. Wu, S.P. Zheng, Y. Chen, Design and operation of a CFB gasification and power generation system for rice husk, Biomass Bioenergy 23 (2002) 181–187.
[31] E. Ferrer, et al., Fluidized bed combustion of refuse-derived fuel in presence of protective coal ash, Fuel Process. Technol. 87 (1) (2005) 33–44.
[32] J.W. Slusser, S.P. Bartlett, A.D. Bixler, Materials experience with the Stockton CFBC, in: Proceedings of Workshop on Materials Issues in Circulating Fluidized Bed Combustor. EPRI Report GS-6747, 1990.
[33] J. Stringer, I.G. Wright, Materials issue in circulating fluidized bed combustors, in: E.J. Anthony (Ed.), Proceedings of 11th International Conference on Fluidized Bed Combustion, ASME Press, New York, 1991.
[34] T.M. Knowlton, S.B.R. Karri, A. Issangya, Scale-up of fluidized-bed hydrodynamics, Powder Technol. 150 (2) (2005) 72–77.
[35] I.P. Boukis, et al., CFB air-blown flash pyrolysis. Part I: Engineering design and cold model performance, Fuel 86 (10–11) (2007) 1372–1386.

Underground gasification of low rank coals

8

Nikolaos Koukouzas
Centre for Research and Technology-Hellas, Athens, Greece

8.1 UCG process

8.1.1 Definition

Within the last hundreds of years, coal mining has given a tremendous push in many aspects evolving our everyday lives. Metals and industrial material production, electrical energy, and transport are some of the uses even nowadays. The continuous improvement of living standards in parallel with population growth leads to predictions supporting that between 2008 and 2035 and world coal demand will be increased by 20% [1]. In the last few decades, a new and ecologically friendlier way to recover carbon from coal seams instead of direct mining gains ground year by year. The Underground Coal Gasification (UCG) method can not only be applied to deep coal deposits that would otherwise remain unmined, but is also related to functional clean coal technologies (CCTs).

UCG is a technological method where the seam of solid carbon is converted into a gaseous phase in situ. In order to achieve this, an injection—often inclined well—supplies oxygen, air, or even stream directly to the coal seam which becomes an underground gasification reactor in high temperatures. Reactions such as pyrolysis convert coal deposits to a synthesis gas known as "syngas". Via other production wells, this gas mainly hydrogen, methane, carbon dioxide, and monoxide travels to the surface for treatment and further use (Fig. 8.1). During this process, the reacted coal and the produced gas are responsible for the formation of cavities in the coal body between injection and production well, which are growing as the process continues. Once gas productivity slows down, new injection or production wells are drilled further, for more unreacted coal to be exposed to the input steam current.

8.1.2 Active worldwide projects

In the last few years, more and more research institutes and government sectors in contact to industry work on worldwide projects that intend to make this technology viable and commercially applicable. Multidisciplinary studies on chemistry, physics, geology, and chemical engineering have to collaborate in tests with different time and length scales in order to better understand technical economic and environmental risks and parameters. Assuming that UCG should efficiently be formed

Low-Rank Coals for Power Generation, Fuel and Chemical Production
DOI: http://dx.doi.org/10.1016/B978-0-08-100895-9.00009-7

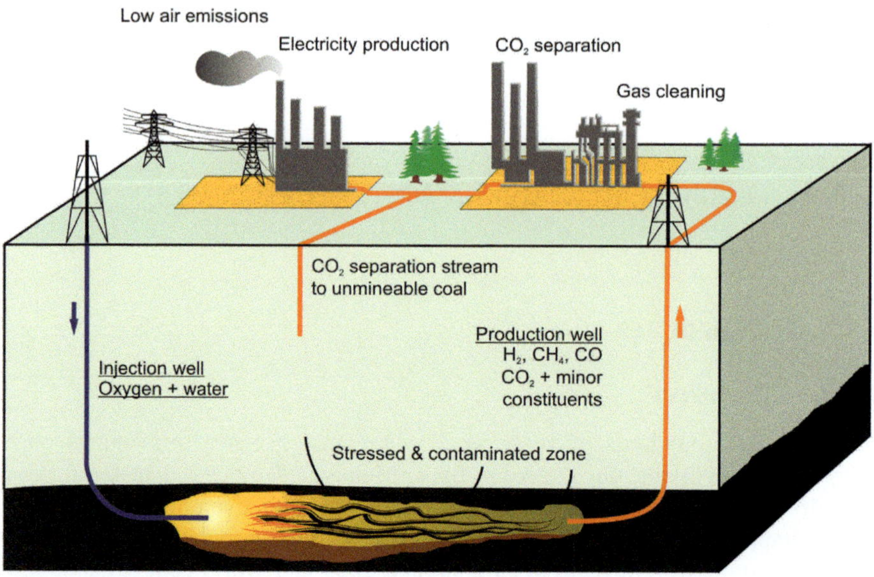

Figure 8.1 Schematic figure on Underground Coal Gasification procedure. *Source*: Modified from UCG Association.

and broadly deployed, it is more likely that the world's coal deposits will be reconsidered greatly upward.

A well-developed large UCG project has been established in Uzbekistan (Yerostigaz), where this is the only commercial facility for more than 50 years (since 1961), providing gasified coal for power production. The total daily production is about 1 million cubic meters of syngas per day. Today, quickly turning into a suitable business development is happening in North America, Australia, South Africa, and China. Additionally, mostly many Eastern European nations such as Poland and Hungary are seriously working for its economic viability (Fig. 8.2).

Notwithstanding the fact that UCG process is based on simple principles, there are obstacles and limitations hindering further commercialization of the technology. These are related to the precision where the coal seam is drilled from both the injection and production wells and the right manipulation for achieving optimal conditions to the underground reactor for a high quality syngas production. In addition, in order to acquire secure findings, the experiments have to be applied in as much as real conditions and pilot scale instead of laboratories, which demand higher amounts of money. There are also environmental issues regarding possible pollution of underground water aquifers, during and after gasification process which tend to local society with a more skeptical attitude. A lot of progress has to be also done on site evaluations, economic studies, and safety in order to convince institutions and investors to push on UCG commercialization [3].

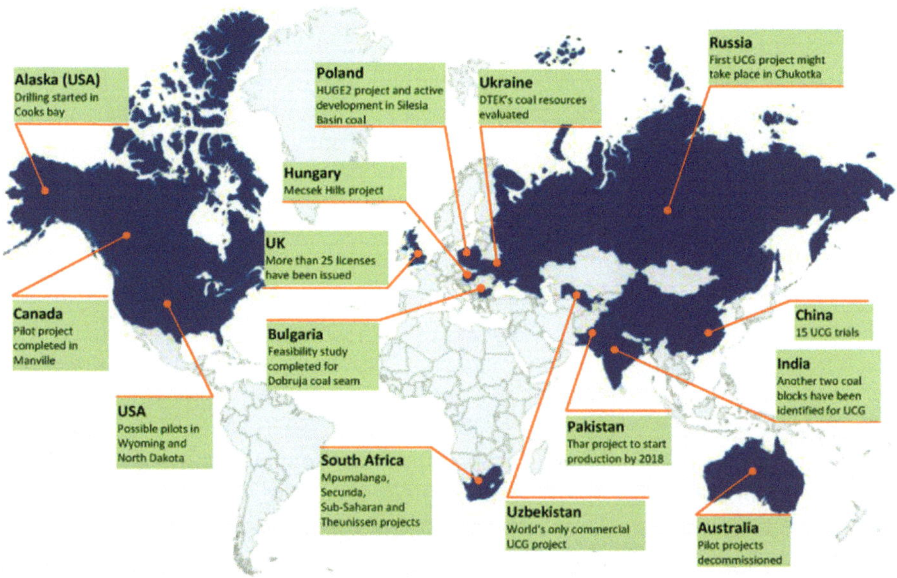

Figure 8.2 Active countries on UCG projects [2].

8.1.2.1 Australia

In the present century, Australia has made significant development on the implementation of UCG projects [4]. Since 1999, a demonstration plant in Queensland was constructed by Linc Energy for UCG purposes. Five UCG reactors were installed followed by the construction of a gas to liquid pilot power plant. Further two UCG reactors were applied by Carbon Energy Ltd., using enough syngas for 1 MW electricity production but for limited time. Other UCG projects were also planned in order to be established, however, these remained inactive due to environmental issues and other related reasons. Despite all these obstacles and difficulties, Australia has demonstrated readiness to attempt projects on power production related to Carbon Capture & Storage (CCS), with parallel concern on collaboration to UCG-CCS involved European research institutes.

8.1.2.2 Asia

According to the confirmed coal reserves, significant amounts of coal remain unmined—especially in China, the buried coal deposits reach 50%. In the last few years, China's government has set new targets for drastic decrease of pollution produced by the coal-fired plants. Thus, the UCG Research Center of China and the University of Mining and Technology (Beijing) (CUMTB) in collaboration with coal industry has run 15 UCG projects. This increased activity on CCTs leads to efforts for enhanced partnerships with countries and companies that have achieved substantial progress on UCG technology in Europe, South Africa, United States, and Australia. Therefore, in 2011, a consortium of the Seamwell International Ltd. (UK) and Chinese Energy Conservation and Environmental Protection Group (CECEP, China) worked to gasify 6 million tons of coal per year in Inner Mongolia's Yi He coalfield, generating 1000 MW of electricity for 25 years.

In India, the Neyveli Lignite Corporation began in 2005, a new project funded by the Ministry of Coal aiming at a suitable lignite block for UCG test, in order to carry out pilot-scale studies and finally to assess the heat value of the gas produced. In addition, the Central Mine Planning & Design Institute Limited (CMPDI) prepared data packages for five potential UCG sites. One of these sites, the Kasta block in Raniganj coalfield, was selected for the UCG experiment. In July 2014 two additional coal blocks, Kaitha Block of Central Coalfields Ltd and Thesgora "C" Block of Western Coalfields Ltd., were identified by Coal India Limited (CIL) for industrial use.

Regarding Pakistan, in December 2009 the Thar Coal and Energy Board (TCEB) joined a partnership with world renowned companies from China, Germany, and South Africa to examine the geological, hydrological, cultural, and environmental impacts in the area after 6.5 MT/y coal mining activity, aiming to generate 1200 MW of power. Subsequently, in the next year a UCG project was started in Block V of Thar Coal deposits. The aim was to gasify 1.4 billion tons of low-grade lignite coal reserves [5], which consequently would set up two pilot 5 MW plants, producing 8000 MW of energy [6]. Throughout the UCG pilot construction phase, the project team dealt with problems related to hole linkage, water incursion, aquifer characterization, and coal ignition. The project will pass to the operational phase producing electricity by 2018.

8.1.2.3 North America

For many decades now, the United States and Canada have shown significant and substantial development not only on UCG modeling but also on field testing. The USA was the UCG leader during the 1980s and early 1990s. In the mid-1990s, as natural gas prices fail to record low levels, activity decreased drastically until 2005 where interest recovered as the US Department of Energy (DOE) funded the technology [7]. Research took up with emphasis on integrating 3D full simulator development in the cavity growth (reaction, geothermal, and hydrology & fluid mechanics) [8].

Many Australian and Canadian Companies (e.g., Linc Energy, Carbon Energy, Ergo Energy, and Laurus) managed the commercial development of UCG in the United States. Linc Energy has undertaken a project in Wyoming and another in Cooks bay Alaska in association with the State Authorities, which have started exploratory drilling since September 2014. Carbon Energy also has coal assets in Wyoming and North Dakota for a possible future pilot [9]. The Canadian Company Laurus Energy has planned to develop a project at Stone Horn Ridge near the Beluga River in southern Alaska in conjunction with Cook Inlet Region Incorporated, a native American-owned corporation in Alaska. The project will be designed and developed with the capability of CCS. The most advanced Canadian UCG development is a pilot project completed by Swan Hills Synfuels with support from the Alberta Energy Research Institute (AERI) in coal seams up to 1400 m deep, the deepest UCG plan ever conducted in the world. This UCG project was agreed to be financed by the Government of Alberta and Swan Hills Synfuels for $285 million to cover the CCS part of the project, but this has not yet accomplished. Laurus Energy Canada Inc. is seeking projects in North America to exploit these large quantities of syngas for power generation, as well as for chemical and fertilizer plants in the Region. Other organizations such as Sherrit Technologies and University of Calgary are involved in clean energy technologies development [10].

8.1.2.4 South Africa

In South Africa the Eskom energy company was the first who initiated UCG investigation in 2007 [11]. Eskom undertaken to construct a 3 MW UCG Pilot Plant next to Majuba Power Station in Mpumalanga and 3 years later the produced syngas was co-fired with coal in the power plant. The associated coal deposit has a bituminous character and the stratigraphy ranges from 1.8 m up to 4.5 m in thickness, lying at depths between 250 m and 380 m. Other Africa's plans include the sub-Saharan project with Linc Energy and Exxaro Resources, and the Theunissen project with Africary Holdings Ltd. [12].

8.1.2.5 Europe

UCG has a deep theoretical and field-based background in countries such as the UK, France, and the Former Soviet Union (FSU) for over 50 years in the latter [13]. Scientific and engineering knowledge on UCG has been continuously developed and has led to several UCG operations. Since 1996 when field work stopped, Russia has been improving the basic structural components and operational parameters of UCG technology. It is anticipated that Russia's first UCG project will take place soon in Chukotka where Clean Energy, a subsidiary of Linc Energy, investigated the coal deposit and concluded at least two appropriate sites for the UCG implementation (The Moscow Times, 2013).

Ukraine continued working on UCG after FSU independence, participating in the first Research Fund for Coal and Steel (RFCS) funded HUGE Project (2007–10), by providing an extensive review of the previous Soviet work on UCG and contributing to the design of the underground gasifier at the Barbra Mine, Katowice. The Ukrainian Technological Academy (UTA) has patented a geo-technology process for obtaining hydrogen by purifying synthesis gas from UCG [14].

In Hungary, Wildhorse Energy is a company which has been activated in UCG, focusing on implementing UCG and developing its prospective uranium deposit in the Mecsek Hills in the Pecs region of Southern Hungary. In July 2012 the Hungarian Government approved UCG as a technology and planned the construction of a 130 MW pilot power plant to demonstrate the ability of UCG dependent on seeking investment partners.

Poland has large coal reserves that can be exploited via UCG technology for power generation. In the 1960s and 1970s, small-scale UCG experiments were carried out, but since 2007 Poland has begun to re-approach its UCG activities through new exploratory and field tests in its territory. An important EU project undertaken by Central Mining Institute (GIG in Polish) is the Hydrogen-Oriented Underground Coal Gasification for Europe project (HUGE, 2007–10), funded under the RFCS program and bringing together 11 partners from seven countries. Its main focus was the theoretical and experimental development of in situ production of hydrogen-rich gas from coal using underground gasification. A follow-up project HUGE2 (2011–14), which was also financed by RFCS, focused on the environmental and safety aspects associated to the UCG process, including underground water contamination and the potential leakage of toxic gases. There is also an ongoing nationally funded UCG project, which is being constructed in an active coalmine in Upper Silesian Basin for further industrial use. In addition, Linc Energy has a joint venture to develop UCG in Poland with an exploration site license in Silesia [12].

In Bulgaria, a comprehensive feasibility study for developing UCG in Dobruja coal deposit has been completed under an RFCS funded project [15]. Geological, geomechanical, cavity, and hydrogeological models have been developed for the adjacent area. Furthermore, engineering, drilling, and completion requirements of wells were investigated. Further study has been implemented for the environmental and economic assessment of the UCG combination and subsequent CO_2 storage.

Since 1950s field trials, the UK did its first steps for the UCG development. Years later, new initiative on UCG (2000–05) led by the UK Coal Authority and supported by the UK Department of Trade and Industry (DTI) examined the feasibility of UCG. The main conclusion was that UCG should be seen primarily as a near-shore and estuary technology and a site was identified in the Firth of Forth for a possible field trial. Since then, more than 25 licenses have been issued for UCG exploration in offshore areas. The current key players are Thornton Energy (Firth of Forth), Five Quarter (Newcastle), and Cluff Natural Resources with eight offshore sites in Firth of Forth, Liverpool, Cumbria, and Durham. Key private sector investors are responsible for the funding of geological evaluation of these license areas. The planning applications are currently underway [16].

The previous review makes clear the fact that UCG technology appears to have strategic developing for the worldwide growing interest related to large-scale syngas production, parallel with the recovery of huge unminable deep coal deposits with distinct cost and environmental benefits. Notwithstanding that we do know much about the control, site selection, and operation of UCG, further knowledge from modeling and field tests must be acquired to ensure that environmental impacts of full commercial deployment are fully understood and assessed. This situation would place coal in a sustainable, secure, and competitive energy mix, allowing its continued contribution to an economical and safe energy production. A complete set of UCG trial project data including coal seam types and thicknesses has been reported in 2007 [17]. The basic point that will lead this technology to further development—making it more attractive commercially in the UCG industry—is the collaboration, the expertise sharing and knowledge among projects and governments with experience in the aspects of environmental impact, planning and regulation of UCG.

The EU provides opportunities by funding projects that set the basis for collaboration between European countries in the field of CCTs like UCG and CO_2 storage solutions. The countries that have shown the greatest interest and most active R&D programs in UCG are China, India, South Africa, United States, Canada, Australia, and certain Member States of the EU. In EU, Poland, Hungary, and the UK are the countries with the greatest progress. The research and development on CCS is active and demonstration scale projects of both capture and CO_2 storage are underway in most countries. The maturity of CCS is of significant importance to facilitate the combination of UCG and CCS. Case studies of UCG-CCS have been carried out in the Powder River basin of Wyoming, United States and the Williston basin, North Dakota, United States. European funds were provided for a pilot investigation of in situ hydrogen production incorporating UCG-CO_2 management in Poland [18].

8.2 UCG research and development

8.2.1 Process modeling

8.2.1.1 Site investigation and geological model construction

The collection of geological information in a study area is the first step before further process. Data digitizing and re-processing well log correlation led to renewed insights into the spatial behavior and geometric characteristics of different formations and laid the basis for further modeling and mapping of data for the subsequent geotechnical and hydrogeological modeling of the study.

The identification of the coal seams of main formations as well as the preparation of basic structural, thickness maps and cross-sections was made using specified digitizing software. The coal properties of different seams are studied (e.g., ash content (A), volatiles (Vs), sulfur (S), and moisture content (W); coal resources of the seams are calculated from seam thickness and actual surface area), in order to have a better understanding of the differences between them.

Structural and tectonic research follows for fault modeling—the existing tectonic units are identified and studied—with calculation of the dipping direction of faults. Petrel is an example of software that provides 3D tectonic model for the deposit, developed by pillar gridding method (Fig. 8.3A), as well as a 3D interpretative geological model (Fig. 8.3B) [15]. The latter included the creation of the key horizons of the model (e.g., the top formation, erosion surface, and top aquifer) followed by a 3D grid editing and zone creation. Investigating the cap rocks shows that the top formation (of all coal bearing formations) had no effect on the UCG process.

The developed 3D interpretative tectonic and geological models for the study area improves our understanding about the geological structure of the basin and the related geomechanical, hydrogeological, and environmental data. These models

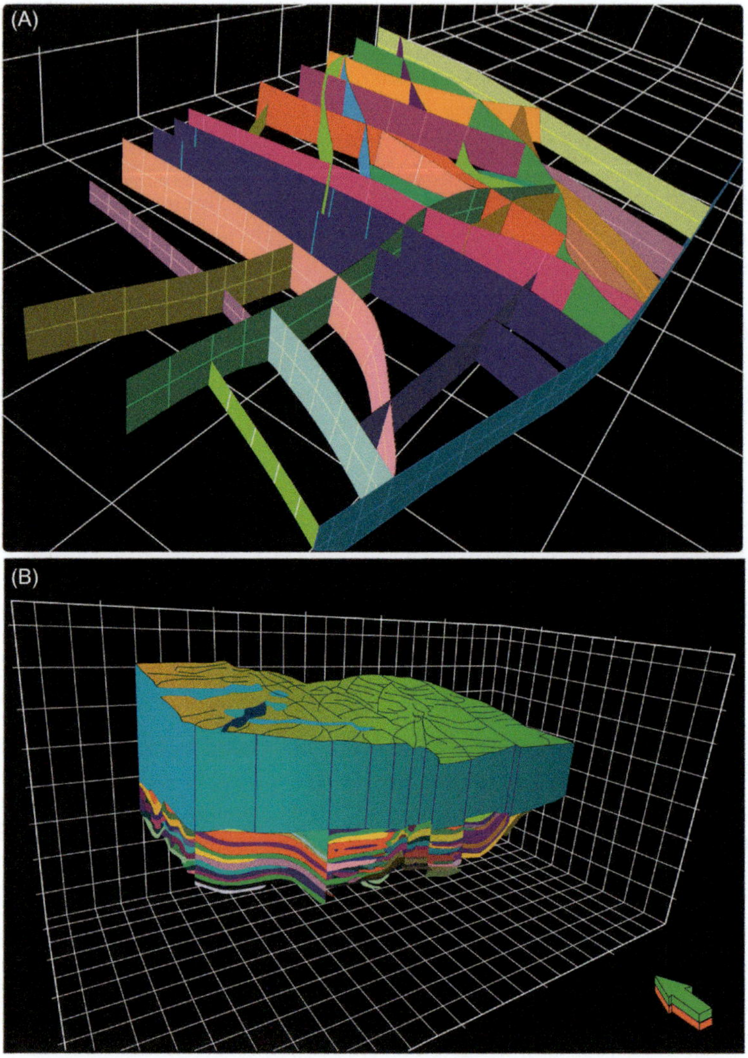

Figure 8.3 (A) Example of the 3D view of fault structure and (B) the geological model of a study area in Bulgaria [15].

constitute a crucial part of the work on assessing the feasibility of the UCG-CCS process for the deep lying coals of the study area. The results arising from the models are used for the development of other state-of-the-art models. A 3D seismic survey around the target should analyze the structure to within 1–5 m, which is less than seam thickness and sufficient to confirm the site for a UCG test. Additional coal and rock sampling is also required in order to clarify the conditions of gasification and the following CO_2 injection at the study area.

Taking into account the world experience, some generalized site selection criteria for UCG have been developed and are additionally assessed and applied to the area that has been proposed for CCS (Tables 8.1–8.4). These tables were formulated based on results from geological, geomechanical, hydrogeological, environmental, economic, and drilling models developed by the data collected in a specific area for feasibility study. Some of these criteria are also supported by general knowledge from previous and common practices.

8.2.1.2 Modeling of the geomechanical and thermal effect of the UCG and CO_2 storage processes

For the purpose of thermo-geomechanical investigation, the first coal seam to gasify is the thickest coal seam in the sequence, which in our example is a seam approximately 10 m thick, located at a distance of 1500 m below ground level [19]. Finite element (FE) software package ABAQUS was used for the thermo-geomechanical modeling. The material properties were calculated by averaging the individual rock properties with the corresponding thickness fraction of each geological section. A detailed library of mechanical and thermal properties of coal and rocks under the effect of heat was established from literature references. The temperature distributions after ignition, captured every 6 hours during 1 day of the gasification process,

Table 8.1 Coal seam bedding depth in the study area in terms of its suitability for UCG [15]

Depth (m)	Suitability
<1200	Not suitable
1200–1500	Highly suitable
1500–1800	Medium level of suitability
>1800	Low level of suitability

Table 8.2 Coal seam thickness in terms of its suitability for UCG [15]

Seam thickness (m)	Suitability
<2	Not suitable
2–4	Medium level of suitability
4–8	Highly suitable
8–12	Medium level of suitability
>12	Low level of suitability

Table 8.3 Thickness of the rocks and their filtration properties [15]

		Thickness of overlaying rocks (m)				
		>200	100–200	50–100	20–50	<20
Filtration coefficient (m/d)	$<1\times10^{-8}$	Optimum	Optimum	Optimum	Possible optimum	Risk
	1×10^{-8}	Optimum	Optimum	Possible optimum	Risk	High risk
	1×10^{-6}–1×10^{-4}	Optimum	Possible optimum	Risk	High risk	Inappropriate
	1×10^{-4}–1×10^{-1}	Possible optimum	Risk	High risk	Inappropriate	Inappropriate
	$<1\times10^{-1}$	Risk	High risk	Inappropriate	Inappropriate	Inappropriate

Color index	Suitability
(dark green)	Optimum
(yellow)	Possible optimum
(light orange)	Risk
(orange)	High risk
(red-orange)	Inappropriate

Table 8.4 Reflectivity of vitrinite in the study area in terms of coal suitability for UCG [15]

Reflectivity of vitrinite (%)	Suitability
0.2–0.5	Medium level of suitability
0.5–0.8	Highly suitable
0.8–1.0	Medium level of suitability
1.0–1.5	Low level of suitability
> 1.5	Not suitable

are shown in Fig. 8.4. The thermal affected area increase for each time increment and the temperature is found to be extremely high in the area nearby the ignition center due to constant heat flux applied at this point.

According to Couch [20], coal would be pyrolized at temperatures around 400°C, so to maintain the heat transfer and model integrity elements with a temperature larger than 400°C heat transfer from the element were allowed. However, in order to realistically represent the mechanical failure of coal after gasification, the elastic modulus of coal was gradually decreased with temperature. The temperature distribution for a period of 3 days after ignition, combining view cuts in the *X*–Y and *Y*–*Z* planes to create a 3D view, is illustrated in Fig. 8.5. As expected, when the ignition point is moved, the position with the highest temperature changes

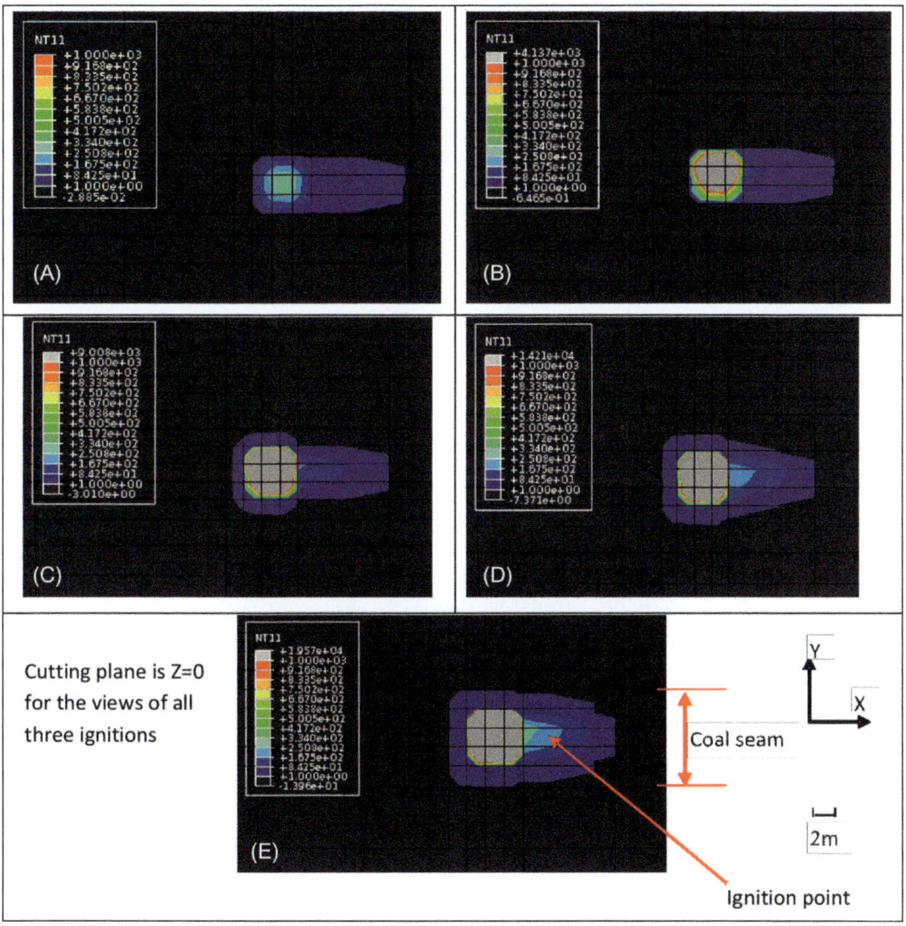

Figure 8.4 Transient temperature transfer under the ignition for 1 day: (A) After 1 h, (B) after 6 h, (C) after 12 h, (D) after 18 h and (E) after 24 h (1 day). *NT*, Node Temperature [15].

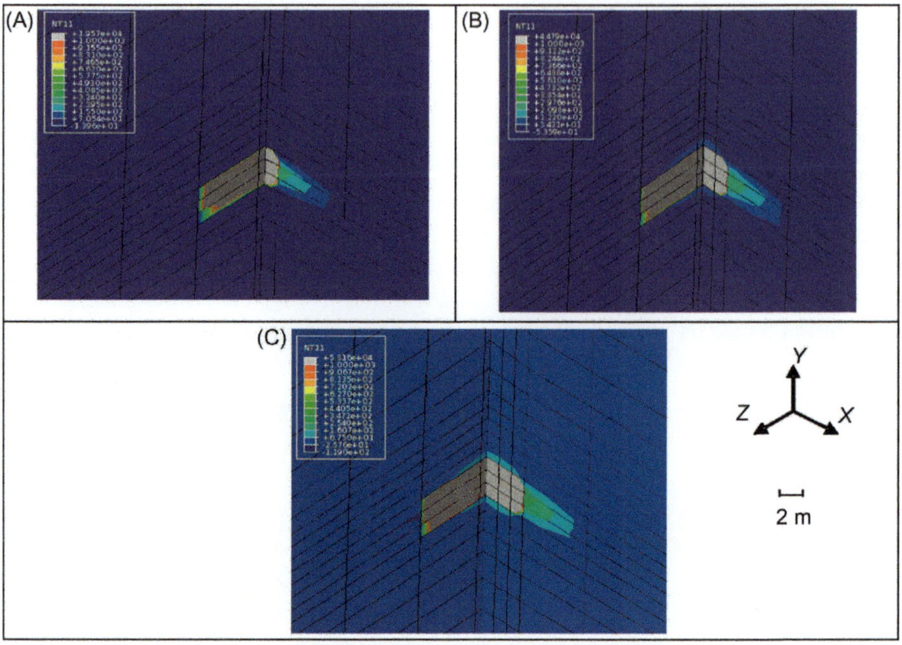

Figure 8.5 Temperature distributions under ignition for 3 days (view in 3D *X*–*Y*–*Z* coordinates): (A) day 1, (B) day 2, and (C) day 3. *NT*, Node Temperature [15].

accordingly. The highest temperature as well as the overall thermal affected area is increased gradually.

The surface subsidence after gasification is shown in Fig. 8.6. Since a geological fault is included in the model, the subsidence curve is not smooth and there is an obvious jump at the location of the fault. It was also found that as the ignition process is taking place, the surface subsidence is slightly increased.

The 3D coupled thermal–mechanical models provide a detailed analysis of the UCG process in terms of heat transfer, cavity growth, and surface subsidence. The 3D models incorporated the real lithological structure from the site and the thermal-dependent material properties from experiments and literature references [21,22] to simulate the real time gasification process. The modeling results showed that a minimum distance of 150 m should be kept between the gasification channels and the nearby faults in the site in order to avoid risky geological interaction which could lead to potential gas or CO_2 leakage.

The developed 3D ABAQUS numerical model was further extended in order to investigate the faults' behavior, the stresses, and permeability in strata around the UCG cavities, as well as the possible roof collapse and the CO_2 storage capacity of the selected site. The model incorporates a detailed geological structure of the site including the positions and depths of the faults as well as the thickness and depth of the coal seam (Fig. 8.7).

A sensitivity analysis on the acceptable cavity distance away from the faults can be assessed by incorporating a series of assessment criteria and assumptions for

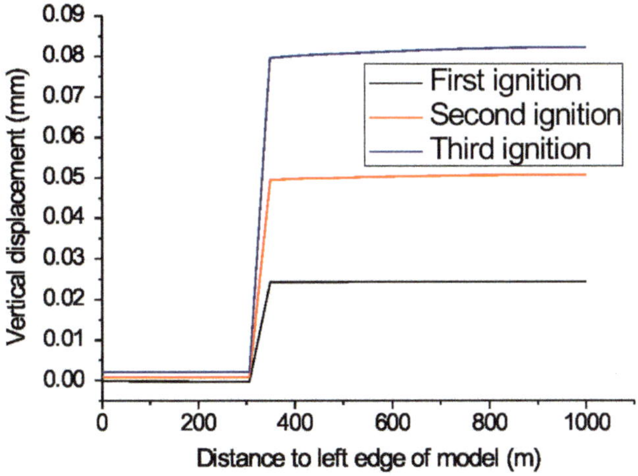

Figure 8.6 3-Day surface subsidence across the cross section [15].

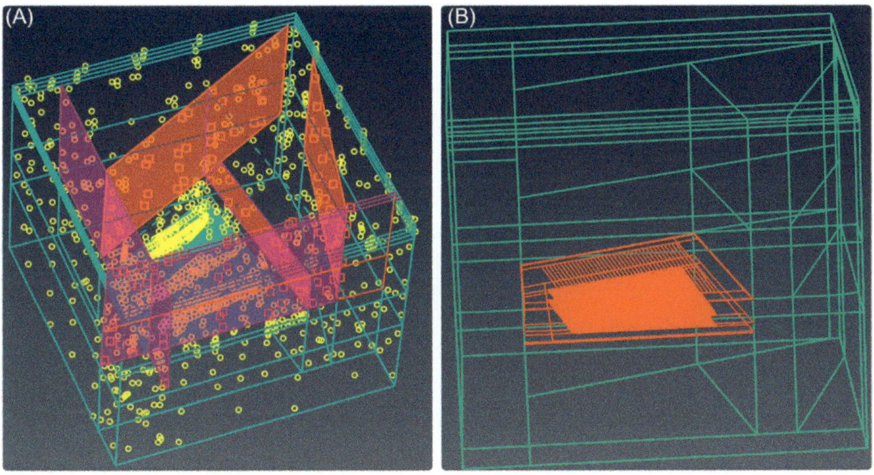

Figure 8.7 3D geomechanical model of the selected site: (A) faults structure in pink and (B) coal seams in red (small yellow circles are the reference points for the surfaces of the parts) [15].

pressure distribution and other parameters (Fig. 8.8). A distance of 150 m away from the fault would also be applicable.

8.2.1.3 Hydrogeological modeling of the selected site

A complete analysis of the hydrogeological information in the target sites is essential. Thus, in our example the most informative reference sources regarding lithological and stratigraphic data from deep wells located in the Varna Artesian Basin

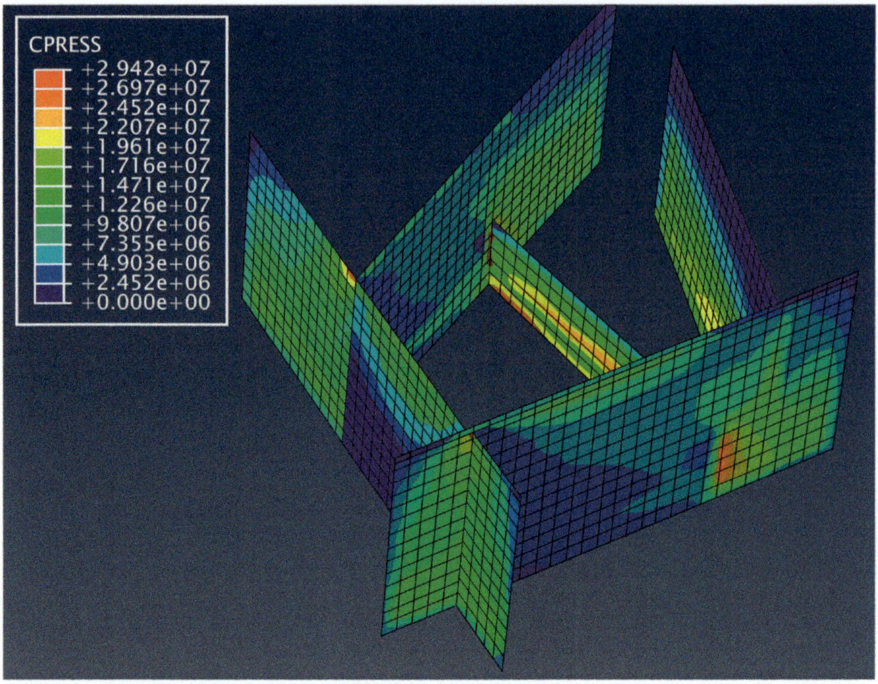

Figure 8.8 Contact pressure distribution on the faults [15].

were collected (including data about available hydrogeological parameters). The regional and local geological and hydrogeological conditions were clarified for their further utilization in the conceptual hydrogeological model development. From a total examination of 344 wells, only 238 of them provided usable information on the regional spreading and parameters of hydrogeological units.

These data provides information about the 3D spatial extent of the existing hydrostratigraphic units, the possible hydraulic connections between them, their hydraulic properties (average estimates), as well as the magnitude and direction of hydraulic gradients. A 3D representation of the top surface and maps of isopachytes were drawn up for the main hydrogeological complexes and aquifers in the area (Fig. 8.9) [15].

The hydrogeological and geothermal data collected allows fixing of the scope of the regional hydrogeological model. The conducted analyses help to clarify the existing regional and local hydrogeological conditions and to create a conceptual model that serves as a base for further development of regional and local-scale numerical models that will be discussed later on.

8.2.2 Laboratory and field tests of UCG

Regardless of the significant number of pilot experimnents on UCG worldwide, only a few of them provide at least limited data on cavity growth in a burning coal

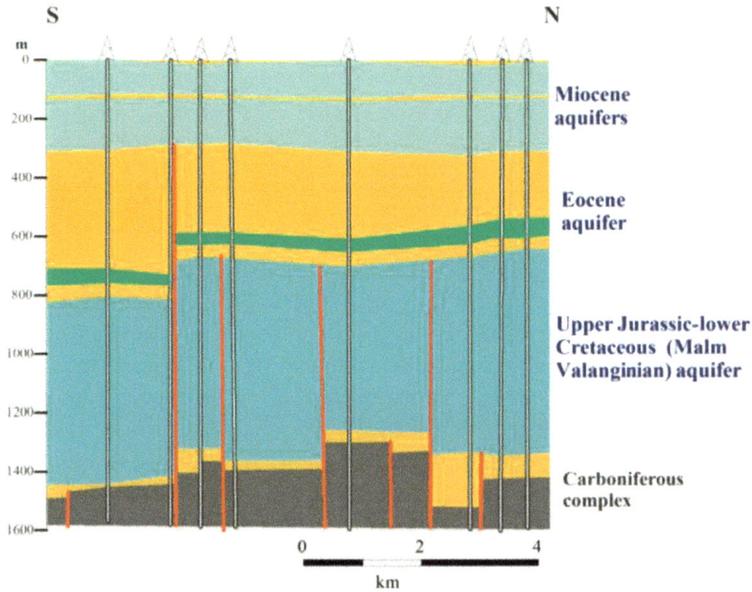

Figure 8.9 Schematic hydrogeological cross-section of the main aquifers above the Carboniferous coal bearing complex [15].

seam. Main obstacle for this is probably the high cost as well as the difficulty of controlling the operating variables. However, a number of laboratory-scale coal block gasification tests have been performed recently under a three-dimensional geometry [23–26]. These experiments are characterized by a drilled borehole through the coal seam; blast gas is injected into the one end of the borehole and the product syngas extracts from the other. These tests have difficult interpretation because they are three-dimensional and the cavity wall temperature changes with time, along the axis of the borehole. Additionally, it is difficult to apply the same or similar hydrostatic stresses on the tested coal/rock samples to represent the real UCG scenario, thus, it is hard to investigate in laboratory conditions the influence of geological parameters on the cavity growth.

A particular field test accomplished by Yang et al. [27] had the specificity of two-stage gasification approach with long channel and big section for hydrogen production in Woniushan Mine, Jiangsu Province, China. This two-stage UCG experiment led to the injection of oxygen-enriched air, stimulating the gasification and increasing the average temperature of the coal seam up to 1000°C. After this first stage, the steam is injected to invoke the reaction between steam and the incandescence for water gas with high content of hydrogen. It can be noticed that the two-stage gasification technology is more proper for large size gasifiers. Another similar two-stage UCG test was later accomplished by Liu et al. [24], in order to investigate both the syngas composition and the field temperature. The main outcome of the test is the effect of oxygen concentration on the time duration of the two gasification stages. During test, there are available data on temperature drop, but not that many regarding the cavity shape.

A semi-industrial test of Enhanced-UCG (EUCG) technology was reported by Wang et al. [28] to gasify the scattering coal seals in Zhong Liang-Shan mine in China for 3 months by making use of the abandoned mine shafts as gasification channels. Due to controlled moving injection points, the operation of this specific EUCG is more controllable in terms of operational pressure and gas flow and meanwhile it seems a feasible technology to "re-mine" the abandoned coal resources, particularly in China.

In the case of HUGE project [26], simulated UCG experiments on lignite and hard coal seams using three different gasification agents, i.e., air, oxygen, and oxygen-enriched air in an ex situ reactor, were carried out. It was found that oxygen was necessary to sustain the gasification process of both lignite and hard coal. Optimal oxygen/air ratios for both types of coal were identified, although the ratio was strongly related to the reactor geometry. Later, Stanczyk et al. [29] also continued with the two-stage gasification approach [24], using large rock and coal samples and applying a vertical weight on the reactor to simulate underground gasification conditions. Even though the production rate of hydrogen remains significant, this experiment gives invaluable information on the temperature profile for future thermodynamic analysis. The electromagnetic technology provides a horizontal profile of the cavity shape on the gasification channel as UCG process continues (Fig. 8.10).

Despite shape of growth of the UCG cavity, rate has great impact on other important phenomena too, such as reactant gas flow patterns, kinetics, and temperature profiles. Daggupati et al. [23] designed a series of laboratory tests to investigate the cavity formation of lignite coal and its dependence on injection flow rate, operation time as well as the distance between injection and production wells (Fig. 8.11). It was reported that the injection flow rate and operation time resulted in monotonic increases in all the dimensions of the cavity, when other factors are kept the same. Keeping specific test conditions, experimental correlations for cavity

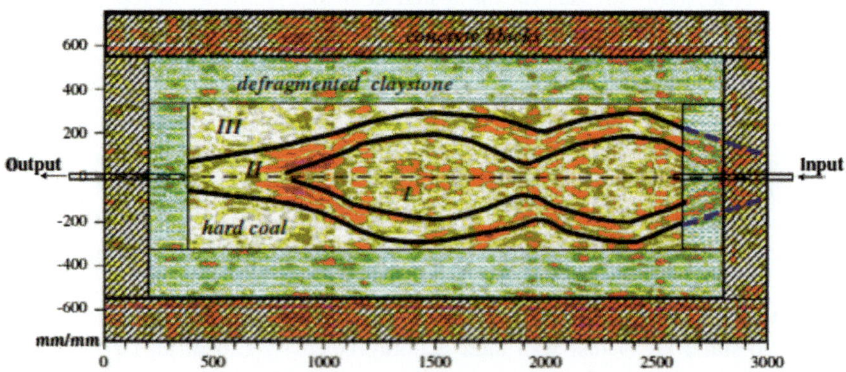

Figure 8.10 Sketch showing a horizontal profile of gasification cavities [29].

growth were proposed in which the correlation coefficients were obtained by fitting the experimental data using statistical software.

Further laboratory work was done by Prabu and Jayanti [25] simulating UCG of high ash coals using pure oxygen or oxygen and steam as the gasifying agents. In their study they probed the effects of coal properties, the feed gas flow rate, and composition on the rate of cavity growth and found out that high oxygen flow rates are required to ensure that the cavity expands in all directions. The high ash content (25 wt%) is responsible for a very fine-grained whitish ash production which was filling the formed cavity (Fig. 8.12).

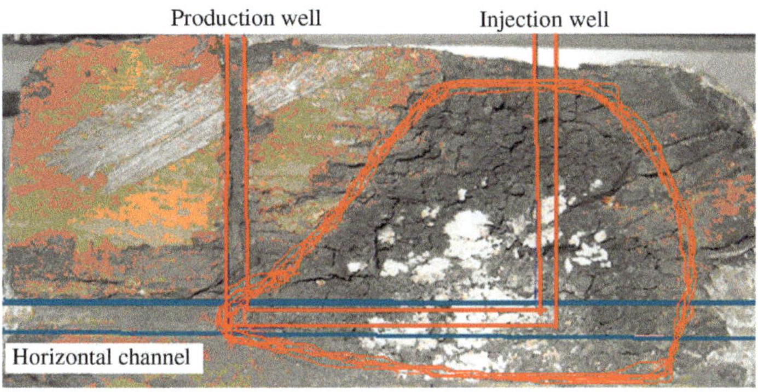

Figure 8.11 Vertical section of the formed gasification cavity [23].

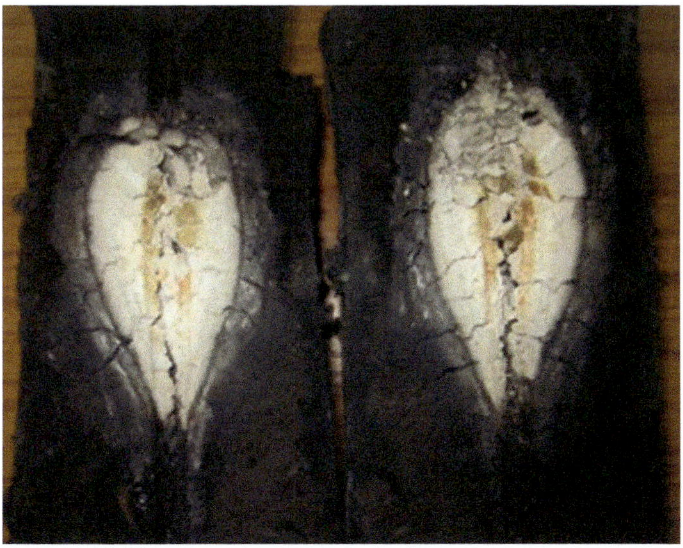

Figure 8.12 The UCG cavity after gasification has a tear drop shape containing fine-sized ash [25].

8.2.3 Numerical modeling

8.2.3.1 Modeling of cavity growth

During the gasification process, the reacted coal gives its position to cavities which consist of coal, char, ash, rubble, and voids; therefore, their shape and size changes as the gasification process proceeds. The cavity size generally is a matter of coal consumption and consequently is associated with the reaction rate that takes place in the cavity reactor. Regarding the cavity shape, this depends on the growth rates in the four directions which is influenced by the gas flow field inside the cavity, and other effects such as thermo-mechanical spalling of the coal, operating conditions, and coal seam thickness. The growth as well as the final form of UCG cavity has direct influence on the economic and environmental factors of the project. Width/height ratio of the UCG cavity controls the resource recovery and the ultimate overall dimensions will dictate the hydrological and geomechanical response of the overburden.

The progress of UCG cavities and the final outline and dimensions of the gasification channel are of great significance for environmental issues and the geological stability of surrounding geological formations [4]. A lot of effort has been made to develop mathematic and/or numerical models, ranging from 1D, to 2D symmetric and to full 3D, to simulate the UCG cavity growth. Perkins and Sahajwalla [30] developed a 1D thermo-mechanical model to test the effects of operating conditions (e.g., temperature, pressure, water influx, and gas composition) and coal properties (e.g., thermo-mechanical spalling behavior, reactivity, and composition) on the rate of local cavity growth and the effectiveness of energy utilization. The UCG cavity growth rate is mostly affected by the thermo-mechanical spalling behavior of coal, the behavior of the ash, and the amount of fixed carbon in coal [16].

Chen et al. [31] developed a 2D thermal model to calculate the temperature distribution in the vertical direction of the UCG cavity and the combustion volume by assuming a circular gasification channel. Yang et al. [2,3] developed a 3D coupled thermo-mechanical FE model to analyze the cavity growth as well as the roof deformation and ground surface subsidence. A coupled temperature–displacement procedure was adopted to simultaneously solve the stress/displacement and temperature fields. Cavity shapes and temperature profile in the coal seam during gasification were investigated by assuming the coal consumption at a specific temperature, e.g., the coal is assumed to be completely gasified once its temperature reaches 1000°C. It should be noted that in this thermo-mechanical modeling the cavity is represented by the elements with thermal and mechanical properties similar to syngas, but the conjunctive heat transfer through gas flow is neglected, thus, the cavity shape asymmetries cannot be anticipated.

Regarding the chemical reaction, the UCG cavity is usually considered as either a packed bed or a free channel. Yang et al. [32] assumed the UCG cavity as a packed bed reactor and used a set of partial differential equations to model the temperature distribution. Furthermore, Khadse et al. [33] developed a 1D transient model for viewing the UCG corridor as a packed bed. The reactor is filled by coal material which goes through the processes of oxidation and gasification in a porous space. The system of equations was divided into two parts, in order to overcome the large differences in the characteristic times of different variables in the model: one consisting of steady-state gas phase and energy balance in only the length domain and the other set of solid balance equations versus time.

Mass/heat-transfer equations combined with chemical processes can give better estimations of temperature distribution in UCG cavity. Perkins and Sahajwalla [34] developed a 2D computational fluid dynamics (CFD) model of the UCG simulating the total effects of heat and mass transport and chemical reaction during the gasification progress. The results demonstrate the importance of transport and reaction on UCG processes. Nourozieh et al. [35] used a comprehensive porous media flow approach to construct a 3D model using of a thin and deep coal seam. The cavity growing is caused by char combustion and gasification reactions, and the rate of this growth depends on the rate of these reactions. Seifi et al. [36] carried out a 3D simulation using the STARS module, which is a process simulator for modeling the flow of three-phase and geomechanical process. This model aims to investigate cavity shape, temperature variation, product gas composition, and flow rates, taking into account heat and mass transport phenomena in conjunction with chemical reactions. The findings of this model in predicting syngas flow rate, cavity shape, and temperature profiles show reliability with those in the literature.

During UCG processes, the formed cavity shape is estimated to present 3D asymmetry [37]. As UCG proceeds, the cavity shape is relevant to gas flow. The buoyancy forces give to the cavity an irregular shape in all three dimensions, and the geometry works as boundary for the gas flow. The characterization of the non-ideal flow routes is an important aspect, as it is more likely to influence the process performance. CFD studies are essential to understand these complex flow patterns within the cavity. Daggupati et al. [37] suggested a model in which they are focusing on cavity and flow patterns. The cavity in this case is split into separate compartments which exchange heat and mass with the cavity roof and the non-carbonaceous rocky floor (Fig. 8.13), providing significant computational savings in predictive modeling for UCG cavity flow patterns. In addition, the injection orientation (vertical or horizontal) and radiation on the reactant gas flow has been examined. Numerical (virtual) tracer experiments have been conducted using CFD techniques defining the residence time distribution (RTD) or exit age distribution.

In 2009, Luo et al. [38] made a model by combining CFD and mathematical cavity growth, showing parameters such as cavity shape and dimensions (Fig. 8.14). Effects of natural and forced convection driving forces on heat and mass transfer in cavity are examined in the 2D cavity growth model which is then expanded into 3D and then integrated with CFD simulation.

8.2.4 Geomechanical and environmental modeling feasibility

Despite the fact that UCG process takes place in deep underground coal seams even below the water table, the produced syngas could travel via tectonic faults and fractures in rock strata to the upper water aquifer, causing contamination. Furthermore, the overlain stratigraphy is driven to spalling and thus unbalances in the hydrostatic pressure of the cavities. This fact consequently produces flow paths where aquifer

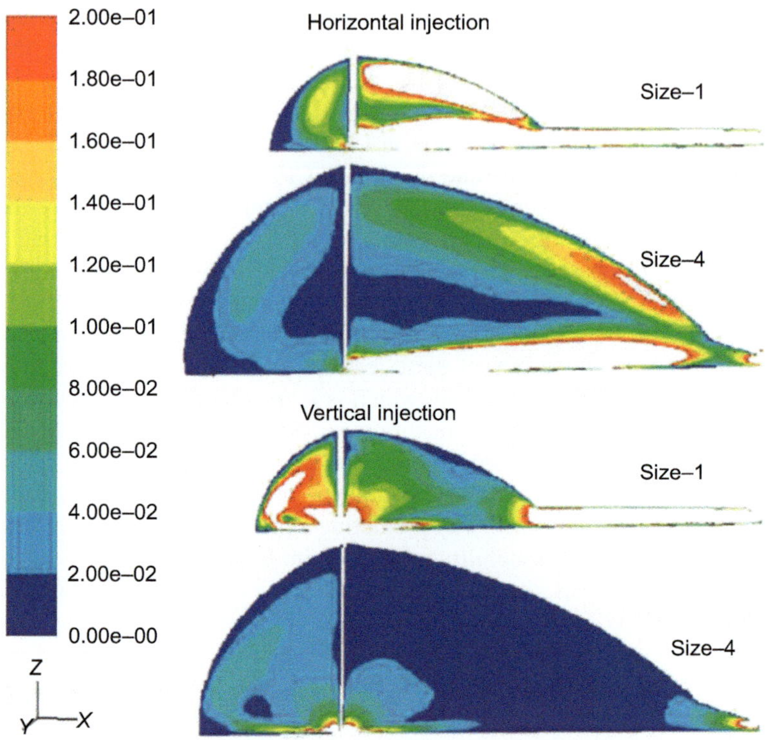

Figure 8.13 CFD flow distribution contours regarding velocity (m/s) on a center plane (width) in a different cavity scale [37].

water migrates into the cavities, and therefore, mixing with the residual chemicals produced in UCG process, and finally causing the contamination of the aquifers. Such an example is the benzene detection in groundwater of Kingaroy project in Queensland, which automatically leads to the discontinuation of the project in 2010 [39].

Another phenomenon that takes place through UCG process is the elevated temperature and permeation of fluid inside the rock pores, which can change the mechanical behavior such as stiffness and strength of strata rock, and therefore, might cause geological instability followed by subsidence. A study by Vorobiev et al. [40] has focused on geomechanical processes in coal and the adjacent rocks. FE and discrete element (DE) are coupled both in two and three dimensions to test various UCG scenarios.

The RFCS recent EU projects focused on the geomechanical response of the tectonic faults in a potential UCG coal deposit [15]. Therefore, the model comprised structural geology including the geometry of faults and the stratigraphic data such as thickness and depth of the coal seams. Additional sensitivity analysis of the cavity away from the faults has been conducted for the final region selection.

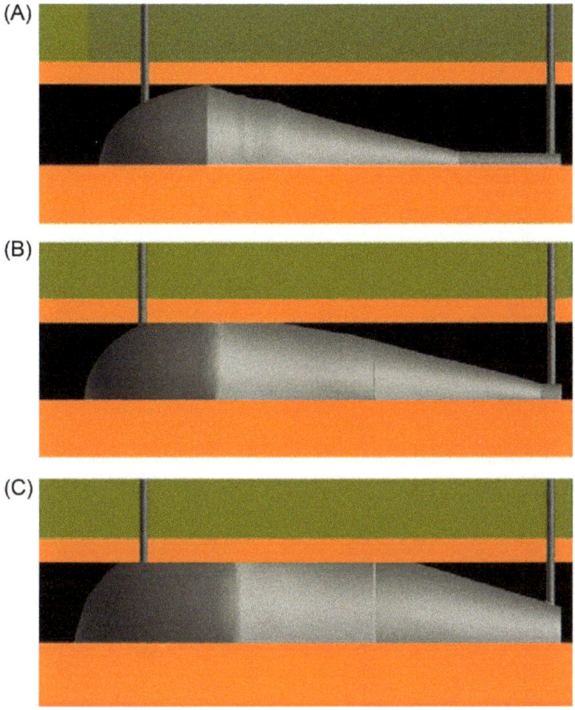

Figure 8.14 Cavity shape versus reaction time [16]. (A) At 2.5 days, the cavity touched the overburden; (B) at 5 days, cavity size and shape; and (C) at 10 days, cavity size and shape.

The UCG cavity growth and gas flow results in subsidence effects of the surrounding rocks as well as in underground water movement. After the end of UCG process we focus on the impact that cavities have to the environment and groundwater. The cavity washed by a net flux of water prevents the dispersion of reaction products into the surroundings in preparation for postburn clean-up treatment which is particularly important when CO_2 is planned to be injected back. Other parameters such as temperature, pressure, porous media properties, and composition of the liquid and gaseous phases in the subsurface after the UCG need to be examined. The 3D software MODFLOW was used in order to develop regional groundwater modeling for non-isothermal groundwater flow and contaminants detection near to a tectonic fault (benzene was considered to be on primary interest) [15]. The results retrieved from the model show that contaminant concentrations would exceed 1 ppb after in a year of UCG operation and in restricted area. As years pass, concentrations are further reduced due to additional chemical degradation (Fig. 8.15).

Dealing with problems such as rock fracturing, aquifer interference, and groundwater flow in coalmine environment, Commonwealth Scientific and Industrial Research Organization (CSIRO) in Australia has developed a dual-phase flow model based on FE method, COSFLOW [41]. In this method, fracturing is

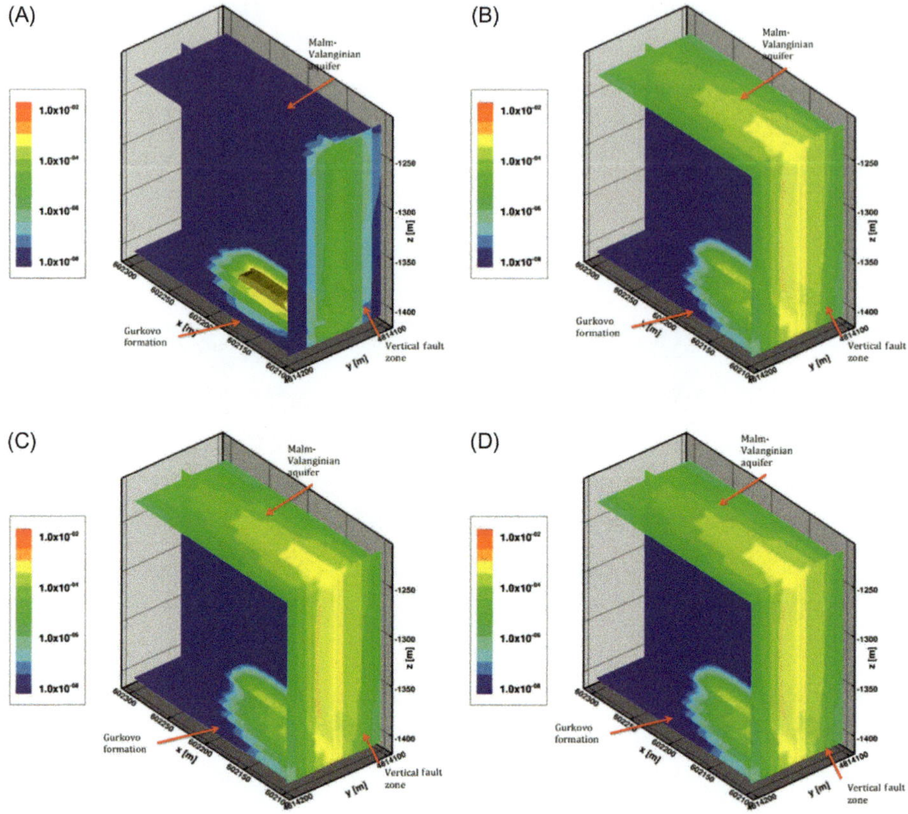

Figure 8.15 Benzene contaminant concentrations in ppm versus time. (A) 1 year, (B) 10 years, (C) 100 years and (D) 1000 years after UCG procedure [16].

estimated from a mechanical sub-model and coupled with the changes of permeability and porosity used in the fluid flow sub-model. The flow in the fracture system is controlled by the pressure gradient and is directed by Darcy's law (Fig. 8.16).

8.3 UCG and CCS

8.3.1 Hydrogeological modeling

8.3.1.1 Selecting the sites using the hydrogeological information

The analyzed archival and published information about the regional and local hydrogeological and geothermal conditions can provide data related to the lithological—stratigraphic data from deep wells. The regional geology and hydrogeological conditions can be clarified for their further utilization in the conceptual hydrogeological model development. The local data offers a large-scale perspective on the expected hydrogeological conditions, but not sufficient information about the local

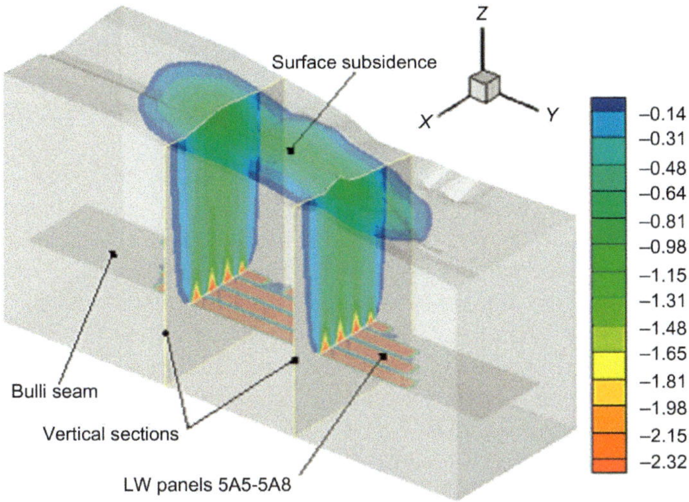

Figure 8.16 COSFLOW modeling on subsidence [16].

site conditions. Thus, the regional data is augmented with additional site-specific data and information precisely for the site-scale hydrogeological model.

All these data give us spatial information about the 3D extent of the existing hydrostratigraphic units and the potential hydraulic interconnections. In such cases there are also estimations for the hydraulic properties and the characteristics (magnitude and direction) of hydraulic gradients.

Water sampling taken from different aquifers can provide information regarding the chemical composition and categorize different water-bearing complexes. They were mainly influenced by the hydrogeological conditions of the water-bearing complexes and less by the formation depth. The data are used for mineral characterization of the cavities and will be formed from the coal seams for potential CO_2 sites. In addition, main characteristics of the geothermal field of the target area such as temperature, geothermal gradient, thermal properties of the rocks, and heat flow have to be studied.

The hydrogeological and geothermal data analysis allowed fixing of the scope of the local hydrogeological model. The analysis of uncertainties and possible alternatives is followed by a conceptual model which is the base of future modeling. The conducted analysis aims to the existing regional and local hydrogeological conditions interpretation and to create a base model for further development of regional numerical models.

8.3.1.2 Feasibility evaluation of UCG and CO_2 storage

Non-isothermal underground water flow and contaminant transport (mainly benzene) is considered for local-scale. It is believed that 1 ton of gasified coal could produce about 36 L of coal tar, and therefore, a relatively small portion (0.1%) of that could be benzene. This UCG contaminant seems to be the key contaminant product for possible migration to the subsurface environment (due to relatively high production during gasification and relatively low detection limits). In such cases, the initial concentration of benzene could reach 425 ppm, hypothesizing a production area with size 20 × 20 × 5 m and porosity around 1, ignoring time for gasification [15].

This model includes physical processes such as fluid, heat, and contaminant flow, whereas the contaminant migration is controlled by advection and dispersion.

The contaminants are estimated to be non-reactive and unaffected by attenuation and decay via geochemical procedures. During the modeling procedure the heat flow may increase. A local-scale hydrogeological model and the performed analyses can indicate that the UCG-CO_2 storage activities at the studied site would have limited impact on the groundwater quality and would not cause unacceptable environmental impacts on the main aquifer in the adjacent area. The injected CO_2 mass is mainly dissolved in the groundwater. Additionally, a number of parameters for UCG with the following CO_2 injection are demonstrated in Table 8.5. These criteria are related to geomechanical, cavity, and

Table 8.5 Parameters that should be taken into account for UCG-CO_2 procedures [15]

Category	Desired value	Comments
Coal thickness (m)	>2 m	Not greater than 30 m. Ideally 5–10 m
Number of seams to be gasified		Avoid seams with overlying coal within 15 m
Thickness variation (% of seam thickness)	<25	Avoid variable thickness seams
Depth (m)	>92	Preferably more than 300 m and not more than 2000 m
Angle of coal seam (degrees)	0–70	Any but steeper is preferred as it may be technically difficult to mine through conventional methods
Variation of the angle of the coal seam (% of average angle)	<2	
Thickness discontinuity (m)	1	Avoid seams with variable partings/discontinuities
Overburden (m)	100	Floor and roof conditions needs to be examined carefully
Coal rank (vitrinite reflectance)	Low rank bituminous	Free swelling index should be low. Sub-bituminous or lower rank, ideally not coking, nonswelling coals
Ash content (wt%)	<50%	
Coal sulfur (wt%)	<1	Volatile matter greater than 10%. Sulfur should be removed along with syngas

(*Continued*)

Table 8.5 **(Continued)**

Category	Desired value	Comments
Coal moisture (wt%)	<35	Preferred 7%–35%. Controlled inflows of water or high moisture contents are desirable especially after ignition
Gross calorific value of coal	>12 MJ/kg	
Thickness of consolidated Overburden	>15	
Seam permeability (mD)	50–150	More permeable greater than 20%. Swelling coals may interrupt the gas circuit. High permeability coals may allow excessive water infiltration causing possible chance of gas leakage and contaminant movement
Porosity of coal seam	>30%	Porous coal seam
Distance to nearest overlying water-bearing unit (m)	100	
Coal aquifer characteristics	Confined	
Available coal resources (106 m^3)	>3.5 Mt	>20 years long operation. Depend upon gas utilization and profitability.
Proximity to faults	>150 m depending on site conditions	If many major faults then site-specific calculation required to be carried out for the accurate estimation of the distance
Distance from active mines (km)	>3.2	
Distance from abandoned mines (km)	>1.6	
Geology-lithology		High UCS, nonporous and impermeable strata
Hydrology	Nonaquifer strata is preferred.	Nonporous strata <30%, Impermeable <5% Moderate water ingress. Avoid potable aquifer and large water bodies
Geotechnical strata properties	Rock strength: Uniaxial compressive strength range 50–250 MPa. Density greater than 2000 kg/m^3	Avoid excessively fractured, faulted and broken rocks as they may cause water inrush or product gas and contaminant leakage

(*Continued*)

Table 8.5 **(Continued)**

Category	Desired value	Comments
Infrastructure availability		Roads, electricity, and power transmission lines
Presence of coal bed methane		Depends upon economics or commercial value of CBM deposit and its interoperability with UCG

Table 8.6 **Parameters for UCG-CO_2 project and their ranges [15]**

Parameter	Minimum	Maximum
Average seam depth (m)	15	1200
Average seam thickness (m)	0.75	18
Seam angle (deg)	0	56
Well spacing (m)	10	210
Coal gasified (t)	0.3	2,781,800
Injection (kmol/h)	7	1223
Injection pressure (kPa)	120	7500
Production pressure (kPa)	110	1000
System pressure (kPa)	207	8000
Total coal wet production (kmol/h)	22	4860
Coal (dry) heat value [MJ/m^3 (dry STP)]	2.57	12.17
Gas recovery (%)	43	110.30
Synthesis gas heat value (MJ/m^3)	2.8	14.60

hydrogeological feasibility research, based on subsurface model elements and mechanical/ physical properties.

8.3.2 Engineering for UCG and CO_2 storage

8.3.2.1 Technical matters on wells for UCG and CO_2 storage

The evaluation of accessible data from other UCG projects gives us the opportunity to proceed with most suitable designs and to evaluate the given parameters based on coal characteristics and oxidant type. The existing literature along with database analysis can provide information about syngas quality which is often influenced by parameters such as initial coal quality, oxidant composition, injection pressure, and coal seam depth (reservoir pressure). In Table 8.6, database selected parameters are displayed. The intersystem relations between the previously mentioned factors and the existing geological data can contribute to estimations for the composition of synthesis gas in the target area.

Reservoir simulations for CO_2 storage in post-UCG cavities can be accomplished. These simulators should be applicable for non-isothermal or quasi-isothermal conditions, or to allow equivalent adjustments to address thermal effects. Another characteristic of these simulations is the capability to simulate CO_2 coal adsorption and residues coal gasification taking into account the surrounding rocks too.

Some of the features that are required for CO_2 storage in a potential UCG site have to do with the appropriate depth, existence of reservoir and cup rocks with adequate porosity, and permeability proportions in each case. Further data are related to the tectonic deformation as well as for trapping capability [15].

In coal seams and the adjacent rocks, directional drilling is a technology that deals with specific technical challenges often observed at conventional hydrocarbon recovery. These are:

- Coal seams are weaker compared to the stratigraphic formations in most oil and gas reservoir fields. Thus, the mechanical stability in-seam holes are a concern. Due to its friability, coal has difficulties during drilling; however, it is reasonably quick process in comparison to massive host rocks.
- In most of the cases, reservoir lithology is thicker than coal horizons. Due to intense brittle—ductile deformation these seams are difficult to be followed laterally.
- In a narrow seam, the drilling precision has to be in higher level than in most hydrocarbon projects. On the other hand the operating depths and pressures are often shallower.
- Coal horizons appear fracture network which is sensitive to blockage from cement or drilling mud.
- After UCG procedure, it is possible that downhole casing equipment is exposed to a severe chemical and thermal environment.

In such a project, at least two boreholes are required for one panel-injection and production well. The actual panel layout consists of a number of injection well branches crossing one deviated production well. In the case of CO_2 injection, only the vertical part of the well(s) would be used because the horizontal part would not be required in view of the cavity development and low permeability will be of minor concern.

The parallel action of conventional drilling of the vertical/cap rock section and the subsequent coiled tubing (CT) for the build-up and horizontal/in-seam section is demonstrated as one of the state-of-the-art practices. The CT drilling technology has the advantage of continuous progress with synchronous hole stabilization. The in-seam leg can be stabilized by a liner. The route of the well has to trace the seam as well as to reach the production and/or ignition wells. After well completion, no significant risk of steel corrosion is expected in drilling equipment since the dry and in supercritical state CO_2 environment keeps corrosion rate at quite low levels.

The operational period of a borehole dual-use during UCG-CCS operations is generally short (2—3 years); however, the performance of materials has to be considered for each stage. The design of the wells has to count on parameters such as the injection rates and pressure, the fluid saturation in the UCG cavity as well as geochemistry, volume, viscosity, and content of CO_2. Specific parameters that are displayed in Table 8.7 have to be studied as a basis for comprehensive preparation of CO_2 storage phase. The final closure of the borehole will follow the general rules that are applied in all CO_2 geological storage sites.

UCG and Carbon Dioxide storage are two technologies that have been passed through long term research, and exhaustive tests, even in some cases they have been applied for industrial purposes. However, there is no site where these two technologies have been combined together. Regardless of the fact that deep located UCG projects are unusual, the rare but valuable information on them helped to identify and evaluate technical concerns related to UCG and CO_2 storage procedures.

For CO_2 storage a minimum depth of 1000 m is required in order to allow storage of it safely in a dense phase. Thus, it requires deep trial gasification sites which in most of the cases do not exist. In these depths, the risk of groundwater contamination is relatively high followed by increased development and operational costs.

Table 8.7 Parameters and conditions for CO_2 storage in a UCG prior site [15]

Parameter	Dual-use post-UCG injection well	Comments
Fluid type in the UCG cavity	Water, CO_2, N_2	CO_2 in supercritical state
Well design life time	About 10 years	1–3 years UCG and up to 1 year of CO_2 injection
Design injection rate	1 kg/s	Based on the overall UCG panel design
Bottom hole temperature	About 100°C after UCG cool down	Cool down process time leads to dissolution of contaminants increasing risk of well corrosion
Well-head pressure	About 170–220 bar at depths of seam sections	Containment loss has to be avoided in the injection zone by not exceeding a pressure threshold
CO_2 purity	95–99.9 wt%	
Corrosion H_2S related and other influence	Exceeding the NACE limits	
H_2O	Assumption that dehydration capacity is exceeded during CO_2 injection process	Related to water back flooding due to shut-ins

8.3.2.2 Borehole planning for UCG and the adjustment for CO_2 injection and storage

The borehole planning and operation for both UCG and CO_2 storage project includes a workover and development program for CO_2 injection. The general and specific parameters correspond to the geological conditions in the area, accounting for the necessary regulatory requirements. The workover and development program have to examine applicable monitoring technologies—respective experiences which can be retrieved through active projects and reviewing previous projects on both UCG and CCS technologies.

The basic setup of the drilling part includes a vertical section which is cased down to the starting point of the horizontal deflection at a vertical and horizontal distance from the planned entry point into the seam. The ultimate diameter of the injection borehole depends on the accepted UCG design followed by detailed cost assessment of seam properties and costs for variable drilling diameters and the relevant pumps [15].

Regarding the injection well, this should be drilled into and along the coal seam, whereas the production well drilled perpendicularly, such that the production well meets the trajectory of the deviated injection well. The proposed circuit (Fig. 8.17) shows the injection wells in red and purple with well sites of the vertical section located in the two corners of the gasification panels.

A few in-seam sections are drilled from a single vertical well, thus, only a limited number of vertical wells will be required. After the vertical drilling achievement the injection wells should be cased from the surface down to the top of the build-up section. The production borehole is expected to be drilled after the deviated injection well and will be cased all the way to the top of the build-up section.

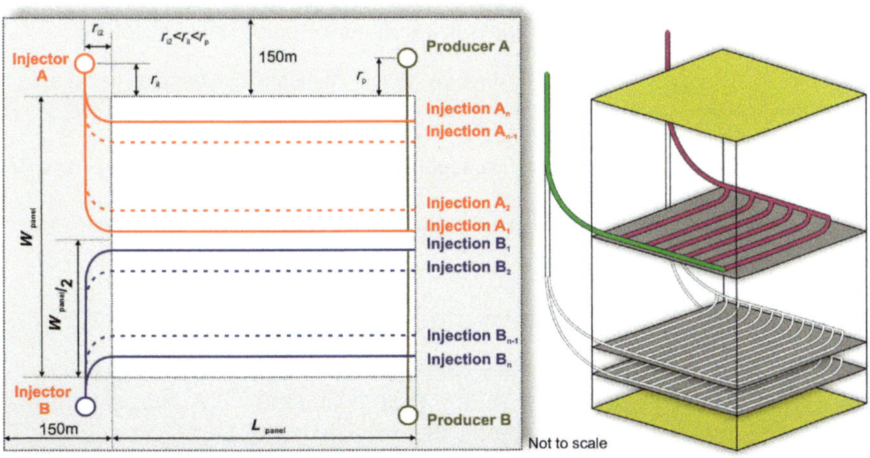

Figure 8.17 Plan view of the proposed panel layout (left); 3D block view with injection and production wells (right) [15].

By the time that this section collaborates with the respective leg of the injection, a liner should be installed in the well, which should be temporarily capped.

When the UCG reactor (the gasification channel or entire UCG panel) is closed down, it is time for shut-down procedures. The aim is to ensure that toxic UCG by-products will not leak, and thus, the freshwater aquifers will be sufficiently protected. The UCG reactor shutdown is undertaken in three main steps [15]. The oxidant injection is reduced and finally stopped. The cool-down process can be supported by water injection into the cavity (being again converted into steam underground) and its subsequent production [2]. In the second step the gasification is finally inactivated synchronously to the related wells. At the end, the water from the extinguished UCG reactor is pumped. The repetition of this process may take place until the concentration of soluble elements is reduced to low percentages. In case of low water permeability of the UCG cavity, the clean-up process can be accomplished by cycling water through the shutdown reactor in order to use it as a solvent for toxins. After the end of UCG procedure and cleaning, water quality is monitored. The national standards at the site are those that specify further treatment at the surface. The postinjection administration of long-term CO_2 storage in UCG sites requires continuous monitoring of the cavity, borehole, and storage space, always considering the local site-specific geological conditions.

8.3.3 Environmental assessment of UCG-CO_2 storage

8.3.3.1 UCG impact on the groundwater environment

As it was already mentioned, groundwater contamination is likely to be a key element for any successful UCG procedure. In order to minimize the risk of groundwater contamination, a comprehensive evaluation of geological and hydrogeological factors at the site is necessary. In these deep environments, the evaluation of the UCG effect on groundwater encompasses a number of parameters as follows:

- Preferential underground pathways through which contaminants migrate to aquifers of economical or ecological significance.
- The influence of anthropogenic actions.
- The migration possibility of significant pollutant amounts further of the permanently unsuitable zone (PUZ).
- The impact of surface activities to groundwater quantity and quality.

8.3.3.2 Underground pathways and human activity

The contaminant migration pathways can be affected by nearby human activities which may increase permeability of the geological formations and even provide new pathways through which contaminants associated with UCG can migrate. The collapse of the UCG reactor, mineral workings, and resource extraction activities such as coal bed methane extraction, and deep site investigation boreholes, can have a great influence on contaminant migration.

8.3.3.3 Geological and hydrogeological evaluation

The evaluation of geological and hydrogeological characteristics requires two main activities: to identify the main potentially transmissive features and links between them that could provide a continuous transmission pathway beyond the permanently unsuitable zone, and determining whether hydrogeological conditions lead to vulnerability. A good understanding of the geology and conceptual hydrogeology at a site is essential. These studies include:

- Detailed host rock examination.
- Permeability and transmissivity probe on Carboniferous rocks and other underlain geological formations.
- Chemical composition analysis of the adjacent aquifers.
- Regional hydrogeological research.
- In the most vulnerable aquifers to future UCG, special work such as thickness distribution mapping, hydrodynamic mapping and data on hydraulic geological factors.

Additional data:

- The role of tectonic deformation in possible migration pathways.
- The technical condition of remaining boreholes.

8.3.3.4 CO_2 storage—Environmental impacts related to UCG

Carbon Capture and Storage is a technique of reducing CO_2 emissions in the atmosphere. After the end of UCG procedure, the abandoned cavities would be accessed by directional drilling or through the existing production wells where CO_2 is injected in supercritical condition for geological storage. For permanent CO_2 trapping, the depth and strata conditions must be the appropriate. The environmental factors considered for the UCG process have also an impact on CO_2 storage.

8.3.3.5 Mathematical modeling for UCG-CO_2 environmental sustainability

This modeling process targets to a reasonable balance between financial, environmental, and social issues during the energy generation. For a scientific contribution to an effective sustainable environmental management, a numerical model can be proposed [expressed in terms of an Environmental Sustainability Index (ESI)] to quantify the environmental sustainability situation of the "in situ" UCG process with CO_2 storage. In a given time and space this parameter allows the definition of an environmental sustainability standard or a minimum permissible level of sustainability for future projects. This method is based on four environmental parameters: (a) atmosphere quality, (b) rock and soil subsidence, (c) groundwater quality, and (d) surface water quality. The main purpose of this index is to establish the required criteria for new UCG projects, as well as for optimization of studies of existing installations.

This ESI model is a function of four component indexes: Subsidence Sustainability Index (SSI), Groundwater Sustainability Index (GWSI), Surface

Water Sustainability Index (SWSI), and Atmosphere Sustainability Index (ASI). The calculation of these indexes considers the condition of sustainability of each pollutant based on threshold limit values, given by the existing standards.

The basic equation for the calculation of the ESI of UCG/CCS is displayed below and the graphic representation of the results is shown in Fig. 8.18:

$$ESI = 1/4(SSI + GWSI + SWSI + ASI)$$

For the calculation of sustainability index (SI) of each component (SSI, GWSI, SWSI, and ASI), the mathematical model uses the condition of sustainability of each element (X and/or X') based on the standard of sustainability or life quality given for the norms. Three sustainability criteria are taken considering the state of the local environmental conditions.

Sustainability criterion: $X' \leq x_i \leq X$ are admissible values, $x_i \geq X$ and $x_i = X_1$ are unsustainable values and $x_i \leq X'$ and $x_i = X'_1$ are unsustainable values (Fig. 8.19).

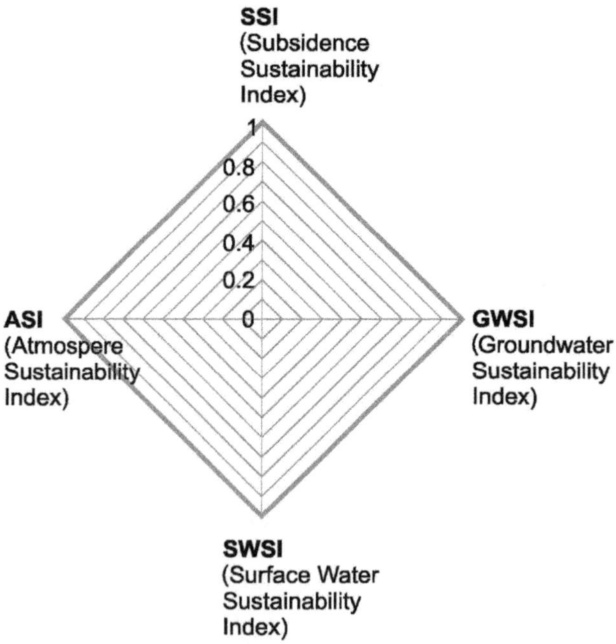

Figure 8.18 Graphic representation of Environmental Sustainability Index of UCG [15].

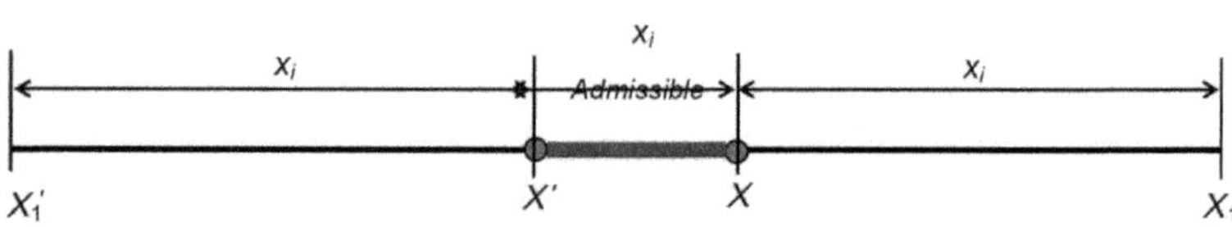

Figure 8.19 Sustainability criterion [15].

Table 8.8 Proposals of ESI_{UCG} and ESI_{CO2} for sustainability conditions [15]

Sustainability level	Color	Conditions based in ESI_{UCG}
Very good		$ESI_{UCG} = 1.00$
Good		$0.75 < ESI_{UCG} \leq 1.00$
Moderate		$0.50 < ESI_{UCG} \leq 0.75$
Low		$0.25 < ESI_{UCG} \leq 0.50$
Very Low		$0.00 < ESI_{UCG} \leq 0.25$

Accepted minimum level of the ESI for UCG and CO_2 storage is proposed in Table 8.8. The four environmental indicators (subsidence, groundwater, surface water, and atmosphere) may vary with time, and thus, the ESI for UCG and CO_2 will also vary.

8.3.3.6 Application of the developed model to possible coal deposits

This ESI model was applied in coal deposits in Bulgaria, the Florina basin in Greece, and the Spanish coal deposit El Tremedal (Teruel). The four environmental indicators (SSI, GWSI, SWSI, and ASI) were calculated for the study area considering the condition of sustainability of each pollutant based on the threshold limit values determined by the existing standards.

8.3.4 Economic assessment

The economic assessment of UCG-CO_2 process is based on a model which used a 308 MW combined cycle gas plant operating on UCG syngas, where CO_2 is separated, compressed, and stored at or near the UCG site (Fig. 8.20). Surface plant cost for CAPEX and OPEX and performance data are obtained from similar projects in the literature. The UCG process assumes CRIP technology. The underground construction costs were calculated in Bulgaria. The individual component and fuel costs such as land acquisition, fees (e.g., concession fee for extraction), piping, measuring, control equipment, drilling, synthesis gas processing as well as oxidizer compression, and injection were brought together in a calculation which takes into account the moderate costs of electricity (COE) and CCS costs [42]. The assumption that the coal—because of depth—has no intrinsic value and royalties payment will not be required has been made.

Considering the thermal regime in the study area, calculations indicate a pressure loss of about 4.2 bar since a well-head pressure (WHP) of about 15.8×103 kPa during the oxidizer injection and of 3.6 bar taking into account the UCG operation at slightly sub-hydrostatic pressure in the target coal seam, which is expected to be

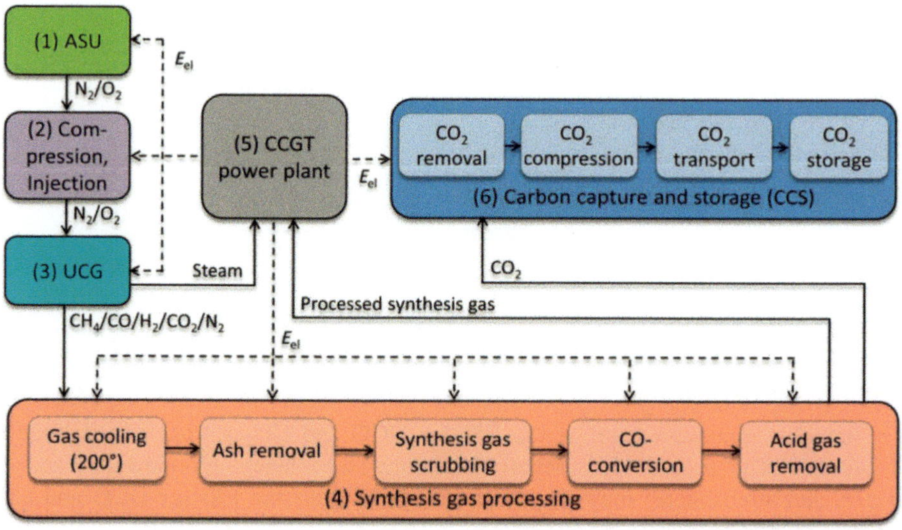

Figure 8.20 Developed model for UCG-CCGT-CCS process [42].

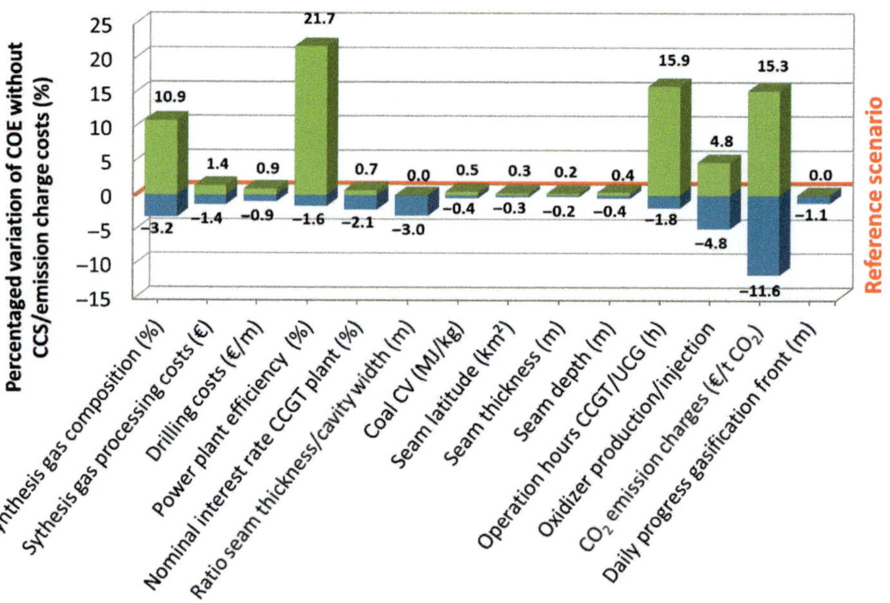

Figure 8.21 Proportional impact of all parameters on COE without CCS/emission storage costs [2].

in an acceptable range for the chosen liner diameters and well dimensions. In this study, it was decided to apply inner liner diameters of about 8.9 cm (3 ½″) and about 12.1 cm (4 ¾″) for the injection and production boreholes, respectively.

After the first gasification reactor depletion, the maximum capture rate from the synthesis gas stream by postcombustion capture will be established according to the available storage capacity of the former UCG reactors (12.6 million m^3) used for CO_2 storage. This refers to 20.5% of the produced CO_2, resulting in a remaining CO_2 amount of 79.5%. As the investigation of alternative storage options for the remaining 79.5% of CO_2 is not within the scope of the UCG-CCS project, this amount is released into the atmosphere applying a fix emission charge of 25 €/tCO_2.

In Fig. 8.21, the designed diagram summarizes the influence of all investigated parameters on a percentage basis. The geological input parameters have a relatively low influence on COE with a maximum of about 3% in the case of the cavity width to seam thickness ratio. However, technical implementation of the CCGT power plant (e.g., operating hours and plant efficiency) may increase COE by up to about 22%, synthesis gas composition accounts for up to 11% and oxidizer production accounts for up to 5%. One reason why the current study geological parameters have a lower influence on COE than technical parameters is that the variation bandwidth was chosen to be narrow (± 8% and 10% against ± 25% for the other cases), otherwise coal resources in the worst-case scenario would be insufficient to provide the daily required fuel for the CCGT plant. Mainly, the low sensitivity of geological parameters on COE is the minor influence of drilling costs (11%) on the overall fuel production costs (5.6 €/GJ) outweighed by the high share of oxidizer production (58%) and synthesis gas processing costs (16%) [2].

References

[1] IEA, World Energy Outlook, 2010. Available from: http://www.worldenergyoutlook.org/media/weo2010.pdf

[2] D. Yang, V. Sarhosis, Y. Sheng, Thermal-mechanical modelling around the cavities of underground coal gasification, J. Energy Inst. 87 (4) (2014) 321–329, ISSN 1743-9671 http://dx.doi.org/10.1016/j.joei.2014.03.029

[3] D. Yang, Y. Sheng, M. Green, UCG: where in the world? TCE: Chem. Eng. 872 (2014) 38–41, ISSN 0302-0797

[4] A.W. Bhutto, A.A. Bazmi, G. Zahedi, Underground coal gasification: from fundamentals to applications, Prog. Energy Combust. Sci. 39 (1) (2013) 189–214.

[5] Khurshid, I., Baig, M.A.S., Choe, J. (2011). Underground coal gasification in thar colliers: encountering the problems and solutions for producing syngas, in: SPE/PAPG Annual Technical Conference, Inslamabad, Pakistan.

[6] Thar Coal, Power Project: Company Expected to Start Production in 2018. <http://tribune.com.pk/story/824866/thar-coal-power-project-company-expected-to-startproduction-in-2018>.

[7] Burton, E., Friedmann, J., Upadhye, R. (2007). Best Practices in Underground Coal Gasification, U.S.

[8] Nitao, J.J., Camp, D.W., Buscheck, T.A. White, J.A., Burton, G.C., Wagoner, J.L., et al., (2011). Progress on a new integrated 3-D UCG simulator and its initial application, in: International Pittsburgh Coal Conference, Pittsburgh, PA, United States.
[9] Carbon Energy Ltd, Carbon Energy Annual Report 2011–2012, Carbon Energy Ltd, 2012. Available from: <http://www.asx.com.au/asxpdf/20120920/pdf/428tsbwmllj8n0>. Quarterly Activities Report for the Period Ended 30 September 2012.
[10] Ergo Exergy, (2014). Available from: <http://www.ergoexergy.com/about_us_ourb_projects.htm>
[11] Eskom, (2013). Co-firing into Majuba Power Station: Project Status. Available from: <http://www.eskom.co.za/c/article/17/project-status/>
[12] Green, M. (2014). Underground Coal Gasification: On the Road to Commercialization. Modern Power Systems. Available from: <http://www.modernpowersystems.com/features/featureunderground-coal-gasification-on-the-road-to-commercialisation-4198939>
[13] E. Shafirovich, A. Varma, Underground coal gasification: a brief review of current status, Ind. Eng. Chem. Res 48 (2009) 7865–7875.
[14] Ukrainian Association for Hydrogen Energy, Hydrogen Production by Underground Coal Gasification, 2010. Available from: <http://uahe.net.ua/index.php?option = com_content&task = view&id = 80&Itemid = 69>
[15] Y. Sheng, A.D. Benderev, Bukolska, K. Imo-Imo Eshiet, C. Dinis da Gama, T. Gorka, et al., Interdisciplinary studies on the technical and economic feasibility of deep underground coal gasification with CO_2 storage in Bulgaria, Mitig. Adapt. Strateg. Glob. Change 21 (4) (2016) 595–627. Available from: http://dx.doi.org/10.1007/s11027-014-9592-1.
[16] D. Yang, N. Koukouzas, M. Green, Y. Sheng, Recent development on underground coal gasification and subsequent CO_2 storage, J. Energy Inst. 89 (4) (2016) 469–484. Available from: http://dx.doi.org/10.1016/j.joei.2015.05.004.
[17] A. Khads, M. Qayyumi, S. Mahajani, P. Aghalayam, Underground coal gasification: a new clean coal utilization technique for India, Energy 32 (11) (2007) 2061–2071.
[18] K.L. Zamzow, Underground Coal Gasification: History, Environmental Issues and the Proposed Project at Beluga, Center for Science in Public Participation, Alaska, 2010.
[19] Overgas Inc. (2011). Geological and geomechanical data for the Dobrudzha coal field area (DCF).
[20] G.R. Couch, Underground Coal Gasification, IEA Clean Coal Centre, London, 2009.
[21] Kim, O.K. (1983). Finite element modelling of themo-mechanical responses associated with underground coal conversion (Ph.D. thesis), Ohio State University.
[22] Lee, J.K., Advani, S.H., Avasthi, J.M., Chen, K.S. (1986). Application of rock mechanics and the finite element method to underground coal gasification process, in: Rock Mechanics: Key to Energy Production, 27th U.S. Symposium on Rock Mechanics (Chapter 101—Oil Sands and Coalbed Methane).
[23] S. Daggupati, R.N. Mandapati, S.M. Mahajani, A. Ganesh, D.K. Mathur, R.K. Sharma, et al., Laboratory studies on combustion cavity growth in lignite coal blocks in the context of underground coal gasification, Energy 35 (6) (2010) 2374–2386.
[24] H. Liu, F. Chen, X. Pan, K. Yao, S. Liu, Method of oxygen-enriched two stage underground coal gasification, Min. Sci. Technol. (China) 21 (2) (2011) 191–196.
[25] V. Prabu, S. Jayanti, Laboratory scale studies on simulated underground coal gasification of high ash coals for carbon-neutral power generation, Energy 46 (1) (2012) 351–358.
[26] K. Stanczyk, N. Howaniec, A. Smolinski, J. Swiadrowsk, K. Kapusta, M. Wiatowski, et al., Gasification of lignite and hard coal with air and oxygen enriched air in a pilot scale ex situ reactor for underground gasification, Fuel 90 (5) (2011) 1953–1962.

[27] L. Yang, X. Zhang, S. Liu, L. Yu, W. Zhang, Field test of large-scale hydrogen manufacturing from underground coal gasification (UCG), Int. J. Hydrog. Energy 33 (4) (2008) 1275–1285.
[28] G.X. Wang, Z.T. Wang, B. Feng, V. Rudolph, J.L. Jiao, Semi-industrial tests on enhanced underground coal gasification at Zhong-Liang-Shan coal mine, Asia Pac. J. Chem. Eng. 4 (5) (2009) 771–779.
[29] K. Stanczyk, K. Kapusta, M. Wiatowski, J. Swiadrowsk, A. Smolinski, J. Rogut, et al., Experimental simulation of hard coal underground gasification for hydrogen production, Fuel 91 (1) (2012) 40–50.
[30] G. Perkins, V. Sahajwalla, A numerical study of the effects of operating conditions and coal properties on cavity growth in underground coal gasification, Energy Fuels 20 (2) (2006) 596–608.
[31] L. Chen, J. Hou, C. Chen, J. Xu, A back analysis of the temperature field in the combustion volume space during underground coal gasification, Min. Sci. Technol. (China) 21 (4) (2011) 581–585.
[32] L. Yang, Study on the model experiment and numerical simulation for underground coal gasification, Fuel 83 (2004) 573–584.
[33] A. Khadse, M. Qayyumi, S.M. Mahajani, P. Aghalayam, Reactor model for the underground coal gasification (UCG) channel, Int. J. Chem. React. Eng. 4 (2006) A37.
[34] G. Perkins, V. Sahajwalla, Modelling of heat and mass transport phenomena and chemical reaction in underground coal gasification, Chem. Eng. Res. Des 85 (2007) 329–343.
[35] H. Nourozieh, M. Kariznovi, Z. Chen, J. Abedi, Simulation study of underground coal gasification in Alberta reservoirs: geological structure and process modeling, Energy Fuels 24 (2010) 3540–3550. Available from: http://dx.doi.org/10.1021/ef9013828.
[36] M. Seifi, Z. Chen, J. Abedi, Numerical simulation of underground coal gasification using the CRIP method, Can. J. Chem. Eng. 89 (6) (2011) 1528–1535.
[37] S. Daggupati, R.N. Mandapati, S.M. Mahajani, A. Ganesh, A.K. Pal, R.K. Sharma, et al., Compartment modeling for flow characterization of underground coal gasification cavity, Ind. Eng. Chem. Res 50 (1) (2011) 277–290.
[38] Luo, Y., Coertzen, M., Dumble, S. Comparison of UCG Cavity Growth with CFD model predictions, in: Proceedings of the 7th International Conference on CFD in the Minerals and Process Industries CSIRO, 9–11 December 2009, Melbourne, Australia.
[39] M. Imran, D. Kumar, N. Kumar, A. Qayyum, A. Saeed, M.S. Bhatti, Environmental concerns of underground coal gasification, Renew. Sustain. Energy Rev (2014) 31600–31610.
[40] Vorobiev, O., Morris, J., Antoun, T., Friedmann, S.J. (2008). Geomechanical simulations related to UCG activities, in: Twenty-Fifth Annual International Pittsburgh Coal Conference, Paper 32-3, Pittsburgh, PA, USA.
[41] Mallett, C. (2006). Managing Ground Deformation in UCG, Underground Coal Gasification Workshop, November 12–15, Kolkata, India.
[42] N. Nakaten, R. Schlüter, R. Azzam, T. Kempka, Development of a techno-economic model for dynamic calculation of COE, energy demand and CO_2 emissions of an integrated UCG-CCS process, Energy 66 (2014) 779–790.

Gasification, DICE, and direct carbon fuel cells for power, fuels, and chemicals production from low rank coals

9

Daniel G. Roberts[1], *Sukhvinder P.S. Badwal*[2], *Louis J. Wibberley*[3] *and Sankar Bhattacharya*[4]
[1]CSIRO Energy, Brisbane, QLD, Australia, [2]CSIRO Energy, Clayton, VIC, Australia, [3]CSIRO Energy, Newcastle, NSW, Australia, [4]Monash University, Clayton, VIC, Australia

9.1 Introduction

Pulverized fuel (pf) combustion in a boiler for power generation is by far the most common approach for the utilization of low rank coals. The properties of low rank coals (in particular moisture and alkali content) mean that such an approach is often very inefficient, leading to very high CO_2 emissions. Low rank coals, however, are inexpensive, and in some regions represent a significant energy resource.

There are advanced technologies suitable for power generation from low rank coals at various levels of technological and commercial readiness: gasification-based systems are demonstrated commercially at large scale; the preparation of low rank coals into a micronized refined carbon (MRC) fuel and the use of this in a modified Diesel engine (direct injection carbon engine, DICE) are nearing commercial-scale demonstration; and the use of low rank coals as feedstocks in direct carbon fuel cells (DCFCs) is an active area of research and development offering the possibility of near maximum possible power generation efficiency. The significant efficiency gains associated with these advanced technologies reduce CO_2 emissions from power generation without the need for CO_2 capture and storage (CCS). For deeper CO_2 emissions' reductions, most of these technologies are also amenable to relatively low cost capture of CO_2.

In addition to efficient, large-scale power generation, state-of-the-art gasification systems also offer the opportunity for low rank coals to be used as feedstocks for the production of chemicals, fertilizers, liquid fuels, explosives, etc. This can add value to a potentially stranded low rank coal resource, with significant economic implications for local economies.

This chapter will discuss these technology concepts, the state of commercial readiness, and the RD&D that is being undertaken to ensure their relevance and applicability to power generation and the production of chemicals and liquid fuels from low rank coals.

Low-Rank Coals for Power Generation, Fuel and Chemical Production
DOI: http://dx.doi.org/10.1016/B978-0-08-100895-9.00010-3

9.2 State-of-the-art: Gasification

Depending on the velocity regime and particle sizes used in the gasifier, gasifiers can be classified into three major types: moving bed, fluidized bed, and entrained flow. There are also variants of these types of gasifiers, in particular within the fluidized and the entrained flow types. Gasifiers also vary in terms of the oxidant used: there are impacts of the use of air or oxygen/steam on feedstock performance and technology selection, and this is particularly significant when considering low rank coals. In this section, we focus mainly on the fluidized bed gasifier and their variants, and entrained flow gasifiers, and discuss some key aspects of the technologies in the context of advanced utilization of low rank coals.

9.2.1 Fluidized bed gasifiers

These gasifiers include the Winkler gasifier, high-temperature Winkler (HTW) gasifier, U-gas gasifier, and transport reactor gasifier.

Fluidized bed gasifiers started as Winkler gasifier nearly 100 years ago as atmospheric pressure bubbling bed (1–2 m/s gas velocity) gasifiers working with coal particles up to 8 mm particle size, and temperature up to 1000°C depending on the softening temperature of the ash. Temperatures being relatively low, large quantities of steam are required to control temperature, in addition to air or oxygen. Steam also acts as a reactant for gasification. Being operated at atmospheric pressure, a simple feeding system could be used for the Winkler gasifier. However, carbon conversion remains low, 90% or below, and few Winkler gasifiers operate these days.

The HTW gasifier was developed to operate around the same temperature as Winkler gasifiers, but at circulating mode and around 10 bar pressure in order to improve the gasification rate and hence the carbon conversion within the same volume. Being pressurized, a different coal feeding system is required, with two lock hoppers switching between atmospheric pressure and the gasifier pressure. In order to enable continuous feeding to a pressurized reactor, moisture content of the coal has to be controlled consistently well to a low level, below 10%. HTW gasifiers operate to achieve similar carbon conversions to Winkler gasifiers, but require smaller volume for the same output. Both Winkler and HTW gasifiers also face problems of agglomeration which limit the bed temperature.

U-gas gasifiers are a variant of Winkler gasifier, developed by the Gas Technology Institute, operating in the bubbling bed regime, but at pressure up to 30 bar and temperature up to 1100°C. The higher temperature promotes agglomeration and the larger particles are selectively removed from the gasifier. Carbon conversion of 95% has been reported from U-gas gasifiers.

Some of the challenges posed by the use of low rank coals in gasification technologies, however, have led to technology developments designed to address specific challenges with their utilization. The Kellogg Brown Root (KBR) Transport Integrated Gasification (TRIG) technology—the so-called transport gasifier—is a high-velocity variant of the circulating fluidized bed concept which was developed to address issues related to carbon conversion commonly associated with more traditional fluidized bed systems. Transport reactor operates as a high-velocity (up to 10 m/s) circulating fluidized bed gasifier with high circulation rate, and around the same temperature and pressure as U-gas gasifiers in both air-blown and oxygen-blown modes. This gasifier was developed by KBR based on the fluid catalytic cracker technology. Combination of high velocity and high circulation rate results in high carbon conversion, reported to be up to 97%. A commercial 550 MWe IGCC plant at Kemper County, Mississippi has been commissioned to operate under air-blown mode designed to capture 65% CO_2. This was developed in the United States and is the gasification technology used in the Kemper County polygeneration project in Kemper County. Although a project incorporating 65% carbon dioxide capture (for Enhanced Oil Recovery) the Kemper County gasifier is an air-blown gasifier, oxygen-blown variants of the TRIG gasifier are under development.

With the use of steam and oxygen, fluidized bed, U-gas, and transport gasifiers do not generate any tar when used with coals. Figs. 9.1–9.3 present schematics of the three fluidized bed variants (www.netl.doe.gov).

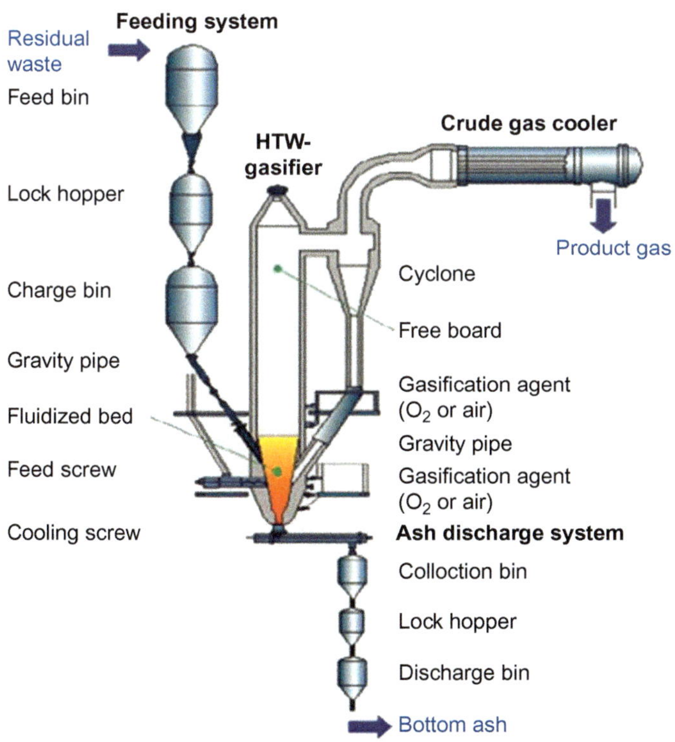

Figure 9.1 High-temperature Winkler gasifier (www.netl.doe.gov).

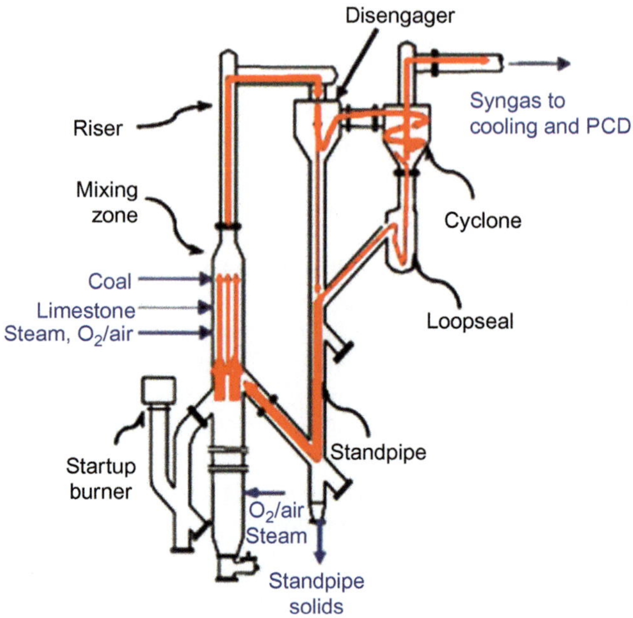

Figure 9.2 Transport reactor gasifier (www.netl.doe.gov).

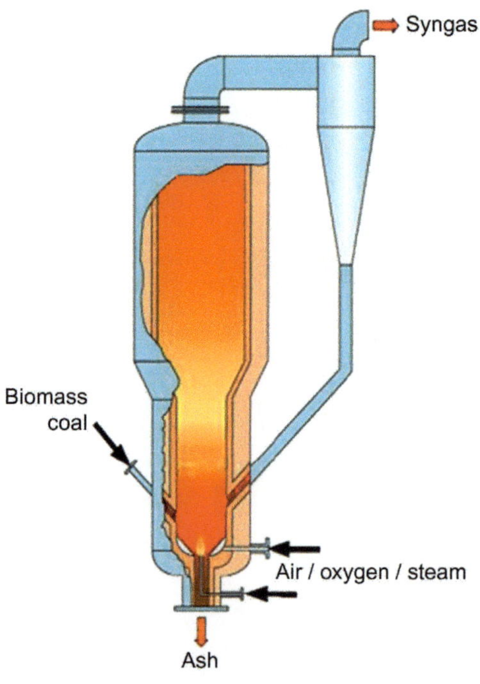

Figure 9.3 U-gas gasifier (www.netl.doe.gov).

All gasifiers in general are of lower capital cost when designed to operate in air-blown mode. This is primarily due to cost associated with the air separation plant, which is of cryogenic type, and required for oxygen-blown gasification.

9.2.2 Large-scale, high efficiency power, chemicals, and fuels

The significance of the low rank coal resource, and the generally low cost of production, makes them attractive as feedstocks for the large-scale production of chemicals, fuels, fertilizers, and other products such as hydrogen. In countries such as Australia, where the low rank coal resource is very high in moisture and associated with low efficiency, high emissions power production, there can be opportunities to develop new, advanced industries using this resource as a feedstock. This has been shown to work for SNG production in the United States. The diverse and often challenging properties of low rank coals, however, mean that the use of low rank coals in gasification systems requires particular consideration of their properties and supporting RD&D. This can be complicated by the geographically specific nature of the challenges: Australian lignites, for example, are characterized by very high moisture ($\sim$60%) and low amounts of alkali-rich mineral matter ($\sim$2% on a dry basis); North Dakotan lignites are also high in sodium but with much less moisture ($\sim$40%); and those from Germany may have less moisture ($\sim$50%) and more ash ($\sim$10%).

A key aspect of the use of any feedstock for nonpower products is the oxygen-blown nature of the gasification system used. As discussed in the previous section, low cost, reliable power generation on lignites is usually undertaken using air-blown gasification systems; however, oxygen-blown gasifiers are more relevant to the more contemporary focus on "carbon capture ready" power generation and the interest in nonpower products. This provides a complication for the use of feedstocks which have been extensively used in fluidized bed technologies: the previous section discussed some issues associated with commercial-scale deployment of fixed bed gasifiers, and there is a further complication associated with the use of oxygen as a gasifying agent and the localized high temperatures that are likely to result. This means that alternative gasifier technology approaches are required, and new insights into coal behavior at higher temperatures and pressures are needed. Entrained flow gasifiers come into consideration in this regard.

The state of the art in commercial-scale applications is high-pressure, high-temperature entrained flow gasification. Such technologies (e.g., Shell, GE, and Siemens) form the large majority of coal-based gasification plant around the world, with considerable deployments in China for the production of liquid fuels and SNG. These technologies have been proven commercially using a range of bituminous (and some sub-bituminous) coals as feedstocks, and we have developed a good understanding of the coal properties that are suitable. There are also newer gasifier technologies being deployed such as the MHI dry-feed air-blown gasifier, the ECUST opposed multi-burner (OMB) oxygen-blown slurry-fed gasifier from China, and the MHPS "EAGLE" oxygen-blown gasifier. The MHI gasifier is a dry-feed, upflow, air-blown gasifier with a 250 MW IGCC plant currently under operation in

Nakoso, Japan, and there are almost 90 ECUST OMB plants handling 3000 tonnes per day (TPD) coal in China. The MHPS "EAGLE" gasifier is designed for integration with fuel cells for power generation with very high efficiencies, and is the proposed gasifier for some emerging H_2 production projects.

We know less about the behavior of low rank coals in the entrained flow gasifier technologies, due to their operation at high temperatures and pressures and the need for a suitable slag layer to be formed with appropriate viscosity and flow properties (Fig. 9.4).

These potential issues, and the relatively high capital cost of high-pressure entrained flow gasifiers, have seen other approaches used to produce high value

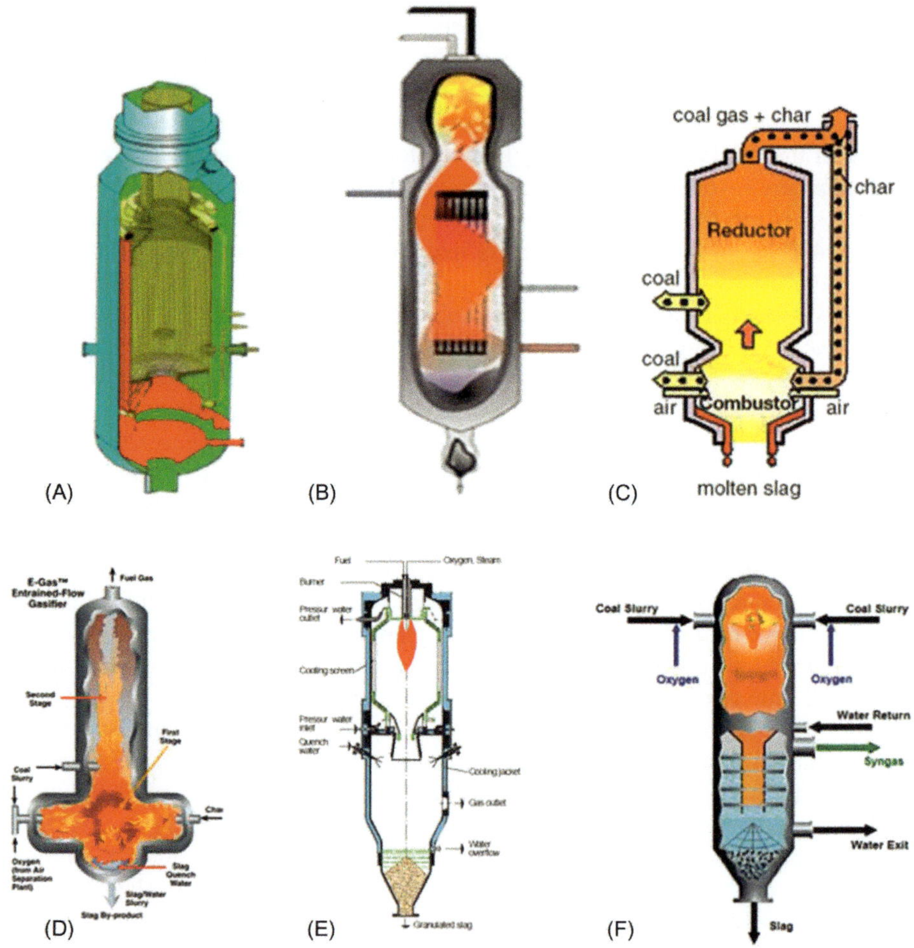

Figure 9.4 Some entrained flow gasifier technologies (taken from M. Gräbner, B. Meyer. Introduction of a ternary diagram for comprehensive evaluation of gasification processes for high ash coals, in: International Conference on Coal Science and Technology, Oviedo, Spain (2011).; and www.netl.doe.gov). (A) Shell, (B) GE, (C) MHI, (D) E-Gas (CB&I), (E) Siemens, and (F) ECUST.

products from low rank coals. Fixed bed gasifiers (such as the Lurgi dry-bottom and BGL Envirotherm slagging gasifiers) produce a methane-rich, often tarry syngas that can be suited to the production of SNG. This, and their relatively low throughput, means they are less suited to large-scale power or FT applications. They are able to handle high moisture coals, but there are some particular requirements of the coal in terms of physical properties, how these change during heating and gasification, and their tendency to form fines that can impact the flow of gas through the bed. Low rank coals that do not form strong, porous beds and which are prone to fragmentation can be troublesome—which has been a limitation to the more widespread use of this technology for this application. There have been successes; however, a particularly successful lignite gasification project is the Great Plains Synfuels plant in North Dakota, which uses a Lurgi dry-bottom gasifier to produce about 153 million cubic feet of SNG annually.

When considering the most suitable gasifier type for the conversion of low rank coals to high value products, clearly a range of coal properties and technology features must be considered. Table 9.1 summarizes the brief review given in this section in the context of gasifying low rank coals for the production of fuels, chemicals, hydrogen, or other nonpower products. This shows that there are

Table 9.1 Summary of the use of different technology types for gasification of low rank coals in nonpower applications

Technology type	Attractive features	Challenges	Comments
Fixed bed	Relatively low cost and ease of operation (for suitable coals)	Specific coal property requirements Syngas quality Fines management	Fixed beds have been shown to be successful in gasification of low rank coals-to-products. This success, however, is strongly dependent on feedstock suitability
Fluidized bed	Low operating temperatures Relatively low cost compared with entrained flow gasifiers	Difficult to operate in O_2-blown configuration Upper pressure limited	Challenges faced by using a typically air-blown technology in a nonpower application may be large
Entrained flow	Feedstock flexibility, high throughput, and syngas quality	Limited experience with low rank coals such as lignites	While EF is a common choice for coal-to-products applications, there remains some uncertainty as to the most suitable approach for low rank coals, in particular lignites

research and development requirements if low rank coals are to be considered as high quality feedstocks for the production of fuels and chemicals; these are discussed further in the next section.

9.2.3 Research needs to enable low rank coals-to-products

As previously mentioned, low rank coals have variable composition in terms of moisture, mineral matter, and mineral composition. These may have high moisture/low ash, low-to-medium moisture/high ash, and variable mineral composition with alkali- or sulfur-rich ash. The following section briefly identifies the research needs linked directly to the above issues.

9.2.3.1 Moisture

High moisture content affects the gasification process in multiple ways; above 10% moisture content or variable moisture content, feeding through pressurized lock hoppers can become erratic affecting the gasification rate and gas composition. Therefore, uniform predrying of high moisture coals to <10% moisture content is a necessity. But such predrying (50%–60% to <10% moisture) has not yet been demonstrated in any coal or biomass gasification plants operating at high pressure.

9.2.3.2 Mineral matter content

The mineral matter content affects the agglomeration of bed materials in fluidized bed gasifiers of all variants and viscosity of slags in entrained flow gasifiers.

Agglomeration in fluidized bed gasifiers takes place due to the formation of low-temperature eutectics between the mineral matter in coal and the bed materials or sintering of the ash particles themselves. The sintering of ash particles followed by agglomeration to larger-sized particles is a function of temperature, composition of the mineral matter in particular the alkali and alkaline earth metals (Na, Ca, and Mg) forming low-melting silicates of these compounds. Reduction in bed temperature below the sintering temperature of the ash coupled with regular bed drainage reduces the problem to some extent. It is therefore important to measure the sintering temperature of the mineral matter content of the coal being used in the gasifier. For U-gas gasifier, the ash requires agglomeration; therefore, the information on the full range of the ash fusion temperature from sintering through to melting point under reducing conditions is necessary.

The role of mineral matter content in entrained flow gasifiers is very important. The mineral matter converted into ash exits the gasifier as molten ash. In order to ensure uninterrupted operation of an entrained flow gasifier, it is of utmost importance to sustain continuous slag flow out of the gasifier. Excessive build-up of slag melt on the gasifier wall reduces heat transfer in a gasifier, causing downtime or shutdown of the gasifier. It is important that the viscosity of the liquid slag is low enough so that slags flow down the gasifier walls and can be drained out from the gasifier as "molten slag." However, if the viscosity of slag is too low, an effective slag layer will not form, and therefore, will not protect the gasifier internal wall. An acceptable industry standard for the viscosity is between 100 Poise and 250 Poise that ensure the satisfactory protection of the gasifier wall and uninterrupted operation of the gasifier. Therefore, information on composition, viscosity, and the change in viscosity with temperature for coal ash slags are essential for efficient operation of an entrained flow gasifier. Such information on ash from low rank coals is scarce in the open literature. In general, calcium and iron content together with the other constituents—silica, alumina, and magnesium oxide—exerts an important effect on the viscosity of slags. Calcium oxide content below 30% is known to keep the viscosity low, but above 30% calcium oxide, the effect reverses. The effect of iron content in ash up to 40% is known to lower the viscosity. Also, in general, high silica and alumina contents, 30% or above, increase the viscosity as these are refractory materials. A comprehensive database of viscosity as a function temperature and mineral matter composition is necessary if low rank coals are to be widely used in entrained flow gasifiers.

9.3 The next big thing: MRC and DICE

DICE, or the direct injection carbon engine, combines the superior thermal efficiency, flexibility, and lower capital cost of the diesel engine with the low cost and availability of coal [1]. The technology dates back to the development of the diesel engine by Rudolf Diesel in 1892, which was originally intended to use pulverized brown coal as fuel, but then the inventor turned to peanut oil after difficulties with the injection equipment. The modern variant of DICE uses finely ground low-ash carbons (e.g., coal, lignite, and biomass) slurried with water—a fuel called MRC. MRC is a liquid fuel (similar in consistency to acrylic paint), which can be used in diesel engines that have been adapted with a slurry fuel injection system and hardened cylinder components. The DICE fuel cycle is depicted in Fig. 9.5.

DICE is based on using adapted diesel engines, which are a mature power generation technology. Such engines play a minor role in baseload power production in today's market due to the high cost of diesel fuel, higher maintenance costs, and, in the past, the relatively small unit size. Currently, the largest diesel engine in commercial production is 76 MW—a size used principally for large ships. If larger capacity engines were required, the manufacturer MAN B&W has a K108 engine design, which in its 18-cylinder form would generate around 120 MW. MAN B&W believes that engines up to 150 MW are technically feasible using current engine and manufacturing technologies [2]. In addition, there are a range of options to optimize large marine engines for land-based power generation to reduce both capital and operating costs.

However, in contrast to conventional coal-fired power plants, which gain economy of scale via increased unit capacity, diesel (and gas) reciprocating engines obtain economy via modularity and multiple units (e.g., a 1000 MW reciprocating gas-engine power plant has been constructed near Salvador using 120 medium-size

Figure 9.5 The DICE fuel cycle.

medium-speed gas engines). It is anticipated that the same modular approach could apply for DICE.

Regarding fuel for DICE power plants, MRC fuel could become a new global commodity for low-emissions power, given that MRC is nonflammable, environmentally benign, and can be safely transported and stored. This option is being actively considered in Victoria, Australia, with the potential to efficiently convert vast reserves of brown coal to export MRC, at around one-third of the cost of LNG. In Australia, MRC fuel also costs less than fuel oil and most gas, and could reduce power generation costs for remote or decentralized generators. This would also enable lower cost electricity for off-grid communities (especially in developing countries) and assist sustainable mining development by providing cost-effective power for mines and communities.

Like other reciprocating engines (gas or diesel), DICE power plants could be modular and can be ramped relatively quickly, and can thus be used for load following and even baseload generation in some circumstances. Development of DICE faltered historically largely because the intended purpose was to use coal as a substitute for oil in shortages that did not materialize. However, today, DICE is being pursued not to replace oil, but as a unique high-efficiency, low-emissions (HELE) coal technology that can support a grid with increased renewables and water constraints, and with decreased emissions.

9.3.1 DICE as a HELE coal technology

DICE could play a role in the transition to a lower emissions energy sector (i.e., HELE, high-efficiency low emissions) as it is a cost-competitive, lower emissions technology in its own right. See Fig. 9.6 for an example based on Victoria,

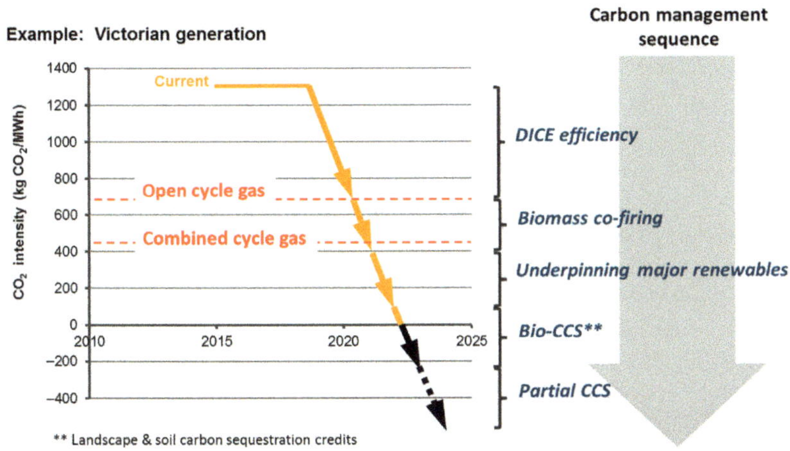

Figure 9.6 Cumulative effects of an integrated carbon management scheme, using Victoria (Australia) as an example.

Australia, showing the cumulative effects of an integrated carbon management scheme.

It is expected that DICE would provide a higher efficiency coal-based generation technology, with a step reduction in CO_2 emissions of 20%–35% for black coals and 30%–50% for brown coals (ranges based on new DICE and new ultra-supercritical pulverized coal, and new DICE and an old subcritical coal-fired power plant in Australia). Greater emissions' reductions can be achieved if MRC composed of a coal/biomass blend is used, where the biomass is likely to be in the form of a char.

Further reductions in CO_2 emissions could be achieved by providing backup and load-following power to enable the cost-effective and increased use of intermittent renewables without the loss in efficiency incurred from load following by conventional thermal power plants. DICE also could assist in the uptake of CCS, delivering a 30%–40% cost advantage (in terms of \$/t CO_2 abated) [3] compared to CCS on conventional coal-fired power generation, through increased efficiency and the ability for waste engine heat to provide at least 65% of the CO_2 stripper heat requirement.

9.3.2 Flexibility: An increasing challenge

Increased flexibility is a challenge for grids—one that is growing in magnitude with the ever-higher penetration of intermittent renewable generation in response to policy initiatives. Continued growth of renewables over the long term will further exacerbate the need to provide more flexible and distributed generation capacity.

Natural gas combined cycle and conventional coal-fired power plants can follow load demand, although their efficiency decreases at lower loads and maintenance requirements increase due to thermal swings. In addition, due to their large unit size, which helps to optimize efficiency, most conventional coal-fired power plants have low ramp rates on the order of about 1%/min, although some plants in Europe have reported ramp rates of about 4%/min. In comparison, DICE provides superior flexibility with ramp rates of 10%/min without maintenance penalties caused by thermal cycling.

The flexibility of DICE is not limited to load following. The technology provides an option for customers seeking new power generation capacity at various sizes. DICE is modular (in 10–100 MW increments). In addition, as DICE is likely to need half the capital investment (US$1.1 million/MW for four-stroke DICE and US $1.6–2 million/MW for low-speed two-stroke marine engines; 2015 estimate) of conventional coal-fired power plants, it could provide an important technology for incremental capacity—either new or replacement.

DICE could also provide users of natural gas with the option of installing multi-fuel gas engines and retrofitting with a DICE conversion kit in the future if natural gas prices become uneconomic. This approach could prevent stranded assets. Such a conversion is not possible for gas turbines, regardless of mineral content of the MRC fuel, due to intolerance to alkalis.

9.3.3 The Water–CO_2 nexus

Cooling water (for noncoastal installations) has become a serious issue for Rankine cycle thermal plants: in many parts of the cost and highest efficiency such plants use around 2000 L of water for condenser cooling for every MWh generated. This level of water consumption is becoming intolerable in some locations (e.g., Australia, India, South Africa, and China), requiring the use of dry cooling. Although this reduces water consumption to around 300 L/MWh, dry cooling increases plant costs, delivered electricity cost, and CO_2 intensity—they last by up to 5%. Thus, dry cooling can result in an additional 1000 kg CO_2 being emitted for each 30 tonnes of water saved.

DICE is much easier to dry cool because the temperature at which heat is rejected is much higher. This enables much lower cost dry-cooling (radiator) systems and completely avoids the need for cooling water. Water is needed for MRC fuel production, but this is at a rate similar to that for a dry-cooled conventional coal-fired power plant. Notably, fuel preparation for DICE can be located remotely from the power plant (i.e., fuel preparation can be co-located with reliable and abundant sources of water).

9.3.4 Natural gas versus DICE

Although natural gas-based generation can be considered an easy option for reducing CO_2 intensity, to compete with coal for baseload electricity natural gas must be a sufficiently low price (without excessive volatility), and have security of supply. The "dash for gas" in Europe has abated due to gas price increases, with recent mothballing of gas generation capacity (both combined and open cycle) in some countries. As examples, GDF Suez has closed gas power plants with a combined capacity of over 7 GW; in the United Kingdom, the Keadby 750-MW power station (commissioned in 1996) was mothballed last year; and RWE is closing its Claus C plant in the Netherlands, just 2 years after it was commissioned. Based on the availability of unconventional natural gas, the United States has expanded natural gas-based generation, although price volatility remains an issue (particularly in winter). As an extreme example, the price in New England in January and February 2014 rose to $17/GJ, whereas the average gas price for the United States for the same period was US$7.4/GJ, and the average coal price was US$2.3/GJ (data from US EIA Electric Power Monthly).

Overall, considering fuel options that can respond to demand should provide new opportunities for coal—providing the technology is highly efficient, highly flexible, and CCS ready. DICE potentially has all of these attributes and is well suited to balance the additions of intermittent renewables to the grid, but does not suffer from the high prices and/or price volatility often associated with natural gas.

9.3.5 Development requirements

Despite many potential advantages, the use of coal in DICE has required addressing a number of technological issues (especially injector nozzle and piston ring wear), which were essentially solved for smaller engines (i.e., <5 MW) in a comprehensive US Department of Energy (DOE) program (1978–92) [4–6]. Since the DOE program ended, a number of significant technology developments have occurred in both coal processing and engine design and materials. These, combined with the use of larger, lower speed engines (and many new drivers), are expected to significantly increase both the technical and the economic viability of DICE [7].

In past programs, the cost of coal processing reduced the cost advantage of DICE, and, together with falling oil prices, was a key factor in the termination of the DOE program. Fuel production now has the advantage of large and efficient mills for micronization (e.g., IsaMill) and improvements in fine-coal cleaning [8]. These technological advances provide a step reduction in processing cost and also allow cost-effective recovery of MRC from tailings. Thus, in Australia, MRC is now estimated to cost Australian dollar (AUD)$2–3/GJ (US$1.4–2.2/GJ) for Victorian brown coal and AUD$4–6/GJ (US$2.9–4.3/GJ) based on bituminous coal. Economic assessments have shown that the increased cost of coal processing is more than offset by increased grade recovery, lower capital cost, and reduced fuel rate of DICE (together with a range of environmental benefits) [9]

The largest challenge facing the development of DICE may be a nontechnical one. At present there is no clear owner or champion for the technology, but many interested parties, including the coal industry, technology providers, engine producers, and power producers. Although DICE needs considerable development and demonstration, it has an enormous advantage via the ability to carry out a near-commercial-scale demonstration at a relatively small size (around 10 MW), both quickly and at relatively low cost. This should enable DICE to leapfrog the usual technology development steps, resulting in a time to first commercial deployment of 5 years, and as low as US$70 million, including fuel processing and logistics. An international umbrella organization, DICEnet (www.dice-net.org), has been established to help coordinate efforts, and a staged, integrated DICE development program has been devised for both black and brown coals with a goal of a large-scale demonstration after 2020. Currently, a number of groups are investigating the DICE fuel cycle in Australia, Europe, China, Japan, the Unites States and South Africa. MAN Diesel and Turbo have engaged with a number of MRC proponents, and are considered the industry leaders in DICE development. MAN has established a staged development program with a specially adapted low-speed 1-MW single-cylinder test engine at Mitsui in Japan. Development efforts will also benefit from recent experience from firing engines with bitumen slurries and residual fuel oils (e.g., Orimulsion and MSAR) [10–12].

9.3.6 Conclusions

Combining the superior thermal efficiency, flexibility, and lower capital cost of the diesel engine with the low cost and availability of coal provides an innovative step change technology that could markedly increase the competitiveness and reduce the environmental footprint of coal-based power generation. Australian R&D over the last 5 years has piggybacked on the extensive R&D undertaken as part of the US DOE program over 1978–92, and focused on "derisking" the fuel cycle. This recent work has confirmed the potential of this alternative technology for coal and other carbons. Internationally there are many opportunities for DICE, including:

- Incremental replacement of less efficient coal generation plants—175 GW of smaller/older capacity identified (e.g., below 300 MW and older than 40 years).
- Smaller incremental capacity for developing nations.
- A cost-effective/CO_2-equivalent technology for replacing open-cycle natural gas turbines used for load following and backup duty.
- New capacity for ancillary services to underpin growth in renewables.
- A replacement technology for high-cost diesel generation in remote mining sites and communities, especially in developing countries.
- Replacement capacity for uneconomic gas generation.

Recent technology assessments have shown no major technical barriers in developing DICE to a commercial scale. Some further development work is required to optimize for a range of coal types and engine technologies, but the development-to-deployment time scale could be as short as five years based on an assessment by MAN Diesel & Turbo [12]. The costs to commercialization (from R&D to demonstration and first commercial plant) would be comparatively low (i.e., 2015 estimate of US$70 million).

Development and deployment of DICE provide an opportunity for the coal industry, the electricity generation sector, and governments to support a breakthrough technology—for the combined benefit of coal and renewables-based electricity systems.

9.4 Toward maximum efficiency: DCFC

9.4.1 Concepts

Fuel cells are reaching a high level of maturity with commercialization under way for number of years, with molten carbonate fuel cell (MCFC), solid oxide fuel cell (SOFC), and polymer electrolyte membrane fuel cell (PEMFC) leading to the market penetration. According to US DOE report, worldwide 50,000 fuel cell units with a total capacity of 185 MW were shipped in 2014 and over 70,000 units with a total capacity of about 340 MW in 2015 [13,14].The total installed capacity for all types of fuel cells now stand at >1 GW. The largest single facility of 59 MW (Posco's Gyeonggi Green Power Plant), based on the MCFC technology, is located in South Korea [14]. Some of the major attractions of fuel cells include high efficiency due to direct electrochemical oxidation of a fuel, the ability to cogenerate both power and heat at distributed sites removing the need for expensive transmission/distribution infrastructure, and low chemical and noise pollution. For most fuel cells, modularity allows economies of upscaling to the required power plant size. A majority of fuel cells use gaseous fuels (H_2/CO) with some exceptions (methanol or ethanol for portable power generation).

The carbon fuel cells (CFCs) refer to a generic name in which the starting fuel is in the form of solid carbon as distinct from other fuel cells. Under this generic name a range of technologies are under development and these have been reviewed in recent review articles [15–18]. These fuel cells have been forecast to have electric efficiency approaching 70% [15,19]. With typical carbon fuel cell operating temperature in the range 600 and 900°C, if the high quality waste heat can be put to good use, then the overall system efficiency can increase to above 85%. Other benefits include low fuel utilization per unit power generated and easy capture of CO_2 for sequestration as highly concentrated stream of CO_2 is produced as the by-product. These efficiency values are highest among various carbon combustion technologies and almost twice those of existing coal-fired power plants (Fig. 9.7).

In a fuel cell, the theoretical efficiency is derived from the fuel oxidation reaction, and is very much dependent on the fuel type and the operating temperature of the fuel cell. For carbon as the fuel, the following oxidation reactions are possible:

$$C + O_2 = CO_2 \tag{9.i}$$

$$C + 0.5O_2 = CO \tag{9.ii}$$

$$CO + 0.5O_2 = CO_2 \tag{9.iii}$$

For direct solid carbon oxidation (reaction (9.i)), the theoretical electric efficiency, as determined by the free energy of the reaction ($\Delta G^\circ = -nE^\circ F$, where n is the number of electrons transferred in the fuel oxidation reaction, E° is the fuel cell open circuit voltage, and F is the Faraday constant), is almost independent of temperature due to near zero entropy term and is slightly above 100% at the fuel cell

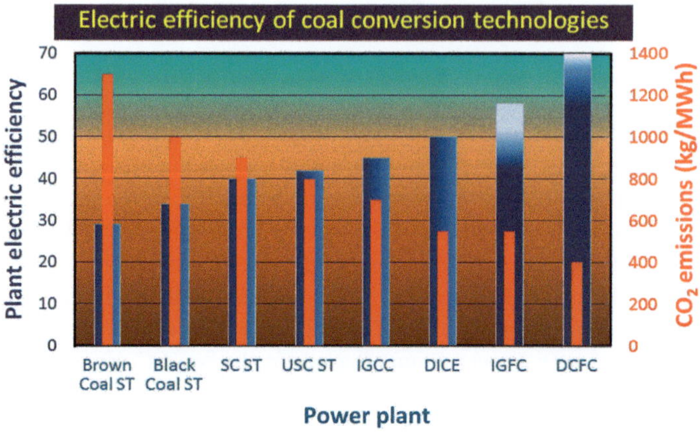

Figure 9.7 Electric efficiency (blue bars) and CO_2 emissions (red bars) for different coal-based power plants. *ST*, steam turbine; *SC*, super critical; *USC*, ultra super critical; *IGCC*, integrated gasification combined cycle; *DICE*, direct injection carbon engine; *IGFC*, integrated gasification carbon fuel cell; *DCFC*, direct carbon fuel cell.

operating temperature of 800°C. Under these conditions, the system would absorb heat from an external source. However, in operation, reactions (9.ii) and (9.iii) are most likely to dominate, thus leading to lowering of the maximum theoretical electric efficiency.

Depending on the combination of these reactions taking place inside a fuel cell at the anode/electrolyte interface, the average maximum or theoretical efficiency will be somewhat lower than 100%. In real fuel cell operation, the practical electric efficiency is given by:

$$\text{Electric Efficiency (operating)} = \text{Th}_{\text{eff}} \times F_u \times V_{\text{eff}} \qquad (9.1)$$

F_u is the fuel utilization factor and for solid fuel is almost 100% (fuel feed and product gas CO_2 are distinct phases—fuel can be separated and recycled), whereas it is 85%–90% for most other fuel cells utilizing gaseous fuels. V_{eff} is the voltage efficiency of a fuel cell once internal resistive and overpotential losses are taken into consideration. The operating electric efficiency is predicted to be around 70% for DCFCs [15,17].

9.4.2 Carbon fuel cell types and fuels

Two major categories of CFCs are (i) DCFCs in which the solid fuel is in contact with the anode of the fuel cell and the chemical energy in the solid carbon are directly converted to electricity either through its direct electrochemical oxidation or the fuel is partially oxidized by the chemical reaction of carbon with CO_2 generated in situ (Boudouard reaction $C + CO_2 = 2CO$); and (ii) indirect CFCs in which an external gasifier converts the solid fuel (coal) to gaseous fuels in an external gasifier by reaction of carbon with CO_2 generated in the fuel cell or with steam and the gaseous fuel, following removal of sulfur, particulates, and other contaminants, is fed to a conventional high-temperature fuel cell [22]. The operating electric efficiency is predicted to be somewhat lower (50%–58%) for indirect CFCs integrated with an external gasifier (integrated gasification carbon fuel cell, IGFC) compared with around 70% for DCFCs [15–17].

The DCFCs are broadly categorized into three main practical systems based on an ion conducting electrolyte: molten hydroxide (KOH/NaOH with OH^- as the conducting specie); molten carbonate ($Li_2CO_3/Na_2CO_3/K_2CO_3$ with CO_3^{2-} as the conducting specie); and solid oxide (Y_2O_3–ZrO_2 (YSZ), Sc_2O_3–ZrO_2 (SCSZ), doped CeO_2 with O^{2-} as the conducting specie). The electrolyte is held between cathode (air electrode) and anode (fuel electrode). Further sub-categories, especially for DCFCs based on a solid oxide electrolyte, include systems where a molten media (salts such as carbonates or molten metals) are employed for solid fuel distribution to the electrode/electrolyte interface. The metal may be reactive (e.g., Sn) participating in the fuel cell reaction or an inert carrier of solid carbon particles (e.g., Ag). The various DCFCs and their operating temperatures are shown in Fig. 9.8 and cell reactions are summarized in Table 9.2 and materials of construction are given in many literature reports [15–19]. Typical operating principle of a carbon fuel cell with solid oxide electrolyte is shown in Fig. 9.9. As most anode

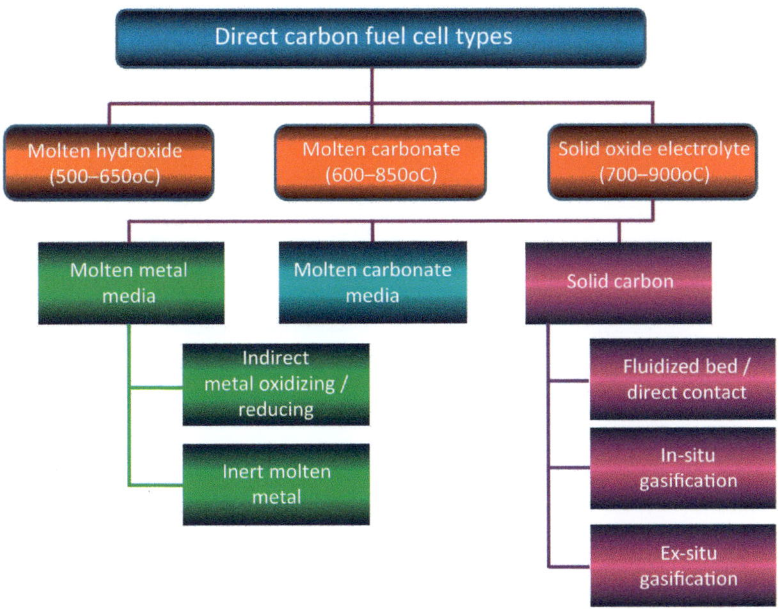

Figure 9.8 Various types of direct carbon fuel cells.
Diagram redrawn from S. Giddey, S.P.S. Badwal, A. Kulkarni, C. Munnings, A comprehensive review of direct carbon fuel cell technology, Prog. Energy Combust. Sci. 38 (2012) 360–399.

Table 9.2 Electrode reactions associated with different types of carbon fuel cells

Types of direct carbon fuel cell	Anode cell reactions	Cathode cell reactions
Molten hydroxide	$C + 4OH^- = 2H_2O + CO_2 + 4e^-$	$O_2 + 2H_2O + 4e^- = 4OH^-$
Molten carbonate	$C + 2CO_3^{2-} = 3CO_2 + 4e^-$	$O_2 + 2CO_2 + 4e^- = 2CO_3^{2-}$
Hybrid carbonate/solid oxide	$C + 2O^{2-} = CO_2 + 4e^-$	$O_2 + 4e^- = 2O^{2-}$
Molten metal—reactive anode	$Sn + 2O^{2-} = SnO_2 + 4e^-$ $SnO_2 + C = Sn + CO_2$	$O_2 + 4e^- = 2O^{2-}$
Molten metal—inert anode	$C + 2O^{2-} = CO_2 + 4e^-$	$O_2 + 4e^- = 2O^{2-}$
IGFC: integrated gasification fuel cell	$C + 2O^{2-} = CO_2 + 4e^-$	$O_2 + 4e^- = 2O^{2-}$

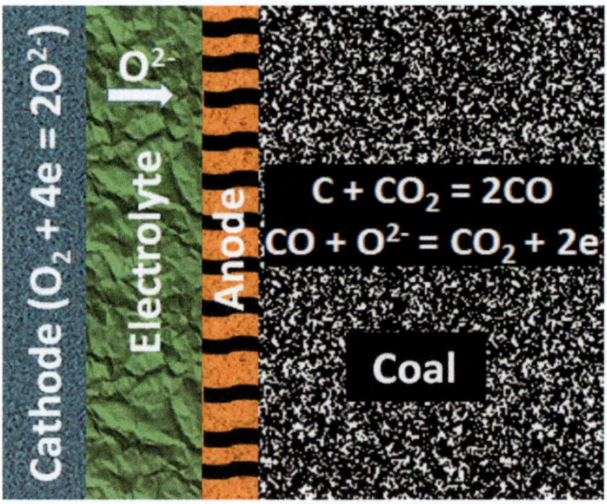

Figure 9.9 Operating principle of a carbon fuel cell with solid oxide electrolyte.

materials have very low tolerance to impurities typically present in coals, DCFCs require pre-processing of the fuel to remove sulfur, volatiles and inorganic (ash), and sulfur impurities, micronization to increase surface area and the use of more resilient anode materials although some inorganic impurities have been shown to assist with the reaction of carbon with CO_2 (Boudouard gasification) and lead to improvement in the performance [20]. A number of articles have discussed in detail the fuel processing requirements for DCFCs [19,21,22].

The external gasification type systems have the benefits in that the impurities in low rank fuels can be removed externally to the fuel cell thus prolonging the life of fuel cell components. This nevertheless occurs at the expense of somewhat lower electric efficiency. However, direct thermal integration of the gasifier with a high-temperature fuel cell such as a solid oxide or MCFC allows higher electric efficiency to be achieved with system optimization. It has been reported that for thermally integrated gasifier indirect CFCs, 50%–58% electric efficiency is possible [16,17] (see Fig. 9.7).

9.4.3 Technology challenges and R&D needs

The CFCs offer many advantages over conventional coal-fired thermal power plants which include high efficiency, less CO_2 and pollutant emissions, and easy capture of CO_2 in concentrated stream and no water management requirements. Although indirect CFCs, where coals are externally cleaned and reformed to produce CO or syngas, have reached a high level of technology maturity partly due to the large effort that has gone into the development of molten carbonate and SOFC, the DCFCs are still at an early stage of development [15,16,22] with a number of unresolved technical challenges. There are many different types of CFCs, each with its own set of technical challenges. Most of the early R&D effort on molten hydroxide and molten carbonate electrolyte based DCFCs has decreased substantially due to low power densities and corrosion/degradation issues [15,16,22]. The major global R&D effort since about 2000 is on CFCs which employ a solid oxide electrolyte. However, one of the key challenge is the continuous supply of solid fuel to the anode/electrolyte interface through the pore structure of the anode.

A number of strategies proposed include the use of a molten carbonate or a molten metal mixed with fine carbon particles to deliver solid fuel to the electrochemically active reaction sites or use of a mixed ion/electron conducting anode to shift carbon electrochemical oxidation reaction from anode/electrolyte interface to anode/fuel interface [15–19,23]. However, these systems suffer from overall system complications and serious corrosion of fuel cell components (reaction of molten carbonates with solid oxide electrolyte) or decomposition of mixed ionic/electron conducting materials in reducing environments in fuel compartment. Thus robust materials and strategies are required to overcome corrosion and degradation issues. Further, optimum continuous fuel feed mechanisms need to be developed for continuous generation of power. Operation of the fuel cell in a batch type fuel delivery mode is more like a battery. Coal preparation to remove volatiles and ash (present in widely varying compositions and physical properties depending on its location) before coal can be fed in a solid or molten delivery system is another challenge along with understanding of carbon conversion and utilization [16,22]. None of the DCFCs have been up-scaled to practical levels (kW or above) and life times of only few hundred hours have been demonstrated, significantly further away from real life times of multi 10,000 hours required for practical devices. At the current level of investment, which is very low and mostly at academic level, compared to other fuel cell types or advanced coal-fired power plants, it will take a long time before DCFCs are likely to be commercialized.

9.4.4 Applications

Coal is forecast to remain a key source for power generation globally for many more decades. With current power plants offering efficiency in the 30%–35%, there is substantial potential for a technology such as CFCs that can deliver twice the efficiency and halve the greenhouse gas emissions.

Indirect fuel cells with thermally integrated gasifier (IGFC) with molten or SOFC technology is likely to be the first technology to be commercialized. Gur has looked at the economic assessment of advanced coal conversion technologies (pulverized coal combustion and IGCC compared with IGFC [22]). The economic assessment compiled the results of many previous studies and took into account the cost of CO_2 capture and sequestration. Three parameters were considered (efficiency, capital cost, and levelized cost of electricity generation in terms of cents per kWh). IGFC technology was clear winner by a long margin in all three categories of comparison. DCFCs, although at a very early stage of development, offer even higher efficiency.

One of the major advantages of fuel cells is that they are modular in construction and the unit size can vary from a few kW for distributed power generation to multi MW for embedded generation to large-scale power plants [12].

Like other high-temperature fuel cells, due to the high operating temperatures, the combined heat and electric efficiency is in the 80%–90% range if heat can be captured and put to good use. However, such systems are best for baseload applications, require thermal insulation, have limited thermal cycling capability, very slow start up and shut down, poor load-following capability, and no power at cold start. Moreover the balance of plant for CFCs may be quite complex.

References

[1] C.M. Jha, F.J. Smit, G.L. Shields, & N. Moro, (1997). Premium Fuel Development, Contract No. DE-AC22-92PC92208, U.S. Department of Energy, National Energy Technology Laboratory, www.netl.doe.gov/publications/proceedings/97/97cl/jha.pdf.

[2] P. Van Dyck, The big one from MAN B&W, Motor Ship 86 (1013) (2005) 8–9.
B. Hsu, Coal-fueled diesel engine development update at GE transportation systems., J. Eng. Gas Turb. Power 114 (3) (1992).

[3] Based on personal communications with Barry Hooper of CO_2CRC in May 2013, Regarding CO_2CRC's Calculations on the Use of the UNO MK3 Capture System for DICE.

[4] T. Ryan, Coal-fueled diesel development: a technical review, J. Eng. Gas Turb. Power 116 (4) (1994) 740–748.

[5] D. Arthur, Little, Inc., Coal-Fuelled Diesel System for Stationary Power Applications—Technology Development, Final Report March 1988–June 1994, Topical Report Contract DE-AC21-88MC25124, For U.S. Department of Energy, Office of Fossil Energy, <www.netl.doe.gov/File%20Library/Research/Coal/major%20demonstrations/cctdp/Round5/M96000600.pdf>, 1995.

[6] K. Nicol, (2014). The Direct Injection Carbon Engine, International Energy Agency Clean Coal Centre, Available at: <www.iea-coal.org/site/2010/publications-section/reports>.

[7] IsaMill. (2015). IsaMillTM Breaking the Bboundaries, <www.isamill.com/EN/Pages/default.aspx>.

[8] L. Wibberley, D. Osborne, (2015). Premium coal fuels with advanced coal beneficiation, Presented at Clearwater Clean Coal Conference, Clearwater, FL, 31 May–4 June, Available at: <jamesoncell.com/EN/Downloads/Pages/TechnicalPapers.aspx>.
[9] Modern Power Systems. (2000). Optimism About Orimulsion?, <www.modernpower-systems.com/news/newsoptimism-about-orimulsion> (21.07.2000).
[10] Wärtsilä. (2005). Annual Report '05 Sustainability Report, <www.euroland.com/arinhtml/sf-wrt/2005/bu_eng_2005/index.htm>.
[11] Quadrise Canada Corporation. (2015). Low Cost Alternative Fuel, <www.quadrisecanada.com/fcs-low-cost.php>.
[12] Personal communications with Larry Silva, Managing Director, MAN Diesel & Turbo Australia—Especially Those Based on an Internal Report (#LDF1-20120014), "A First Plan for the Development of a Stationary Low-Speed Two-Stroke Engine to be Operated on Coal Water Slurries," Issued by MAN Diesel & Turbo on 29 June 2012.
[13] S. Curtin, J. Gangi, Fuel Cell Technologies Market Report 2014, US Department of Energy, <http://energy.gov/sites/prod/files/2015/10/f27/fcto_2014_market_report.pdf> (cited 2.06.16).
[14] The Fuel Cell Industry Review 2015, E4tech, <www.FuelCellIndustryReview.com>, November 2015 (cited 10.06.16).
[15] S. Giddey, S.P.S. Badwal, A. Kulkarni, C. Munnings, A comprehensive review of direct carbon fuel cell technology, Prog. Energy Combust. Sci. 38 (2012) 360–399.
[16] T.M. Gür, Progress in carbon fuel cells for clean coal technology pipeline, Int. J. Energy Res. 40 (2016) 13–29. Available from: http://dx.doi.org/10.1002/er.3288.
[17] L. Deleebeeck, K.K. Hansen, Hybrid direct carbon fuel cells and their reaction mechanisms-a review, J. Solid State Electrochem. 18 (2014) 861–882. Available from: http://dx.doi.org/10.1007/s10008-013-2258-1.
[18] P. Desclaux, M. Rzepka, U. Stimming, R. Hempelmann, Actual State of Technology in Direct Carbon Fuel Cells, Z. Phys. Chem. 227 (2013) 627–649. Available from: http://dx.doi.org/10.1524/zpch.2012.0253.
[19] J.F. Cooper, in: S. Basu (Ed.), Direct Conversion of Coal Derived Carbon in Fuel Cells in Recent Trends in Fuel Cell Science and Technology, Anamaya Publishers, New Delhi, India, 2007, pp. 248–266.
[20] A.C. Rady, S. Giddey, A. Kulkarni, S.P.S. Badwal, S. Bhattacharya, Catalytic gasification of carbon in direct carbon fuel cells, Fuel 180 (2016) 270–277. Available from: http://dx.doi.org/10.1016/j.fuel.2016.04.047.
[21] A. rady, S. Giddey, S.P.S. Badwal, B.P. Ladewig, S. Bhattacharya, Review of fuels for direct carbon fuel cells, Energy Fuels 26 (2012) 1471–1488.
[22] T.M. Gür, Critical review of carbon conversion in carbon fuel cells, Chem. Rev. 113 (2013) 6179–6206.
[23] A.K. Kulkarni, F.T. Ciacchi, S. Giddey, C. Munnings, S.P.S. Badwal, J. Kimpton, et al., Mixed conducting anode for direct carbon fuel cell, Int. J. Hydrog. Energy 37 (2012) 19092–19102.

Further reading

E.J. Carlson, EPRI Report: Assessment of a Novel Direct Coal Conversion Fuel Cell Technology for Electric Utility Markets, EPRI, Palo Alto, 2006.

Part III

Other applications of low rank coal and carbon capture and storage

Coal-to-liquids and polygeneration using low rank coals

10

Zhongyang Luo
Zhejiang University, Hangzhou, PR China

10.1 DCL processes

The reaction process of coal liquefaction can be divided into two steps [1]: the first step is to break the bridge bonds between the structural units of coal by heating, and form a single structural unit as the main body of the free radicals; the second step is to maintain free radicals' stability in the solvent by hydrogenation in the role of the catalyst, so the solvent should have good heavy aromatic solubility, and can provide hydrogen to free radicals to prevent them polymerizing. In addition, hydrogenation can also make the structural units continue to remove oxygen, nitrogen, sulfur and other miscellaneous atoms, and the structural unit of further cracking, the aromatic partially saturated to reduce the relative molecular mass and increase the hydrogen carbon atomic ratio, thus gaining low relative molecular mass oil, very similar to petroleum fractions. Coal liquefaction oil products contain more heteroatoms and aromatics, generally have to go through hydro-refining or hydrocracking process, then can be qualified oil.

DCL (short for direct coal liquefaction) is one of the coal liquefaction technologies and its detailed description is given below.

10.1.1 The principle of DCL processes

The treatment of coal, under high temperature and pressure, done by the DCL process involves a catalytic reaction treated with hydrogen which is a treatment that converts it into liquid fuel, including gasoline and diesel among other types of fuel and into some chemicals. This process [2], under high temperature and pressure atmospheric conditions, breaks down the organic structure of coal into fuel. This does not involve the intermediate gasification step that is normally the prime constituent of this reaction, which is completed with use of a few solvents and/or catalysts. Even though the liquid hydrocarbons consist of a higher hydrogen–carbon molar ratios than coals, other appropriate steps or process are engaged in the DCL process to achieve the required result or fuels such as hydrogenation of carbon-rejection.

The whole liquefaction process can be divided into the following four main units [3]: (1) coal slurry preparation unit: the raw coal is broken to 0.2 mm below, and made to be coal slurry with solvent, and catalyst; (2) reaction unit: carrying out

Low-Rank Coals for Power Generation, Fuel and Chemical Production
DOI: http://dx.doi.org/10.1016/B978-0-08-100895-9.00011-5

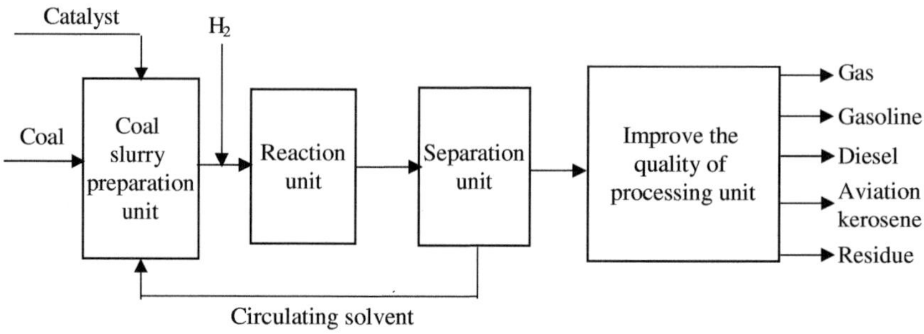

Figure 10.1 Process flow diagram.

catalytic hydrogenation reaction under high temperature and high pressure, and getting liquefied crude products; (3) separation unit: separating the residue, liquefied oil, and gaseous products of the reaction, heavy oil as a circulating solvent for coal slurry; (4) improve the quality of processing unit: processing the liquefied crude oil into gasoline, diesel and aviation kerosene, and other products. The simple DCL process is illustrated in Fig. 10.1.

10.1.2 Typical DCL processes

At present, the world's typical coal direct liquefaction technologies mainly include the German integrated gross oil refining (IGOR$^+$) process, the Japanese NEDOL process, and the US H-COAL and HTI processes.

1. German IGOR$^+$ process:

   IGOR$^+$ process is a new generation of coal liquefaction technology developed on the basis of the original IG process in Germany. The liquefaction process reduced the reaction pressure from 70 MPa to 30 MPa and increased the amount of coal processed by 50%. In addition, in the treatment of residues, a modern distillation method was used instead of a mechanical separation method which was deficient both in terms of environmental protection and technology [4].

   The characteristics of this process [5] are as follows: (1) the reaction conditions are more stringent, the reaction temperature of 470°C and the reaction pressure of 30 MPa; (2) using the aluminum industry waste (red mud) as the catalyst; (3) solid–liquid separation using vacuum distillation method; (4) liquefaction and liquefied oil upgrading process are operated in the same high-pressure system, and can get a very low content of heteroatoms of refined oil liquefaction. The liquefied oil can be distilled to obtain diesel oil which cetane number greater than 45, gasoline fraction then by reforming can be high-octane gasoline [6]; (5) circulating solvent for coal slurry uses hydrogenated oil, good hydrogen performance, and high conversion rate of coal liquefaction.

2. Japanese NEDOL process [3,6]:

   Japan NEDOL process consists of coal pretreatment unit, liquefaction reaction unit, liquefied petroleum distillation unit, solvent hydrogenation unit, and so on, mainly dealing with sub-bituminous coal and low-quality coal. Using the prehydrogenated medium and heavy oil solvent to prepare coal slurry with coal and catalyst, preheating together with the hydrogen at a certain temperature and pressure to make them react and liquefy, then the liquefied crude oil is separated, refined, and modified. After hydrogenation as hydrogen-donating solvents, the middle oils and heavy fractions are recycled by the NEDOL process.

The characteristics of this process [7] are as follows: (1) the reaction temperature is 430–465°C and the reaction pressure is 17–19 MPa; (2) catalyst is synthetic iron sulfide or natural pyrite; (3) solid–liquid separation using vacuum distillation; (4) circulating solvents hydrogenate separately to improve the hydrogen supply ability of the solvent. In the process, the advantages of three Bituminous coal liquefaction methods, i.e., direct hydrogenation, solvent extraction, and solvent decomposition, are summarized.

3. US H-COAL and HTI processes [4,6]:

   The H-COAL process was developed by HRI Company from the fluidized-bed heavy oil hydrocracking process (H-Oil). The process utilizes a fluidized bed catalytic reactor which has the reaction temperature range of 425 to 455°C, having the pressure of 20 MPa where the catalyst is aluminate extruded bar with alumina as the carrier. The inner part of the reactor is provided with a circulation pipe and the bottom part is equipped with a circulation pump. The bed of the granular catalyst maintains boiling state mainly by the upward flow of circulating material conveyed by the circulating pump. The volume of the boiling catalyst bed is larger than the volume of the initial packed catalyst bed by 30%–40%. Applying the appropriate coal, the total conversion of the H-COAL process can exceed 95%.

   The HTI process was modified to a two-stage catalytic liquefaction process (TSCL) based on the H-COAL process, using a suspended bed reactor and HTI's patented iron-based catalyst. The main technical characteristics of the process are as follows: (1) using colloidal highly dispersed nano-scale iron-based catalyst reduces the catalyst cost and increases the activity; (2) using external circulation full back mixed three-phase bubble column reactor enhances the reactor processing ability; (3) on-line hydrocracking of liquefied crude oil further improves the quality of distillates; (4) reaction conditions are mild, temperature is 440–450°C, pressure is 17 MPa, high yield, low hydrogen consumption; (5) solid–liquid separation using supercritical solvent extraction method, the yield of oil was increased by 5%.

4. Russia low pressure hydrogenation liquefaction process [8]:

   This process is a direct hydrogenation liquefaction process developed according to the characteristics of coal in Russia at the 20th century 1970s and 1980s. The process special points: using molybdenum catalyst with high hydrogenation activity, and centrifugal solvent recycle and incineration were used for recovery; the liquefaction reaction gas pressure is low, the liquefaction pressure of lignite hydrogenation is 6.0 MPa, bituminous coal is 10.0 MPa, which is helpful to reduce the investment and operation cost; using instantaneous vortex chamber coal drying technology can increase the ratio of raw coal surface area and pore solvent, reduce coal particle size, which favors hydrogenation and liquefaction reaction; using semi-offline fixed-bed catalytic reactor for hydrogen purification of the liquefied crude oil, easy to operate.

5. China Shenhua coal direct liquefaction process [9,10]:

   Based on the analysis of DCL technologies at home and abroad, Shenhua Group completed the development of Shenhua coal direct liquefaction technology through technical innovation and technical integration in accordance with the needs of its own industry development. At the end of 2007, a million-ton demonstration project was completed. Since the end of 2008 opened up the whole process, until mid-September 2010, the cumulative run time was more than 5000 hours. Continuous operation has exceeded 2900 hours, the average load of 75% or more. The main products are liquefied petroleum gas, naphtha, and diesel. This indicates that China has become the only country in the world to master key technologies for direct liquefaction of coal with a capacity of one million tons. The main process features are as follows: (1) the catalyst is the domestic developed "cheap disposable type" of the "863" synthesis catalyst; (2) using two-stage reactions, according to Ref. [9], the reaction temperature is 455°C and pressure is 19 MPa; (3) coal slurry preparation process taking recycling hydrogen solvent and coal premixing and first-cycle mixing process; (4) coal liquefaction part uses two-stage series of full-return suspended bed reactor technology; (5) solid–liquid separation of the reaction product using vacuum distillation; (6) all recycled hydrogenated solvents and liquefied petroleum products use forced circulation fluidized bed reactor, the catalyst can be daily online replaced, leading to good performance of the hydrogen supply solvent after hydrogenation and stable product properties.

Although a lot of coal liquefaction technologies have been explored and applied to the practical application, there are still defects such as serious wear and tear of critical equipment, harsh operating conditions, high investment, and high oil residue. Therefore, in the new process and its catalyst development, economic and environmental evaluation, it is necessary to do plenty of in-depth and detailed research.

10.2 ICL processes

Indirect coal liquefaction (ICL) consists of two major steps and three main subprocesses, including Fischer–Tropsch (F-T) synthesis, Methanol synthesis, and methanation. First, coal is converted into a syngas. Second, the CO and H_2 in syngas are converted into a range of hydrocarbon fuels/products such as gasoline, diesel, methanol, and chemicals using the abovementioned subprocesses, here the methanol synthesis is used to do the conversion to gasoline and/or petrochemicals [2].

Relying on ICL technology, not only gasoline, diesel, kerosene, and other common petroleum products can be extracted from coal, but also aviation fuel, lubricating oil and other high-quality petroleum products and olefins, paraffin and other high value-added products can be extracted.

10.2.1 Fischer–Tropsch synthesis

F-T process, developed by Franz Fischer and Hans Tropsch, is among the most popular and widely used ICL processes to acquire liquid hydrocarbons from CO and H_2 mix. It is a collection of chemical reactions, developed in 1925 at the Kaiser Wilhelm Institute, Germany and was commercially used to produce liquid fuels such as gasoline and oil in the early 1930s. F-T process is recently been used in the industry in South Africa by Sasol to make liquid fuels [11].

Syngas is produced by gasifying coal in the presence of oxygen and steam and later the syngas is delivered to the F-T reactors after cleaning it of process impurities. Fig. 10.2 shows this process in a simple block flow diagram. The syngas after cleaning is converted to desired liquid fuel consisting of hydrocarbon rich constituents. This rich fuel is used to make ultra clean diesel and in some cases even into jet fuels. Additional refining might be required to achieve the desired result.

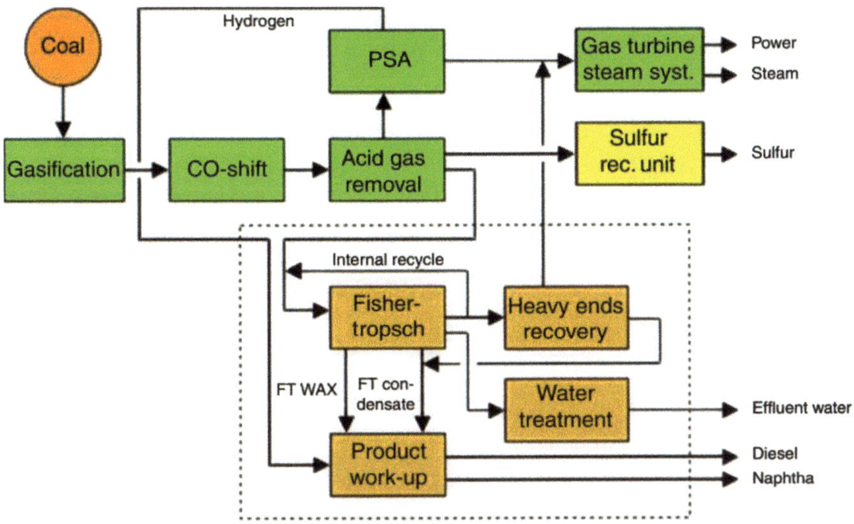

Figure 10.2 Fischer–Tropsch coal to liquids process. From K. Radtke, M. Heinritz-Adrian, C. Marsico, New wave of coal-to-liquids. An opportunity to decrease dependency on oil and gas imports and an appropriate approach to a partial revival of domestic coal industries, VGB PowerTech 86(5) (2006) 78–84, Figure courtesy: Uhde Corporation of America.

Furthermore, some of the syngas, which might not be a part of refinement, is recycled or combusted in a combined cycle power plant to maximize the efficiency of fuel production. Here a gas turbine is used to generate power [12].

1. Reaction:

As mentioned previously in this chapter, the F-T process is constituted by a particular set of reactions, resulting in similar hydrocarbons such as C_nH_{2n+2}. Among these sets of chemical reactions, the important ones are as follows:

$$(2n+1)H_2 + nCO \rightarrow C_nH_{2n+2} + nH_2O$$
$$\Delta \approx -38\ \text{kcal/mol CO}$$

To produce alkanes here, the n is typically 10–20, while methane formation having $n = 1$ is deemed undesired. This might be due to the straight chain production of the fuel which can be categorized as diesel, in case of such kind of alkanes and even similar reactions give alkenes as a product including some other alcohols which might also include a small amount of oxygenated hydrocarbons [13].

Another major reaction, olefin synthesis:

$$nCO + (2n)H_2 \rightarrow C_nH_{2n} + nH_2O$$

Other side effects:
Methane: $CO + 3H_2 \rightarrow CH_4 + H_2O$
Methanol: $CO + 2H_2 \rightarrow CH_3OH$

Ethanol: $2CO + 4H_2 \rightarrow C_2H_5OH + H_2O$
Carbonization reaction: $2CO = C + CO_2$

In addition, there are other side reactions of the formation of higher alcohols, aldehydes, ketones, acids, esters, and other oxygen-containing compounds.

2. Process conditions:

There is a strong effect of temperature of the each particular reaction on the distribution and chemistry of the desired product. The catalyst metal also has similar effect on the abovementioned conditions. In this scenario, the F-T operating temperature should be between 150°C and 300°C. The higher the temperatures the faster are the reactions, and consequently, the conversion rates lean toward the greater side and thus prefer production of methane. Therefore the reaction temperature is balanced somewhere between the middle part of the temperature range, sometimes even lower than that, depending upon results obtained. Another factor, pressure, also plays a similar role but more favorable as it produces longer chained alkanes and also improves the conversion rates. Typical pressures range from one to tens of atmospheres. Furthermore, even though increasing the pressure tenfold might bear better results, the cost and catalyst deactivation during the coke formation leaves it unfavorable to realize gaining this benefit.

Synthetic catalyst is mainly made of Co, Fe, Ni, Ru, and other metals of Group VIII of the periodic table. In order to improve the activity, stability, and selectivity of catalyst, some auxiliary components are added such as metal oxide or salt class. Most catalysts require a support, such as alumina, silica, kaolin, or diatomaceous earth. Synthetic catalyst needs to be activated by reduction of $CO + H_2$ or H_2. At present, the use of the world's more mature indirect liquefaction catalysts is mainly iron- and cobalt-based two major categories, Sasol mainly used iron-based catalyst. In the Sasol fixed bed and slurry bed reactors, precipitated iron catalysts are used. In the fluidized bed reactors, molten iron catalysts are used.

3. Product distribution:

The Anderson–Schulz–Flory distribution [14] gives us an improved hydrocarbons distribution by considering the weight fractions, which can be expressed as:

$$W_n/n = (1-\alpha)^2\alpha^{n-1}$$

W_n of the resultant hydrocarbons by analyzing the carbon atoms, n, contained in them where the chain growth probability, α, can be used to describe the probability of a molecule continuing the reaction in forming a longer chain if α in this case can claim to be observed relating to the catalyst and our required specific process conditions.

10.2.2 Methanol synthesis with subsequent conversion to gasoline or petrochemicals

Methanol or methyl alcohol (CH_3OH) is a common domestic and industrial solvent, which can be produced by reaction of CO with H_2 in the presence of a catalyst, widely used are copper and zinc oxide mixtures, owing to the fact that it is a colorless liquid and can be readily mixed with different organic liquids and even water. Due to its low boiling temperature of 65°C, it is commonly used in methylated spirits domestically. In the industry, many of the common chemicals such as formaldehyde, dimethyl terephthalate, methylamines and methyl halides, methyl methacrylate, acetic acid, and gasoline are used as a raw material.

The copper and zinc oxide constituted catalysts were used in the beginning by ICI in 1966 supplemented by alumina. At 5–10 MPa and 250°C, the reaction is characterized by high selectivity (>99.8%):

$$CO + 2H_2 \rightarrow CH_3OH \quad \Delta H_{298} = -90.84\ kJ/mol$$

During the production of syngas, one mole H_2 is extra and H_2O is a byproduct as methane produces a single mole of hydrogen by each CO mole and only two of the H_2 moles are consumed. Here H_2O can be easily recycled by relying on the water-gas shift reaction and the H_2 mole is dealt with by letting CO_2 into the synthesis reactor. This way, the extra mole is also converted to form methanol by the equation [15]

$$CO_2 + 3H_2 \rightarrow CH_3OH + H_2O \quad \Delta H_{298} = -49.57\ kJ/mol$$

where the H_2O byproduct is recycled via the water–gas shift reaction

$$CO + H_2O \rightarrow CO_2 + H_2$$

This gives an overall reaction, which is the same as listed above:

$$CO + 2H_2 \rightarrow CH_3OH$$

Further on, a zeolite catalyst is used to polymerize the methanol produced by syngas in methanol synthesis process. This zeolite catalytic conversion, developed and used by Mobil in the early 1970s, is known as the moniker MTG, methanol to gasoline, process which is now strongly used and developed in China to make coal-to-chemicals industry. Olefins, MEG, DME, and some aromatics are its products. A demonstration plan by Jincheng Anthracite Mining Group (JAMG) in Shanxi, China is working on these principles.

Based on this methanol synthesis, China has also developed a strong coal-to-chemicals industry, with outputs such as olefins, MEG, DME, and aromatics.

1. MTG:

   Methanol to Gasoline is the first process we will discuss in this section. It has already been mentioned above that with the MTG process, methanol has been converted to gasoline for quite a number of years already, even though it is already been used as a fuel and can be blended easily with gasoline [16]. Methanol synthesis as the first stage in the indirect liquefaction affords 100% selectivity for gasoline, much higher than that obtainable in the F-T synthesis with its wide range of hydrocarbon products.

   Catalysts have been core in methanol conversion and realization on the industry and the recent popularity in China. In the case of Mobil MTG, zeolite ZSM-5 has been used to convert methanol to dimethyl ether (DME) in the first step by dehydrating methanol. Consequently in the same scenario, it is then converted to those hydrocarbons that are in the boiling range of gasoline including LPG and fuel gas. The energy released, which depends on the distribution of the resultant products, for this reaction ranges from 1.5 to 1.7 kJ/g and the complex sequence of this reaction can be represented as given below [17]:

$$2CH_3OH \rightarrow (CH_3)_2O + H_2O$$
$$(CH_3)_2O \rightarrow (CH_2)_2 + H_2O$$

Light olefins → Heavier olefins

Heavy olefins → Aromatics; alkanes; cycloalkanes

To further enhance the MTG process, Mobil has been working to develop a set of plant configurations that will ultimately help it to realize zero waste by recycling into an equilibrium mixture. This is done by two reactors in series where the first, adiabatic fixed-bed, is used for dehydrating methanol as previously mentioned and the other, a fluidized bed process, from which recycled gas containing light hydrocarbons is used to be converted over by using zeolite catalyst. The methanol dehydration is done to an equilibrium mixture of DME and water and is reacted with the recycle gas, which also aids in removing some heat from the second stage of this process. The catalyst in the MTG reactor requires regeneration for about every 20 days.

In 1980, an international project was initiated to construct and operate the fluidized bed MTG process on a 16 m^3/d demonstration scale at Wesseling, Federal Republic of Germany. Participants included Mobil, Uhde, and Union Rheinische Braunkohlen Kraftstoff (URBK).

2. Methanol to petrochemicals:

Methane is also commonly converted to other petrochemicals, such as olefins, by a process known as methanol to olefins or in short, MTO and similarly methanol to propylene commonly known as MTP [18].

a. MTO:

MTO refers to the direct conversion of methanol to lower olefins (ethylene, propylene) technology. MTO was first proposed by Mobil, followed by BASF, Exxon, UOP, and Hydro. Representative MTO process technologies are UOP, UOP/Hydro, Exxon Mobil and DMTO, DMTO-II process (Technology of Dalian Institute of Chemical Physics, China).

MTO reaction mechanism is that first methanol dehydrates to produce DME, then the DME continues to dehydrate and be converted to ethylene, propylene-based low-carbon olefins, a small amount of C_2-C_5. The low-olefins are further reacted by cyclization, dehydrogenation, hydrogen transfer, condensation, and alkylation to produce saturated hydrocarbons, aromatic hydrocarbons, C_6 olefins, and cokes of different molecular weights.

b. MTP:

MTP refers to the process in which methanol is first converted to DME and then DME is directly converted to propylene. The difference from MTO is that the catalyst used has a higher selectivity for propylene. The reactor is mainly in fixed bed form. Tsinghua University had a successful trial of the current fluidized bed methanol to propylene (FMTP). They mainly used Germany Lurgi MTP process and intercooled adiabatic fixed-bed reactor. South Chemical Company provided the special zeolite catalyst.

MTP process usually uses fixed-bed reactor with some features, such as easy to enlarge, few risks, and less investment. And it can maximize the product selectivity.

At present, there are three sets of MTO/MTP demonstration device under construction in China. Shenhua 600,000 t/a methanol-to-olefins plant had a successful trial to produce qualified ethylene and propylene products in Baotou in 2010.

10.2.3 Methanation [19,20]

Methanation is a chemical reaction that converts COx to methane through hydrogenation. The process for converting CO_2 to CH_4 across the 'Sabatier' reaction (10.1) was proposed by Paul Sabatier and J.B. Sendersens in 1902. Because of the up-to-date good availability and low enough price of natural gas, units of large-scale methane production on the basis of carbon dioxide have not been widely established, while just a few plants on the basis of carbon monoxide. With the rapid development of renewable electricity, Power-to-Gas technology as an energy-efficient energy storage solution, methanation is being more and more important and gains a lot of attention.

$$CO_2 + 4H_2 \leftrightarrow CH_4 + 2H_2O \quad \Delta H_{298} = -165\ \mathrm{kJ/mol} \tag{10.1}$$

This reaction thermodynamically favored taking place at relatively low temperatures and high pressures. CO to CH_4 recaton is described by 10.2:

$$CO + 3H_2 \leftrightarrow CH_4 + H_2O \quad \Delta H_{298} = -206\ \mathrm{kJ/mol} \tag{10.2}$$

that is the reverse process of Steam Reforming reaction. In effect, Equation 10.1 can be seen as the sum of Equation 10.2 with the reverse Water Gas Shift reaction, 10.3.:

$$CO_2 + H_2 \leftrightarrow CO + H_2O \quad \Delta H_{298} = 41\ \mathrm{kJ/mol} \tag{10.3}$$

$$CO + 3H_2 \rightarrow CH_4 + H_2O$$

The Great Plains Gasification Plant in the US producing 160 million cubic feet per day of SNG is the only large-scale commercial coal-based SNG plant out of China, while the SNG projects just have been experiencing a slow start in India, South Korea, etc. [21,22].

10.3 Polygeneration of power, liquid, and gas: through pyrolysis

Polygeneration of power, liquid, and gas based on heat carrier pyrolysis is a kind of coal gasification technology which utilizes heat carrier to provide the heat in the pyrolysis process. This technology combines coal pyrolysis, gasification, and combustion together and produces heat, power, and coal gas at the same time. This technology is appropriate for medium and small towns as well as factories without special oxygen generating equipment and wishing to have a lower operating cost [23]. According to the characteristics of the pyrolyser and heat carrier, it mainly includes technology based on coal pyrolysis on fluidized bed or moving bed, and that based on coal pyrolysis by semi-coke heat carrier.

10.3.1 Polygeneration of power, liquid, and gas based on fluidized bed

The main process characteristic of this technology based on fluidized bed is to utilize circulating hot ash or semi-coke of circulating fluidized bed boiler as the heat source for coal pyrolysis and partial gasification. Coal pyrolysis and partial gasification produce coal gas with medium heat value, and then the generated gas is output after the dust removal, and semi-coke, with used circulating ash together, is sent back to the circulating fluidized bed boiler. Superheated steam generated in the process of semi-coke combustion, which can be used for power generation and heat supply. The schematic of this technology is shown in Fig. 10.3.

The polygeneration technology based on fluidized bed has the following characteristics:

1. High thermal efficiency: In the polygeneration system, the heat for coal flash pyrolysis and partial gasification in the pyrolyser at atmospheric pressure comes from circulation ash of circulating fluidized bed. Hot semi-coke generated in the pyrolyser is directly used as a circulating bed boiler fuel, thus greatly improving the fuel efficiency to make the overall thermal efficiency of whole polygeneration system up to over 85%, which is higher than the efficiency of general coal-fired industrial boilers and gasifier.

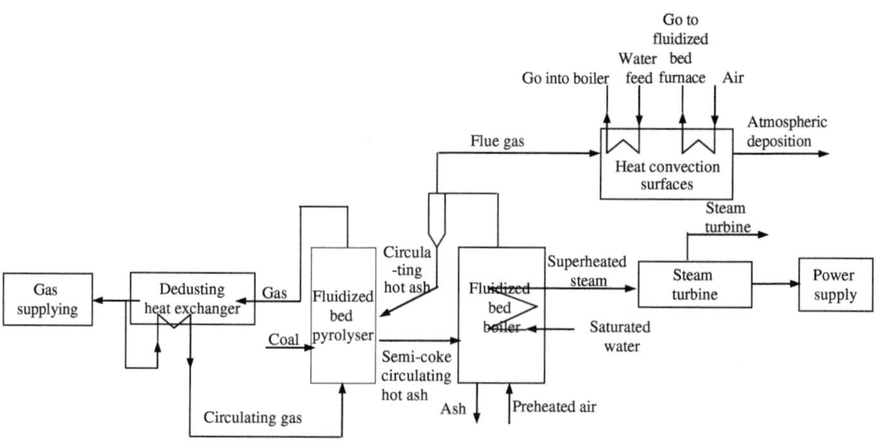

Figure 10.3 Schematic of polygeneration of power, liquid, and gas based on fluidized bed.

2. Low pollutant emissions: Since the boiler and pyrolyser of polygeneration technology employ fluidized bed technology, it has the ability which exists in fluidized bed technology to control SO_x and NO_x emissions in combustion and pyrolysis processes. In addition, the combination of combustion and pyrolysis processes makes the system have a lower pollutant emission than the general single fluidized bed boiler.
3. Wide coal adaptability: The boiler and pyrolyser of the polygeneration system are fluidized bed, and the coal adaptability of fluidized bed combustion is better. This technology is applicable to both cohesive and noncohesive bituminous and lignite with a volatility of greater than 20%.
4. Meet the city gas quality requirements: Because the coal pyrolysis process of polygeneration is dominated by pyrolysis of coal and partly gasification, both the CO content in coal gas and the heat value of gas can meet the quality requirements of the Ministry of Construction on urban gas.

10.3.2 Polygeneration of power, liquid, and gas based on moving bed

The basic principle of polygeneration technology based on moving bed is basically the same as that based on fluidized bed, and the main difference lies in the pyrolyser. The pyrolyser of the polygeneration process based on the moving bed is a moving bed, while that is a fluidized bed.

Fig. 10.4 shows the process of this technology from Institute of Engineering Thermophysics, Chinese Academy of Science. Coal is heated by flue gas in the dryer to remove moisture. After separation in the separator, the dried coal enters the mixer to mix with hot ash, which is from the boiler, and enters the reactor. Circulating fluidized bed boilers can provide a sufficient number of hot ash at 900°C as a solid heat carrier. In the case of good mixing, coal is quickly heated

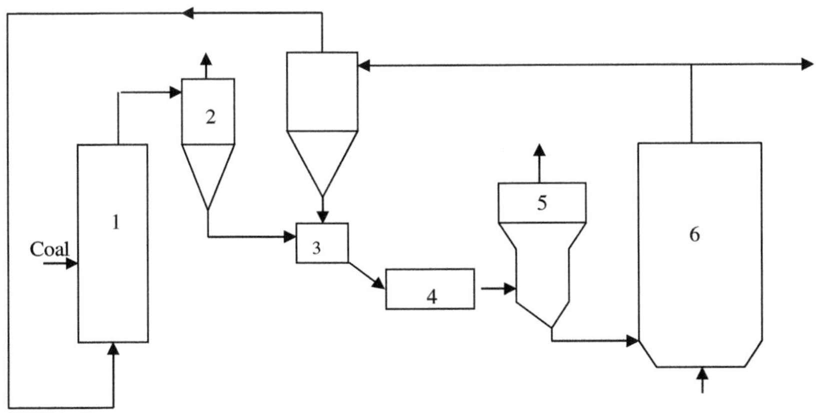

Figure 10.4 Schematic of polygeneration of power, liquid, and gas based on moving bed. 1, coal dryer; 2, separator; 3, mixer; 4, reactor; 5, separator; and 6, circulating fluidized bed boiler.

in the reactor to the desired temperature for pyrolysis to produce coal gas, tar, steam, and semi-coke, and the time and temperature required for pyrolysis vary with coal type. The solids are separated from the gas in separator. The gas after purification can be used as urban gas. The semi-coke and ash mixture is sent back to the circulating fluidized bed boiler to be used as fuel.

This process has the following characteristics:

1. Wide coal adaptability: Bituminous coal, lignite, and other coal with high volatility can be used in this process. There is no special requirement for coal adhesion and coking property.
2. Most of the other existing process cannot use coal powder as raw material, but there are lots of coal powder in China. This process is not subjected to this restriction.
3. High gas quality: Its heat value is higher than that of the other process gas, and CO content is low. Gas quality is in line with national standards.
4. The semi-coke is directly sent back to boiler, so it has less heat loss and high thermal efficiency. It does not need to consider coke selling problems and does not need another waste heat boiler to recover heat.
5. The special structure of mixer and reactor can ensure good mixing. Its size is small because pyrolysis time is only one-tenth of the coke and vertical furnaces.
6. No combustion equipment compared to coke and vertical furnaces. Other auxiliary equipment, such as flue gas processing, deslagging, etc., are also combined with boiler. Equipment of coal storage and preparation is shared with the original boiler which just increase capacity slightly, thus the investment can be greatly reduced.
7. The combination of pyrolyser with circulating fluidized bed boiler as a whole makes the total plant area reduce a lot. Operation and management people are reduced accordingly.
8. Good adjustment, gas quality variable, and fast response: Gas production can be reduced when less gas is used at night. Gas production of coke oven and vertical furnace cannot be adjusted.
9. Can improve the working conditions of the operator because of low pollution emissions.

10.3.3 Polygeneration of power, liquid, and gas based on semi-coke heat carrier

Core of polygeneration technology based on coke heat carrier is setting semi-coke as solid heat carrier, material and heat transport is organized according to the heat required of pyrolysis process through fluidized.

Dalian University of Technology is the first research unit to carry out pyrolysis technology with solid heat carrier for lignite. The process is shown in Fig. 10.5, lignite is sent from hopper to mixer and mixed with hot semi-coke. Due to mixing enhancing and fine coal particles dispersing uniformly, heat exchange between the coal and semi-coke is fast, which result in flash pyrolysis. The mixture of coal and semi-coke is sent from the mixer to reactor where the carbonization reaction is completed and volatile products are extracted. Semi-coke is sent from the reactor to the lower part of the riser to combust partially with air or is heated to promote by hot flue gas. Hot semi-coke goes back to the collection tank and then to mixer so that recycling could be completed.

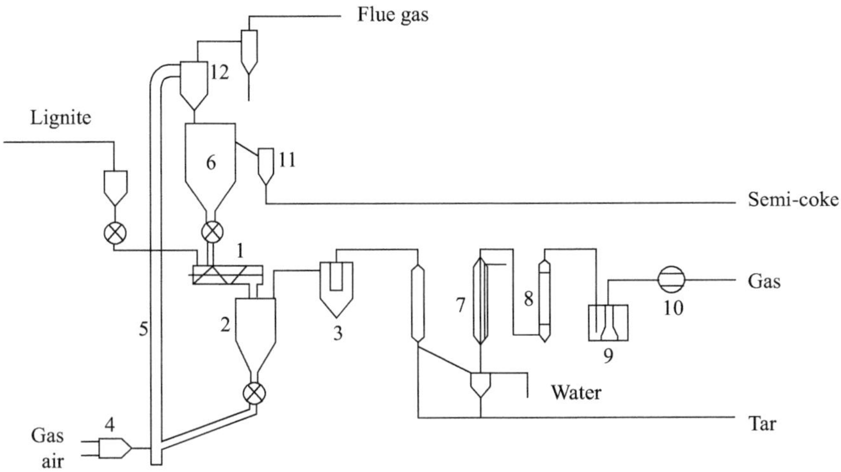

Figure 10.5 Process of new dry distillation method with solid heat carrier for lignite.
1, mixer; 2, reactor; 3, filter; 4, ombustion furnace; 5, riser; 6, assembly tank; 7, cooler; 8, drier; 9, freezer; 10, evacuator; 11, tank; and 12, cyclone.

Carbonized volatile products are cooled and condensed to produce tar and condensate when going through gas cooler and dust separator. Gas goes through drying process to separate water and then extract light oil by cool air at around −30°C. Gas is transported by pump.

This process has the following characteristics:

1. Using solid heat carrier to supply heat for coal pyrolysis can obtain high heating speed, large capacity of unit production, high tar yield, and high gas heating value.
2. The heating and circulation of the hot semi-coke powder is completed in the heating riser with gas–solid two-phase flowing. This process is separated from the coal pyrolyser to prevent the flue gas used to heat semi-coke powder from being mixed with coal gas. As a result, coal gas with high heating value can be obtained.
3. The semi-coke heat carrier and fine particles of raw coal completely mix directly. Due to the large surface area of powder particles, a large number of heat transfer surface are provided so that the heat exchange process can be completed within a few seconds, which make coal pyrolysis rapid to transfer into gas, tar, and semi-coke.
4. Mild reaction conditions, atmospheric production, no pure oxygen required, open and stop flexibly.
5. Wide coal adaptability: Lignite, long flame coal, weakly caking coal, nonstick coal, oil shale, and oil sands can be processed using this technology.
6. Suitable for large-scale with fluidized operation, internal heat, and high capacity of unit equipment.
7. High thermal efficiency, lower energy consumption and cost of the production.
8. The reliable operation of high-temperature dual-screw mixer is the key to the technology.

10.4 Polygeneration of power, liquid, and gas: through partial gasification

10.4.1 Introduction

Coal gasification technology is the main way to achieve the clean and efficient utilization of coal resources. In the 21st century national key technology programs for United States and the "clean coal technology (CCT) plan" of the US Department of Energy, the key technologies for the future of the EC and Japan's future plans focused on coal gasification research and development, resulting in a number of advanced coal gasification technology, such as the United States Texaco, Destec, and KRW, Germany Prenflo and HTW, German and British joint development of BGL/Lurgi, Netherlands Shell and other technologies, as well as the research and development of high-temperature gas purification technology. These coal gasification technologies can effectively improve coal utilization efficiency and reduce SO_x, NO_x, and CO_2 emissions. However, this technology treats the coal as a raw material to convert it to syngas in a single step that requires extremely high temperature, high pressure, and long residence time. Meanwhile, the vast majority of phosphorus, chlorine, alkali metals, and other pollutants transferred from coal into the gas; therefore, the coal gas must be strictly purified under high temperature before using it.

Coal is organic matter mixed with minerals and kinds of impurities, formed from the plant residue after hundreds of millions of years of geological evolution. On the one hand, due to the characteristics of composition, structure, and solid state of coal, the conversion rate of solid particles is slowed down with the degree of transformation in coal gasification process. Therefore, in a single gasification process to obtain high or even complete conversion, the operating conditions must be high temperature, high pressure, and long resident time, which not only increased the technical difficulty, but also increased production costs. On the other hand, according to the intrinsic characteristics of coal itself, for the more active components, volatiles and part of the fixed carbon, pyrolysis reaction can be conducted under relatively mild conditions, and for its nonactivated components, combustion is a violent chemical reaction, thus the rate is much higher than the coal gasification.

Based on the properties of coal, the coal pyrolysis, gasification, and combustion processes are coupled together to convert the more active components of coal into gaseous products, while the nonactive components are used for combustion, reducing the requirements of system's technical conditions and production cost, realizing the grading of coal utilization. At the same time, owing to the generally low reaction temperature, the coal partial gasification technology can combine with the current relatively mature low-temperature gas purification technology, to achieve the effective purification of gas. In addition, most of the sulfur and nitrogen elements in coal are converted into H_2S and NH_3 during the gasification process, which can be removed in the process of gasification or gas purification, and the ash removal is relatively easy. Chlorine and alkali metals and other pollutants in the semi-coke

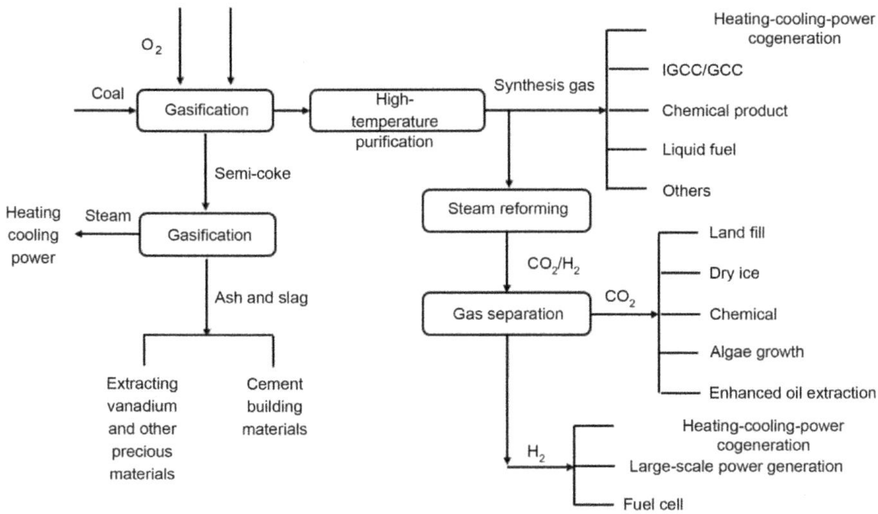

Figure 10.6 Schematic of staged conversion polygeneration based on coal partial gasification. *Modifed from Luo, Z., et al., Technology and engineering example of poly-generation of heat, power and gas from coal. 1st ed. 2004, Beijing: Chemical Industry Press* [25].

greatly reduced compared to the original coal, so the cost of the pollutant removal after combustion is also greatly reduced [24].

Coal partial gasification combustion technology is developed based on the above ideas. The core idea is to achieve different use for different components of the coal. First, in the gasifier, coal partial gasification produces gas, which can be used in many ways, such as gas–steam combined cycle power generation, as fuel gas, or as the raw materials in chemical industry. Then, the semi-coke residue transferred into boiler combusting to product steam, which is used to generate power or heat supply, and the fixed carbon no longer needs mandatory conversion by water vapor gasification and other means. A schematic of staged conversion polygeneration based on coal partial gasification is shown in Fig. 10.6.

In addition to the advantages the traditional coal gasification technology has, it also has the following advantages:

1. The technology is simple and low cost.
2. In the partial gasification process, raw coal containing phosphorus, chlorine, alkali metals, and other pollutants are converted into the gas with volatile mostly. Because of low temperature of the partial gasification technology and the relatively mature low-temperature gas purification technology, the pollutants in gas are easy to remove, which are conducive to environment protection.
3. As the product of coal partial gasification, semi-coke is difficult to burn with low volatile content, so it is convenient for storage and transportation. At the same time, the residue of phosphorus, chlorine, alkali metals, and other pollutants in semi-coke greatly reduced compared to coal, thus burning up is relatively clean.

10.4.2 Development status at home and abroad

After years of development, the current clean coal technologies based on the partial gasification in foreign countries mainly are gasification combined with combustion technology, combined cycle technology, and polygeneration technology. Compared with other advanced technologies (IGGC, etc.), this kind of technology has the advantages of simple system, small investment, and wide coal applicability. It has gained the attention of governments and scholars and has been developed rapidly in recent years.

In the overseas study, the advanced coal-fired power generation technology based on partial gasification is mainly represented by the second-generation pressurized circulating fluidized bed combined cycle (2G-PFBC or APFBC) developed by Foster Wheeler Company [26] and the air blow gasification cycle (ABGC) developed by Babcock Company [27]. In addition, there are Foster Wheeler's high-performance coal-fired power generation system (HIPPS) [28], Japan's self-developed second generation of pressurized fluidized bed combined cycle (APFBC), and pressurized internal circulating fluidized bed combined cycle (PICEB) [29].

2G-PFBC was developed on the basis of the first generation of pressurized circulating fluidized bed combined cycle (1G-PFBC) to increase the inlet temperature of the gas turbine and then increase its efficiency and the gas turbine/steam turbine power ratio. The theoretical net efficiency of the 1G-PFBC system is 39%−41%, while that of the 2G-PFBC power generation system can reach 44%−47% [30]. Schematic of second-generation PFB combined cycle is shown in Fig. 10.7. In 2G-PFBC system, coal is partially gasified in the pressurized gasifier, producing low heat value gas and semi-coke, and the gas combust in gas turbine

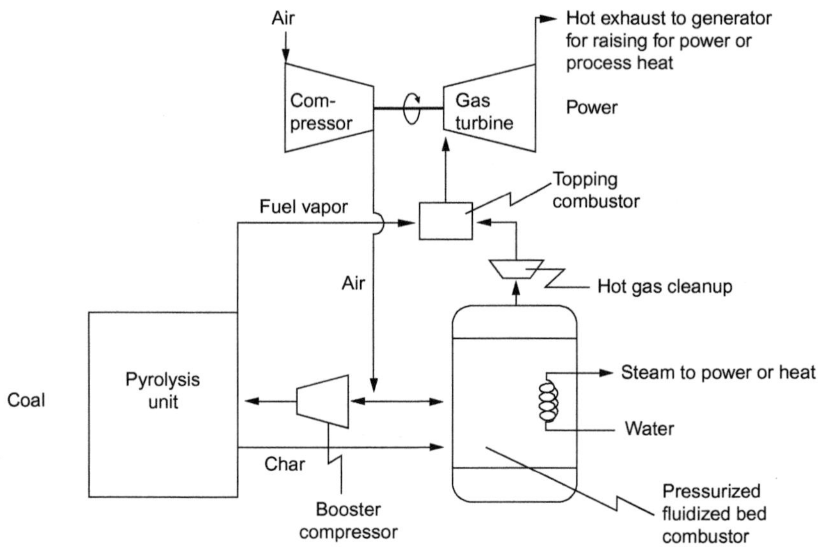

Figure 10.7 Schematic of second-generation PFB combined cycle [31].

precombustion chamber after removing dust in the high-temperature filter, and high temperature gas combusts to drive the gas turbine. The semi-coke not fully gasified in the gasifier combust in the pressurized circulating fluidized bed, part of the flue gas after combustion is used as the oxidant of the gas turbine precombustion chamber, the left transfers use to generate steam through the fluidized bed heat exchanger, driving the steam turbine for power generation. Gas turbine exhaust gas is sent to the waste heat boiler for steam generation, and generated steam flows into the same steam turbine for power generation as well.

The efficiency of such a system is much higher than that of 1G-PFBC, and the whole system is simpler than IGCC system based on complete coal gasification. In addition, by injecting limestone into the gasifier as the adsorbent of H_2S, H_2S can be absorbed to be CaS, which can effectively remove the pollutants in the coal utilization and simplify the system equipment. The technical difficulty of the system is that the gas produced by the gasifier and the part of the flue gas produced by the pressurized circulating fluidized bed combustion need to be cleaned in high temperature before entering the combustion chamber of the gas turbine. Therefore, the designed operating temperature of ceramic filter is so high (970°C or so) that is easy to cause blockage.

Apart from the 2G-PFBC, the coal-fired HIPPS is one of the US Department of Energy's "2000 Combustion Program" [28]. Coal is partially gasified in the gasifier, produced semi-coke burns out in the pulverized coal combustion furnace to heat the compressed air and generate steam after fine grinding. The cleaned gas enters the gas turbine and burns out using heated compressed air. The steam generated by the pulverized coal boiler and the waste heat boiler is supplied to the steam cycle power generation equipment. The efficiency of this program is relatively high, but the technical difficulty lies in material problems of the high temperature air heater.

Different from the 2G-PFBC technology, the ABGC technology of the Mitsui Babcock gets its semi-coke produced by partial gasification sent into the atmospheric pressure circulating fluidized bed boiler for combustion, and the generated steam drives the steam turbine to generate power. Because the heat of semi-coke combustion is only for the production of steam, the gas turbine/steam turbine power output is low, resulting in its overall system efficiency being less than 2G-PFBC and HIPPS technology. However, the gas produced by the gasification in the system is cooled to 400–600°C and fed to the gas turbine combustor, and the gas generated from the combustion of the semi-coke is not fed into the gas turbine combustor. Thus, the whole system requires only one high temperature dust removal equipment, the operating temperature is 400–600°C, which is far below compared to the 2G-PFBC [27]. In general, the filter device is of longer life, technically simple, and more feasible to achieve industrialization. A schematic of ABGC system is shown in Fig. 10.8.

Japan develops the PFBC power station technology and more advanced partial gasification combined cycle power generation technology quickly in recent years [29].

Through the introduction of foreign technology and the combination of development by itself, Japan has accumulated a lot of experience. At the same time, it also

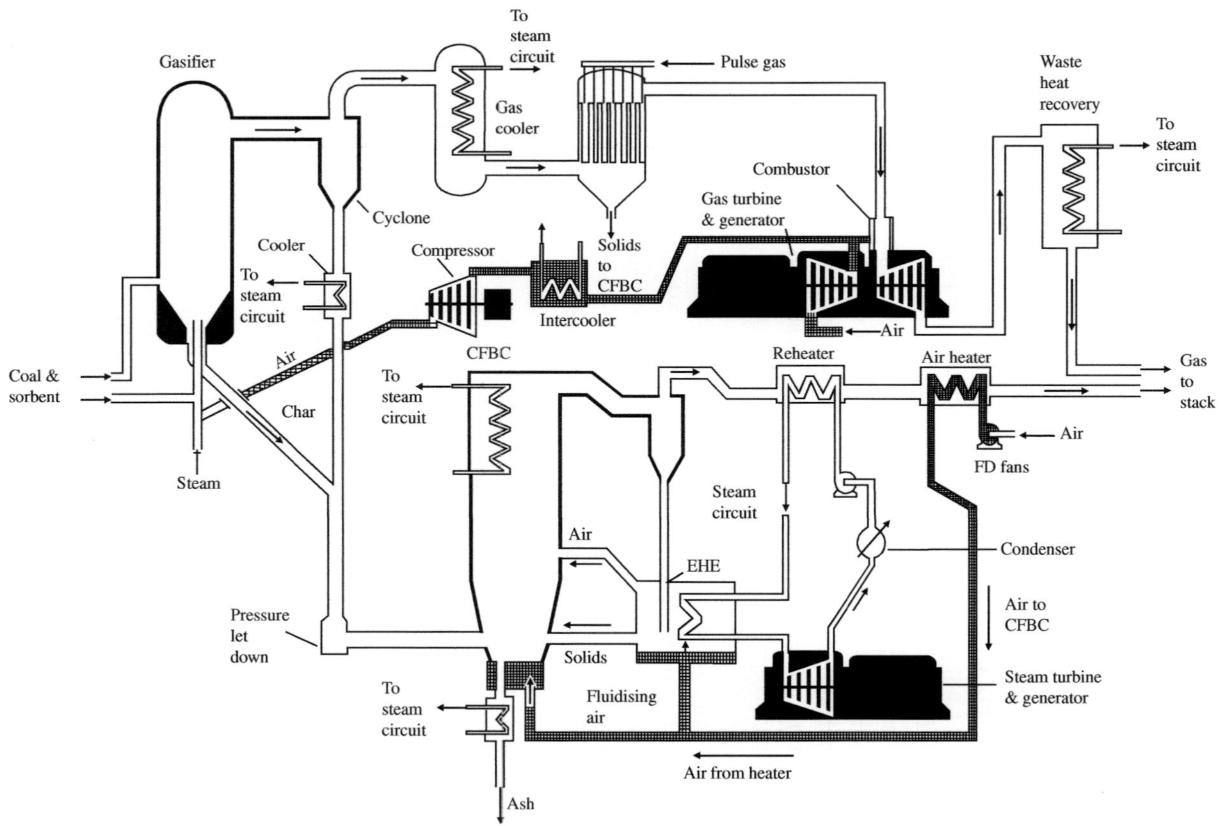

Figure 10.8 Schematic of ABGC system [27].

made significant progress in the whole high temperature ceramic filter technology, reheat steam cycle PFBC boiler, power plant efficiency, and environmental indicators. Development of advanced coal-fired power generation technology based on the partial gasification has made rapid progress, it exploited the advanced pressurized circulating fluidized bed combined cycle system (APFBC) test device, as well as the pressurized internal cycle fluidized bed gasification (PICFG) technology.

Comparison of several foreign coal-fired power generation technology is shown in Table 10.1.

In the domestic research, due to the present situation of "rich coal, lean oil, and less gas," the heat and power co-generation technology based on coal partial gasification is very important in our country [33]. Compared to the United States and other developed countries, China's research in integrated utilization technology based on partial gasification started late. Under the support of the National Basic Research and Development Program "Fundamental Research on Coal Pyrolysis, Gasification and High Temperature Purification Process," the main research units of Zhejiang University, Institute of Coal Chemistry, Chinese Academy of Sciences, and Southeast University are, respectively, responsible for research and development of atmospheric gasification and combustion, pressurized gasification and atmospheric pressure combustion, and pressurized gasification and combustion integration utilization technology.

At present, the Institute of Thermal Engineering, Zhejiang University, put forward the atmospheric pressure coal partial gasification combined cycle power generation technology, designed and constructed a 1 MWe coal pyrolysis, gasification, and combustion staged conversion test facility [34]. The technology is coupled with coal gasification technology, atmospheric fluidized bed combustion technology, and gas–steam combined cycle technology. On the basis of many verification tests of polygeneration technology at 1 MW test facility, in cooperation with the Huainan Mining (Group) Co., Ltd., a 12 MW demonstration project designed by Zhejiang University had been erected and operated successfully in August 2007. Many cold and hot tests had been conducted on this system. Results show that it could maintain stable and continuous operation and produced not only high yield tar and high quality gas, but also heat and power [35]. A schematic diagram of the 12 MW gas, tar, and power polygeneration pilot plant is shown in Fig. 10.9.

The Thermoenergy Engineering Research Institute, Southeast University and Institute of Coal Chemistry, Chinese Academy of Sciences, and other units basing on foreign advanced power generation system technology, combined with China's coal resources and power generation device conditions, have also proposed some of the coal partial gasification combustion power generation technology programs [36]. At present, the integrated verification system of pressurized coal gasification and combustion has been constructed and some pressurized gasification tests have been completed, having got a lot of results through the design and operation of the 1 MWe grade PFBC test device and Jiawang 15 MW PFBC pilot power plant. Based on the successful building of the first 15 MW 1G PFBC-CC pilot power plant in Jiawang power plant in China, Thermo-energy Engineering Research Institute of Southeast University constructed a 2G-PFBC pilot-scale experimental

Table 10.1 Comparison of several foreign coal-fired power generation technology [32]

Items	IGCC	2G-PFBC	ABGC	HIPPS
Gasification method	Spout fluid bed, complete gasification	PFB, partial gasification	PFB, partial gasification	PFB, partial gasification
Agents	Oxygen	Air	Air	Air
Gasifier structure	Complex	Simple	Simple	Simple
Semi-coke combustion	None	PFBB	AFBB	Powdered coal combustion chambe
Coal pretreatment	Pulverize/pulp, complex	Crush, relatively simple	Crush, relatively simple	Crush, relatively simple
Gas purifying	Complex:dedust/desulfuration	Relatively simple: dedust/purification	Relatively simple: dedust/purification	Relatively simple: dedust/purification
Flue gas purifying	None	Relatively complex:high-temperature purification	Simple:dedust	Relatively complex: dedust/desulfuration
Air heater	None	Flue gas of PFB	None	Medium/high temperature heater
Gas turbine/steam turbine power ratio	High	Relatively high	Low	Relatively high
System composition	Complex	Relatively simple	Simple	Relatively simple
Cycle thermal efficiency	Relatively high	High	Relatively high	High
Power consumption rate	High	Relatively high	Low	Relatively low
Loop net efficiency	Relatively high	High	Relatively high	High

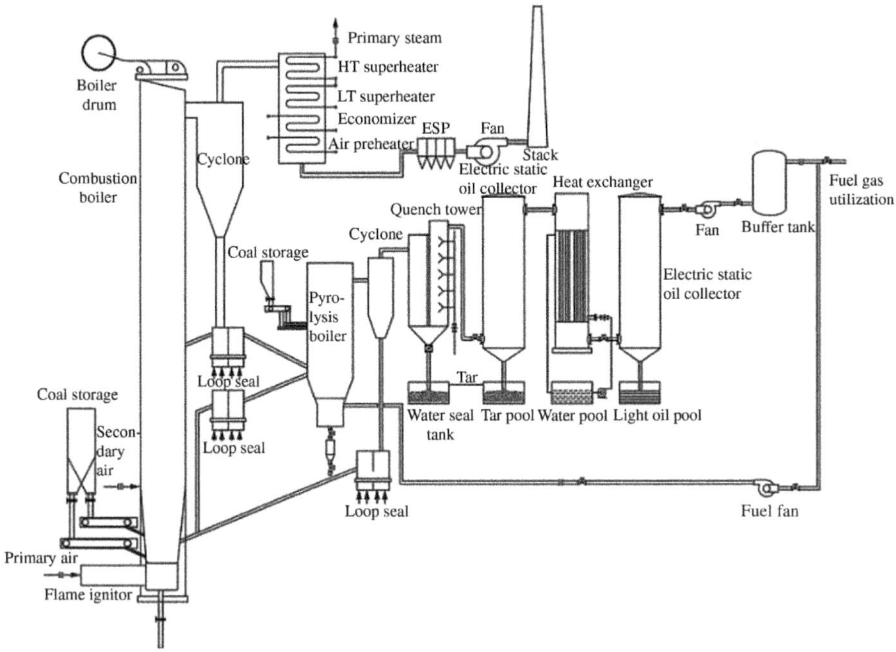

Figure 10.9 The 12 MW gas, tar, and power polygeneration pilot plant [35].

system, including a 2MWth pressurized spout-fluid bed partial gasifier. A long-time-run performance test was carried out to study the different operating parameter effects on coal partial gasification. The results showed the feasibility of gasifier to provide suitable fuel gas and residual char for downstream system of 2G PFBC-CC [37]. The schematic of 2 MW pressurized spout-fluid bed partial gasifier system is shown in Fig. 10.10.

10.5 Polygeneration of power, liquid, and gas: through gasification

Polygeneration system based on coal integrated gasification is the focused research in coal use fields. Many countries have formulated the related research and development plan, and invested a lot of money for the research. In "Vision 21" plan, which on behalf of the American's important research direction in the field of energy production in the 21st century, polygeneration system based on coal integrated gasification is one of the focuses. At present the plan has carried out a lot of research work, and even some primary demonstration plant is under construction.

Many research institutions in China are also studying on the polygeneration system based on coal integrated gasification positively. The Chinese Academy of

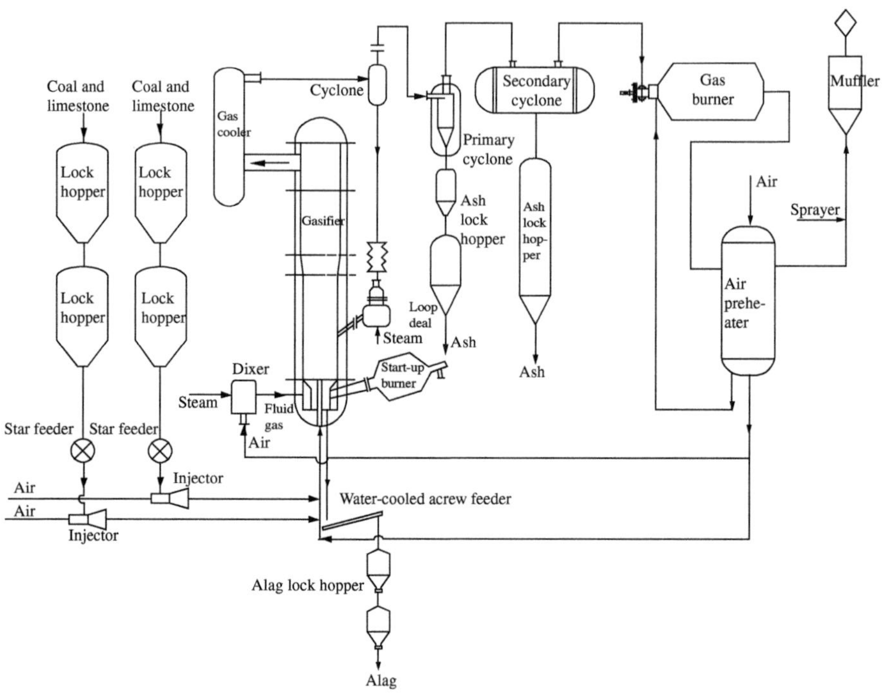

Figure 10.10 Schematic of 2 MW pressurized spout-fluid bed partial gasifier system. [37].

Engineering Thermal Physics is researching the all-round system of IGCC, and building a polygeneration demonstration project cooperating with Yankuang Group. Tsinghua University, Zhejiang University, and other research institutions also have built integrated gasification polygeneration systems which has its own characteristics.

Polygeneration of power, liquid, and gas system with the core of coal gasification is an organic coupling of a lot of technology module. From the feeding of coal and other resources to the output of steam, power, fuel, and other products, there involves many processes such as coal gasification, syngas purification, pollution controlling, and syngas conversion. In the polygeneration system, there are some key technologies such as fuel gasification, gas cleaning, pollution controlling, gas turbine, syngas conversion, gas separation technologies, and the performance of these subunits has extremely important impact on the polygeneration system.

10.5.1 The syngas convert into clean fuel and high quality chemical products

It is extremely important for polygeneration factory to use syngas to produce clean fuel and chemical products rather than generate power. The key link of the polygeneration technology to obtain high benefit is that the syngas can be used to produce many high value-added chemical products or fuels.

At present, although a lot of use of syngas as raw material production process has been applied, such as F-T synthesis process, methanol and downstream products production, DME synthesis, and some technologies have been mature, these technologies often purse high conversion efficiency in their own production process, which would make the process of the production more complicate. It is not consistent with the goal of polygeneration technology. Therefore, one of the key links in the development of polygeneration technology is to research and develop the technology of synthetic gas into clean fuel and high quality chemical products.

10.5.2 The development of gas turbine technology [38]

Combined cycle power generation module is an indispensable unit in the polygeneration system with the core of coal gasification, and this unit uses the tail gas from chemical synthesis process and part of the synthesis gas as fuel to produce power. In some polygeneration systems, the unit also has the effect of supplying steam to the gasifier. As the core equipment of combined cycle power generation module, the performance of large gas turbine directly determines the overall performance of the module, and then takes effects on the whole polygeneration system.

As the third generation power machine with the most market application prospect in the 21st century, gas turbine is the core technology of power system. Gas turbine technology is the integration of a variety of high-tech technology, and the level of technology has become an important indicator of overall national strength, such as national science and technology level, military strength.

Although the gas turbine technology has been widely used, and its technical level has been allowed to apply to polygeneration systems, in order to obtain higher performance, research and development of gas turbine is still very active currently. Among the many gas turbine development projects, the most representative development projects are the Advanced Turbine Power System (ATS) and IHPTET, the Advanced Gas Turbine Cooperation Program (CAGT) jointly implemented by the United States and Europe, the European EC-ATS program, Japan's "New Daylight Program" and "coal gasification combined cycle power system" program, etc.

From the point of current development projects and new products, gas turbine technology has undergone development in several major areas: new energy systems and systems design technology, new cooling technology, new materials and thermal insulation coating, aerodynamic thermodynamics design technology, and low emission combustion technology.

10.5.3 Fuel cell technology [39]

Fuel cell uses fuel (natural gas, syngas, hydrogen, methanol, etc.) as the anode reactant, and uses oxygen in the air as cathode reactant, so the anode is also known as the fuel electrode, the cathode is also known as the air electrode. It uses electrolytes (acids, alkalis, solid oxides, etc.) to separate anode and cathode, and charged ions was circulated through the electrolyte and potential difference was produced, causing the flow of electrons in the external circuit and the formation of low-voltage direct current, and producing water and carbon dioxide simultaneously. If the battery has continuously obtained fuel and oxidant from outside, it can continuously generate power. Compared with the conventional combustion power generation, the fuel cell has the following advantages: (1) high power generation efficiency; (2) less environmental pollution; (3) fast construction and wide application; (4) good load response characteristics.

In the polygeneration system, the high purity hydrogen can be obtained from the syngas through transformation reaction and gas separation. The obtained hydrogen can be used as the raw material of the fuel cell for power generation, and can greatly improve the energy conversion efficiency of the system. At present, the goal of obtaining more than 60% energy conversion efficiency in polygeneration system is based on the integrated system of integrated fuel cell power generation and gas turbine power generation. It can be seen that the fuel cell power generation technology suitable for polygeneration system is one of the key technologies of polygeneration system.

10.5.4 High temperature gas dust removal and desulfurization technology [40,41]

The main pollutants in syngas are solid particles and inorganic sulfide. In addition, there are some but harmful pollutants that cannot be ignored, such as organic sulfur, NH_3, HCl, coal tar, and alkali metal compounds. These substances not only cause downstream equipment damage and affect the normal operation of the process, but also cause serious environmental pollution. Therefore, coal gas purification is one of the key technologies of coal gasification. The two main parts of coal gas purification are the dust removal and desulfurization.

At present, the equipment that can effectively remove dust particles in high temperature gas are mainly high-performance mechanical dust collector (such as cyclone), electrostatic precipitator, acoustic agglomeration dust collector placed before the mechanical dust collector, and high-performance barrier filter. The operating cost of the cyclone is the lowest, but it is generally used as a predust removal equipment. Dust removal efficiency of electrostatic precipitator can reach 90%–99.6%, but there are still some problems with corona stability, material stability, and thermal expansion. Acoustic agglomeration dust collector can be used to reunite dust in front of the cyclone to increase the dust removal capacity of cyclone, but it has not been applied to industrial experiments. High-performance barrier filter can be divided into ceramic filter, particle bed filter, and metal felt filter. The dust removal efficiency of particle bed filter is more than 99%, but the bed is easy to jam when it is running at high temperature. The metal felt filter is not suitable for dust removal under the condition of high temperature gas. The dust removal efficiency of ceramic filter is over 99%, and it can remove particles above 5 μm.

Gas desulfurization can be carried out using wet and dry methods. Wet desulfurization technology is more mature, but it needs to be carried out at room temperature. It is bound to waste part of the sensible heat after gas heat exchange, and the dry desulfurization has the following advantages compared to conventional wet desulfurization:

1. Recovers the sensible heat so that improve system efficiency.
2. Does not need to remove water vapor and CO in the hot gas.
3. Saves the gas heat exchanger, reduces the plant equipment investment, and reduces the production cost.
4. Makes sulfur recovery flexible, and adjusts the production of sulfur or sulfuric acid according to the market supply and demand situation.
5. Tar and other impurities in the gas are to block the system due to condensation.

Therefore, high-temperature dry gas desulfurization technology is one of the most effective purification technologies.

10.5.5 Gas separation technology

There are two processes which comprise the gas separation technology in the polygeneration system: (1) separate the production of oxygen from the air for the coal gasification; (2) separate the gas from the syngas to produce hydrogen for the fuel cell and separate CO_2 for use or disposal.

Traditional cryogenic air separation technology for oxygen making has been very mature, but its high energy consumption makes the efficiency of the polygeneration system or power generation system reduce greatly. Ion transport membrane (ITM) oxygen separation device is one of the promising applications in the non-cryogenic air separation technology. This technology can be combined with the production process in the polygeneration system, such as gas turbine power generation, to separate the high purity oxygen from the high temperature air, and the energy consumption is very low.

The separation of hydrogen from syngas is the key process to realize the fuel cell power generation. At present, there are many hydrogen separation technologies, such as porous membrane separation, separation of palladium membrane, methanation, nitrogen washing, and so on, the cost and the purity of hydrogen need to be further improved.

10.5.6 System integration theory and tools [42]

Polygeneration system is composed of multiple production processes and there is no fixed pattern. The design and construction of each system includes a fuel composition, product, system structure, control of pollutants, the scale, and so on, which are based on the local actual situation, market demand, resources, and economic conditions. The plan of polygeneration system is to meet the needs of the product, as far as possible to improve energy efficiency, minimize the discharge of pollutants. The rigorous system analysis in the early work and the optimization of the system integration in the plant construction are required. The process of system analysis includes market analysis, economic analysis, energy plant process definition, process definition, and so on. Through the process of system analysis, we can decide the choice of the technology module and subsystem of the whole energy plant. Polygeneration system contains a lot of advanced technology modules, so that the system integration is relatively more complex. So the system integration is the crucial first step to ensure the system to achieve high efficiency, low pollution, and high benefits.

References

[1] H.H. Lowry, M.A. Elliott, Chemistry of Coal Utilization, John Wiley & Sons Inc, New York, 1981.

[2] Coal Liquefaction, <https://en.wikipedia.org/wiki/Coal_liquefaction> (11.11.16).

[3] G. Shu, Z. Xu, The technology of direct coal liquefaction, China Coal 23 (10) (1997) 21–23.

[4] S. Xu, Coal Chemical Technology, 1st ed., Dalian University of Technology Press, Dalian, 2016.

[5] Technology Status Report—Coal Liquefaction, <https://www.netl.doe.gov/File%20Library/Research/Coal/energy%20systems/gasification/technology-status-coal-liquefaction.pdf> (16.08.16).

[6] H. Li, Technology progress and prospect analysis of direct coal liquefaction process, Coal Chem. Ind. 30 (3) (2002) 8–12.

[7] J. Xu, et al., Advances in the direct coal liquefaction, Chem. Ind. Eng. Prog. 31 (S1) (2012) 119–123.

[8] J. Li, D. Wang, The discussion of coal direct liquefaction process, Clean Coal Technol. 11 (3) (2005) 49–52.

[9] X. Li, et al., Progress and Development Direction of Shenhua Coal Direct Liquefaction Demonstration Project, Coal Chemical Industry, 43(4), 2015, 12–15.

[10] Zhang, Y., et al., Process for Direct Coal Liquefaction, US7763167, 2010.

[11] J. Bae, et al., Economic evaluations of direct, indirect and hybrid coal liquefaction, Korean J. Chem. Eng. 29 (7) (2012) 868–875.

[12] R. Malhotra, Fossil Energy: Selected Entries from the Encyclopedia of Sustainability Science and Technology, Springer, New York, 2012.

[13] T. Kaneko, F. Drebyshire, E. Makino, et al., Coal liquefaction, in: Ullmann's Encyclopedia of Industrial Chemistry, Wiley-VCH Verlag GmbH & Co. KGaA, Weinheim, 2012, pp. 311–389.

[14] Spath, P.L. and D.C. Dayton, Preliminary Screening—Technical and Economic Assessment of Synthesis Gas to Fuels and Chemicals With Emphasis on the Potential for Biomass-Derived Syngas, 2003.

[15] O. Deutschmann, et al., Heterogeneous Catalysis and Solid Catalysts, 3. Industrial Applications. In: Ullmann's Encyclopedia of Industrial Chemistry, Wiley-VCH Verlag GmbH & Co. KGaA; Weinheim, 2000.

[16] J.H. Edwards, N.R. Foster, The potential for methanol production from natural gas by direct catalytic partial oxidation, Fuel Sci. Technol. Int. 4 (4) (1986) 365–390.

[17] J.C.W. Kuo, Gasification and Indirect Liquefaction, Springer, USA, 1984, pp. 163–230.

[18] Y. Cao, The development of methanol to gasoline or olefines, Coal Chem. Ind. 04 (2010) 25–27.

[19] Methanation, <https://en.wikipedia.org/wiki/Methanation> (11.11.16).

[20] Methanation process, <http://www.helmeth.eu/index.php/technologies/methanation-process> (11.11.16).

[21] C. Yang, et al., China's synthetic natural gas revolution, Nature Climate Change 3, 2013, 852–854.

[22] Y. Han, et al., Should China continue developing the coal-based synthetic natural gas? Energy Sources Part B: Econ. Plan. Policy (2017) 1–8.

[23] X. Qingxiu, et al., Circulating fluidized bed gas–steam coproduction furnace, Furnace Technol. 08 (1994) 11–15.

[24] Y. Li, Basic Research of Part Pyrolysis and Gasification Technology Based on Coal (Doctoral dissertation), Zhejiang University, Hangzhou, 2015, p. 172.

[25] Z. Luo, et al., Technology and Engineering Example of Poly-generation of Heat, Power and Gas from Coal, 1st ed., Chemical Industry Press, Beijing, 2004.

[26] J.M. Wheeldon, Commissioning of the circulating PFBC in the Foster Wheeler advanced PFBC train at the PSDF, Fuel Energy Abstr. 43 (4) (2002). p. 260.

[27] N. Paterson, Fuel behaviour studies in the air-blown gasification cycle, Fuel, 76 (13) (1997) 1319–1325.

[28] Robertson, A., Development of Foster Wheeler's Vision 21 Partial Gasification Module, Office of Scientific & Technical Information Technical Reports, 2001.

[29] H. Sasatsu, et al., CTF development at Wakamatsu 71 MWe PFBC combined cycle power plant, J. Jpn. Inst. Energy 81 (1) (2002) 25–37.

[30] A. Robertson, Second-generation PFB plant performance with W501G gas turbine, Fuel Energy Abstr. 43 (4) (2002), 282.

[31] J.M. Beér, High efficiency electric power generation: the environmental role, Prog. Energy Combust. Sci. 33 (2) (2007) 107–134.

[32] R.R. Mckinsey, An assessment of advanced pressurized fluidized-bed combustion power plants, Fuel Energy Abstr. 3 (1995) 160.

[33] W. Chen, R. Xu, Clean coal technology development in China, Energy Policy 38 (5) (2010) 2123–2130.

[34] J. Wang, System Integration Research on Coal Partial Gasification and Semicoke Combustion, Zhejiang University, Hangzhou, 2008.

[35] M. Ni, et al., Research on coal staged conversion poly-generation system based on fluidized bed, Int. J. Coal Sci. Technol. 1 (1) (2014) 39–45.

[36] N. Cai, et al., Thermal performance study for the coal-fired combined cycle with partial gasification and fluidized bed combustion, Proc. Inst. Mech. Eng. Part A J. Power Energy 215 (4) (2001) 421–427.

[37] X. Rui, et al., Coal gasification characteristics in a 2 MWth second-generation PFB gasifier, 18th International Conference on Fluidized Bed Combustion, The American Society of Mechanical Engineers, Toronto, 2005.

[38] P.C. Chen, H.M. Chiu, Y.P. Chyou, in: P. Varbanov, et al. (Eds.), Effect of Different Gas Turbine on Integrated Gasification Poly-Generation Plant with Methanol and Power Generation, in Chemical Engineering Transactions, Aidic Servizi Srl, Milano, 2013, pp. 271–276.

[39] P. Margalef, et al., Efficiency of poly-generating high temperature fuel cells, J. Power Sources 196 (4SI) (2011) 2055–2060.

[40] J. Yu, et al., Correlation of H_2S and COS in the hot coal gas stream and its importance for high temperature desulfurization, Korean J. Chem. Eng. 28 (4) (2011) 1054–1057.

[41] S.D. Sharma, et al., A critical review of syngas cleaning technologies—fundamental limitations and practical problems, Powder Technol. 180 (1–2) (2008) 115–121.

[42] A. Buttler, C. Kunze, H. Spliethoff, IGCC-EPI: decentralized concept of a highly load-flexible IGCC power plant for excess power integration, Appl. Energy 104 (2013) 869–879.

Coke production from low rank coals

11

Herbert Richter[2], *Jozo Mamić*[1], *Mohammed Morsy*[1], *Klaus Jähnig*[3], *Christian Trautmann*[1] *and André Werner*[1]
[1]ARCUS Technologie GmbH & Co. KG, Cottbus, Germany, [2]International Gasunion (IGU), IL, United States, [3]DBI-EWI GmbH, Freiberg, Germany

List of abbreviations

Abbreviations	Description
adb	air dried basis
arb	as received basis
ARCUS	ARCUS Technologie GmbH & Co. GTL Projek KG, Germany
BHT(-Koks)	Ger.: Braunkohlenhochtemperatur(-koks), Engl.: lignite high-temperature (coke)
CEA	Coal Energy Australia Ltd.
CFD	Computational fluid dynamics
GDR	German Democratic Republic
LEAG	Lausitz Energie Bergbau AG (Lausitz Energy Mining Corporation)
VEB	Ger.: Volkseigener Betrieb, Engl.: Publicly Owned Operation
wf	water free
wt%	weight percent

11.1 Introduction

Coal pyrolysis has been considered an attractive means of using coal, especially for low rank coal/lignite and with the beginning of the worldwide oil crisis of the 1970s, researchers started again significant effort on studying coal pyrolysis due to the importance in the gasification and combustion of coal. Increasing the efficiency of the technological conversion methods and better coal utilization are normally achieved through a more complete understanding of the initial pyrolysis step including the rate and extent of thermal decomposition of coal particles.

The serious environmental problems, which accompany energy production from coal, resulted in a paradigm shift for coal exclusive usage for energy production. Coal's consideration as a potential source of energy will diminish with time as a result of the advances in renewable energy production technologies.

Cokes that are produced from lignite on its pyrolysis are considered to be one of the important raw materials in steel industry and can act as an adsorbing agent for the purification processes in different industries. Aliphatic and aromatic hydrocarbons

Low-Rank Coals for Power Generation, Fuel and Chemical Production
DOI: http://dx.doi.org/10.1016/B978-0-08-100895-9.00012-7

beside hydrogen gas are also used in many areas of the chemical industry and their production from coal using the pyrolysis process does not produce the same amount of greenhouse gases and is a lot less polluting if compared with the direct burning of coal to produce energy. As a result, the production of cokes and valuable gases from the pyrolysis process is a very attractive way of using coal, particularly when considering lignite with a high volatile content and a low calorific value.

11.2 Characteristic of lignite

Coal consists of a variety of substances with different chemical structures. The coal composition depends on the plant-based origin of the raw material and the coalification process. Hence, lignite can have variety of different properties. Nevertheless, the knowledge about the properties of lignite is essential to determine the behavior during refining processes. The most important parameters to characterize coal are the moisture, ash content, volatiles content, and the elemental composition of carbon. The analysis can be carried out in the so-called proximate and ultimate analysis. The proximate analysis test includes typically:

- Ash content
- Calorific value
- Sulfur content
- Moisture
- Volatile matter
- Fixed carbon

In addition, the ultimate analysis includes the elemental composition of the coal. For example, Table 11.1 shows the elemental composition of raw lignite for three mining areas in East Germany.

The moisture content of raw lignite is usually around 50% but can vary greatly depending on the location. The total water content, for instance, in Reichwalde is about 56.2%, while in Schleenhain it is about 51%.

11.3 Pyrolysis of low rank coals

Pyrolysis or devolatilization refers to the destructive distillation of low rank coals/lignite. Volatiles are the primary gaseous decomposition products, some of which

Table 11.1 Ultimate analysis of some lignite areas in East Germany

Open-cast mining	C (wt% wf.)	H (wt% wf.)	N (wt% wf.)	O (wt% wf.)	S (wt% wf.)
Welzow-Süd	60.07	4.36	0.74	23.36	1.48
Reichwalde	57.71	4.55	0.75	18.93	2.79
Schleenhain	62.94	5.63	0.59	13.39	3.16

will be liquids or solids at ambient temperature and pressure. Coke is the residual solid matter, which may or may not yield further volatiles if the heating process is continued. Should the coke be heated in the presence of air, all of the combustible material will be consumed, leaving a residual mineral matter called ash.

Even under the restrictions of small particles, the pyrolysis of lignite is a very complex process. This is partly due to the fact that coal is not a homogeneous material; different portions (on both microscopic and macroscopic scales) of a single coal sample exhibit widely differing chemical compositions and physical properties. In addition, samples from different portions of a mine will not be identical.

Furthermore, pyrolysis of coal is generally a complicated process as there is a broad range of coal ranks (bituminous, lignite, etc.), each of which decomposes in a slightly different manner. Thus, it is possible to characterize only the "average" pyrolysis of coal particles of a given rank in terms of the products produced as a function of the experimental variables. Variables known to be important to the pyrolysis process are pulverized coal rank, composition, and particle size. Also, the thermal history of the particles including heating rate, final temperature, duration of heating, and the cooling process are of great influence. In addition, the composition and pressure of the gas used for the pyrolysis reactor inertization play a significant role.

11.4 Technical solutions for the refining of nonbaking lignite in Central Germany and Lusatia region

11.4.1 The Lurgi low-temperature carbonization furnace process

In the following sections the history and the development of producing coke from lignite in the Lusatia region in Germany are described. In the 1930s large-scale low-temperature carbonization plants were built mainly in Central Germany and the Sudetenland. The low-temperature carbonization plant in Espenhain, Germany, was built in 1935 (Fig. 11.1). All plants, including the one in Espenhain, were equipped with Lurgi furnaces. As feedstock, lignite briquettes were used. The lignite briquettes were micrograined with water content less than 15 wt%, the so-called F2–F4 briquettes.

After the Second World War, the low-temperature carbonization plant in Espenhain was heavily destroyed. Due to the rapid development of coal refining planned by the Rammler [2] and Bilkenroth lignite-fired high-temperature coking plants—known as BHT-coking (Ger.: BHT = Braunkohlehochtemperatur) in Lauchhammer and Schwarze Pumpe, Germany—little financial resources were made available for the preservation of the low-temperature carbonization plant in Espenhain and Böhlen. In the mid-1960s it was decided to shut down the plants. Nevertheless, it was still operating. In spite of a further decommissioning decision, the operation of the largely worn-out facilities, including visible environmental effects, was continued until the end of the 1980s.

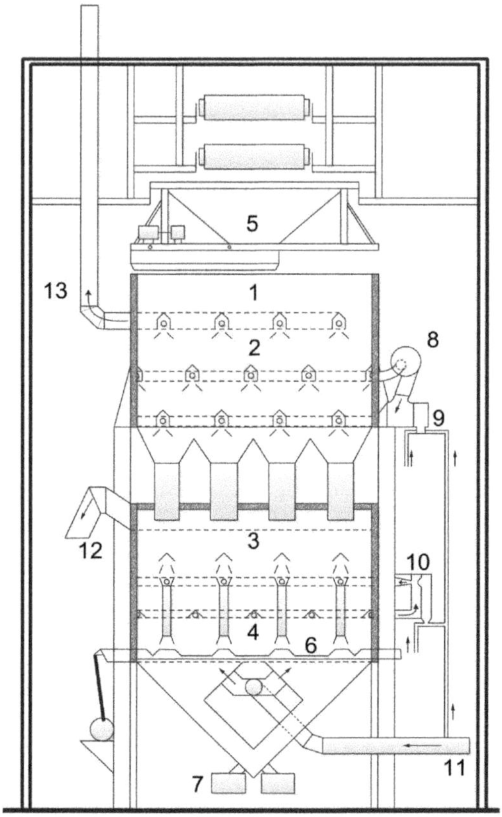

Figure 11.1 Basic construction of the Lurgi-flushing gas low-temperature carbonization unit in Espenhain, Germany, based on [1]. 1. Coal hopper, 2. drying zone, 3. carbonization zone, 4. coke cooling zone, 5. coal discharging car, 6. coke discharge machanism, 7. coke discharge packets, 8. circulation fan for drying zone, 9. combustion chamber for drying zone, 10. combustion chamber for carboniztion zone, 11. admission of purified circulating gas, 12. offtake from mixture of circulation gas and carbonization gas, and 13. offtake for waste gas from drying zone.

The economic importance of the low-temperature carbonization products was underestimated. Long-chain hydrocarbons, such as the tars, were the basis of the East German detergent industry. In the German Democratic Republic (GDR) it was not possible to build up any petroleum-based solution for the detergent industry.

At the end of the 1970s, the plant in Espenhain was assigned to the gas combination Schwarze Pumpe without making any economic decisions for the necessary effort. Little could be done to maintain the operation. In 1988, the plant in Espenhain reached the following results as shown in Tables 11.2 and 11.3.

The operating costs for one year were 1045 million GDR marks. The plant was unprofitable. There was a loss of 35 million GDR marks per year.

Table 11.2 Indicators of the 1988 annual report, production indicators

Product	Quantity	Unit
Lignite briquettes	4.3	mt/a
Coke	2.0	mt/a
Tar	73,845	t/a
Light oil	42,945	t/a
Crude phenol	15,780	t/a
Crude pyridine	605	t/a
Sulfur	23,251	t/a
Gasoline	6795	t/a
Methanol	1940	t/a

Table 11.3 Indicators of the 1988 annual report, specific indices

Parameter	Quantity	Unit
Briquetting input factor	1.814	$t_{raw\ lignite}/t_{briquette}$
Briquetting per furnace	462	t/d
No. of furnaces	32	–
Water content of raw lignite	51.8	wt%
Water content of briquettes	14.7	wt%
Coke yield	0.464	$t_{coke}/t_{briquette}$
Fischer yield	75.9	%
Tar yield	84.3	$kg_{tar}/t_{briquette}$
Raw gas yield	368.8	$m^3{}_{gas}/t_{briquette}$
Light oil yield	19.4	$kg_{oil}/t_{briquette}$

At the end of the 1980s a comprehensive research and investigation program for the development of a new, economic, and environmentally friendly low-temperature carbonization furnace was carried out. The German fuel institute in Freiberg, the project management company pyrolysis kinetic models (PKM) Leipzig, and the carbonization department in Espenhain were involved in the research. After reliable research and experimental results, the project for the new furnace was completed in 1991. The research and experimental results of that time are still completely available. It is certainly of great interest for today's coal refining processes. It contains several decades of carbonization process experience. The briquettes input for a single low-temperature furnace was improved to 540 t/d. A significantly reduction of the carbonization temperature was achieved. The main target of the designed plant was to obtain a high tar yield. The residual volatiles in the coke were less than 2 wt%. The Lurgi low-temperature carbonization process is also applicable for other materials, such as hard brown coal.

After the fall of the Berlin wall there was a totally different commodity market for the chemical and detergent industry. For the worn-out plant in Espenhain only one right decision could be made: a final shut down. Today, the industrial area has been modernized and a 5 MW solar power plant is located next to the former carbonization plant.

11.4.2 Lignite high-temperature coking process by Rammler and Bilkenroth

Research and development, in a laboratory scale, for the production of cokes from lignite for blast furnace application were carried out in Germany at the lignite research institute "Braunkohlenforschungsinstitut" at the "Bergakadamie" in Freiberg from the beginning of 1925. However, no large-scale plant was in operation. After the Second World War, it was clear that in the area of the former GDR the demand for raw materials such as raw iron, crude steel, metallurgical coke, and coking coal could not be ensured within the territory of the GDR.

For example, only around 3% of hard coal production and around 5% of coke production capacity of the former German Reich was located in the territory of the GDR. Coking coal and iron ore deposits were almost non-existent. However, the GDR was equipped with large lignite coal deposits of varying quality. But, by that time around the world no process for the production of coke for blast furnaces from lignite was known. The objective for the GDR was to develop process for the production of raw iron from native iron poor and silicic acid-rich ores under utilization of coke from lignite. Nonetheless, the coking behavior of the lignite is fundamentally different from that of the hard coal. While hard coal is baking, lignite has, due to its colloidal structure, no baking properties at all and does not soften during heating procedures. The colloidal structure is destroyed upon heating by decomposition and evaporation of the colloidally bounded water. Lignite tends to disintegrate during any thermal treatment. To produce a lumpy and structured coke from lignite, other processes in comparison with hard coal needed to be developed. Lignite needs to be preprocessed to achieve a mechanical strength and lumpiness that remains even under heating processes. One way to achieve these properties is to apply a binderless briquetting in which between the lignite grains cohesive forces and bounding bridges are built by the water molecules. During the degassing, a volume and mass reduction of the briquettes occur as shown in Fig. 11.2.

Under the supervision of Rammler and Bilkenroth several test series were carried out. For Rammler and Bilkenroth, the coking tests in the discontinuously operated horizontal chamber furnaces of Espenhain and August Bebel were only necessary intermediate stages. For both, it was clear that the technical solution had to be a continuous vertical column furnace with indirect heating (Fig. 11.3).

It was decided that in Lauchhammer the future lignite high-temperature coking plant shall be located. In Lauchhammer a screening device was installed that allowed the fabrication of finest grain briquettes. In a fifth test series, extremely fine grain briquettes from Lauchhammer were used in the vertical chamber furnace

Figure 11.2 Comparison of size and structure of fine grain briquettes (F4) of hard coal and lignite to coke [3].

Figure 11.3 Prof. Rammler and Dr. Bilkenroth and the government representatives walking to the first lignite high-temperature coking plant in June 14, 1952 [4].

at the Dresden (district of Reick) gasworks for coking. Also, after the not very favorable experiences at the Dresden (Reick) gasworks with regard to the achieved coke quality, the vertical chamber furnace was still technically the most favorable furnace type for the lignite high-temperature coking (Fig. 11.4).

As a result of the experience gained, the following additional procedural requirements were established:

1. A separate briquette dryer needs to be installed ahead of the coking unit.
2. The heating of the vertical chamber oven must be from the bottom to the top so that the temperature decreases to the top heaters.

Figure 11.4 Official opening of the coking plant in Lauchhammer in June 14, 1952 [4].

3. For the coke discharge from the vertical chamber furnace, an air-tight chamber closure has to be developed.

In 1951, the development of the low-shaft furnaces in Donauwörth (Kleiner Max) and in Unterwellenborn (Maxhütte) had been so far advanced that there was an urgent need to use larger quantities of BHT-coke. For this reason the gaswork in Delitzsch, that was already shutdown, was converted into a test facility. The plant consists of vertical chambers that were batchwise in operation. Some chambers were used for the predrying of the briquettes, which were then transferred into the coking chambers.

In May 1951, the "Deutsche Ofenbaugesellschaft" received the order to design and configure a coking technology and coke drying unit. In the beginning of November 1951, in order to secure the special planning and execution as well as the special construction on the large construction site, all available specialists and companies, especially in the field of refractory construction, were integrated in the "VEB Projektierung, Konstruktion, Montage Kohleverarbeitung Leipzig/Berlin," later known as the "VEB PKM Anlagenbau Leipzig." The newly established "VEB PKM Anlagenbau Leipzig" was the general contractor for the construction of 24 coking oven units, for the dry-coke coolers, the condensation units, and all other utilities of the large coking facility in Lauchhammer.

On June 14, 1952, after a construction period of only 8 months, the worldwide first lignite high-temperature coke production was officially set in operation in Lauchhammer.

This was the starting point for 40 years of lignite high-temperature coking in Lauchhammer. Fourteen years later also coking plants were running in Schwarze Pumpe. Fig. 11.5 shows the technological concept of the pyrolysis oven.

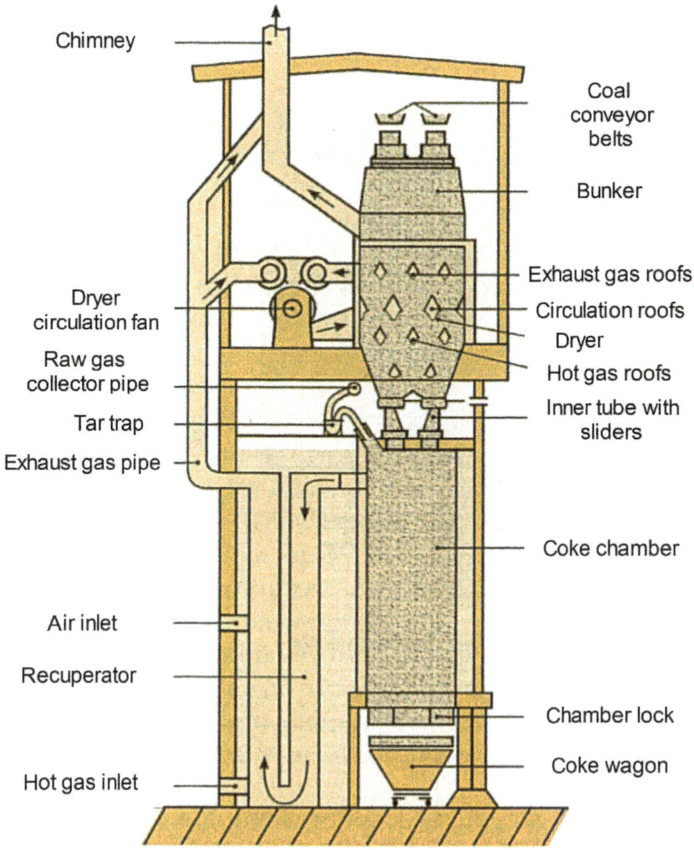

Figure 11.5 Section of the vertical chamber oven in Lauchhammer, based on [3]. Left: heating chamber with preheating; right: coke chamber with flue gas cooling.

The commissioning and operation of further furnace units was carried out on the basis of the completed design documents, the parallel research and development work, and the evaluation of the operating results. After more than 3 years of operation on October 2, 1955 more than 1 million BHT-coke were produced. In 1957 the coking facility in Lauchhammer was fully established with 24 furnaces in operation and more than 1 million tons of BHT-coke were produced per year. In Fig. 11.6 a picture of the fully established plant in Lauchhammer is shown.

The facility consists of the following main units:

24 furnace units in two lines, 16 coke coolers, 1 coke screening plant, 1 coke loading zone, 6 condensation units, 1 light oil extraction plant, 1 tank storage unit, 1 dephenoling unit (phenosolvane extraction), 1 biological sewage treatment unit, 1 dry desulfurization unit, 1 compressor station with piston compressor, 1 rough and fine water treatment unit, 1 feed water treatment unit, and 1 air separation plant.

Figure 11.6 Picture of the coking plant in Lauchhammer [5]. In the center the gas scrubbing units and the liquid production are located surrounded by two furnace lines to the left and right.

11.4.3 Development of the BHT-coking facility in Schwarze Pumpe

For the lignite, refinements combine Schwarze Pumpe (later known as gas combine Schwarze Pumpe) and three high-temperature coking plants were planned according to the original overall concept for the site. The three coking plants were named by their geographical location: east, west, and center. For each coking plant, a separated briquette factory was assigned.

The construction of the coking plant "Center" started in 1964. On September 1, 1969, the first furnace unit started its operation in Schwarze Pumpe (Fig. 11.7).

Within the following 2 years and 5 months, the remaining nine furnace units had been put into operation. In Fig. 11.8 a picture of coke plant "West" is presented.

Fig. 11.9 shows the central control room and the coke loading station at Schwarze Pumpe.

Due to the fact that not enough experienced workforce was available, it was necessary to train and qualify the workforce in parallel with the commissioning of the coking plants.

Compared with the high-temperature coking plant in Lauchhammer the coking plant "Center" in Schwarze Pumpe was a significant further development and became the centerpiece of the third-generation of coking plants in the GDR.

In Fig. 11.10 a schematic sketch of the coking furnace with built-in cooling zone and two-stage drying system is presented.

Figure 11.7 Picture of the coking plant "Center" in Schwarze Pumpe [5].

Figure 11.8 Picture of the coking plant "West" in Schwarze Pumpe [5].

The significant changes were mainly:

- A two-stage dryer for a more gentle heating and drying of the briquettes.
- A new heating system for the coke chambers to improve the residual drying and heating conditions.
- An undercooling zone to reduce environmental impact.

(A)

(B)

Figure 11.9 Picture of the in central control room (A) and the coke loading station (B) at Schwarze Pumpe [6].

With the new technical innovations, there was also a great hope that the coke quality can be improved significantly. In particular, a higher abrasion resistance and larger quantity of particles with size above 30 mm were desirable. However, only small improvements for the abrasion resistance and particle size could be achieved in comparison to the Lauchhammer coking plant. In addition, the coke produced in Schwarze Pumpe had a significantly higher reactivity compared to Lauchhammer. The spontaneous combustion could be controlled relatively well by applying spent sulfite liquor, but another problem was the appearance of so-called burning

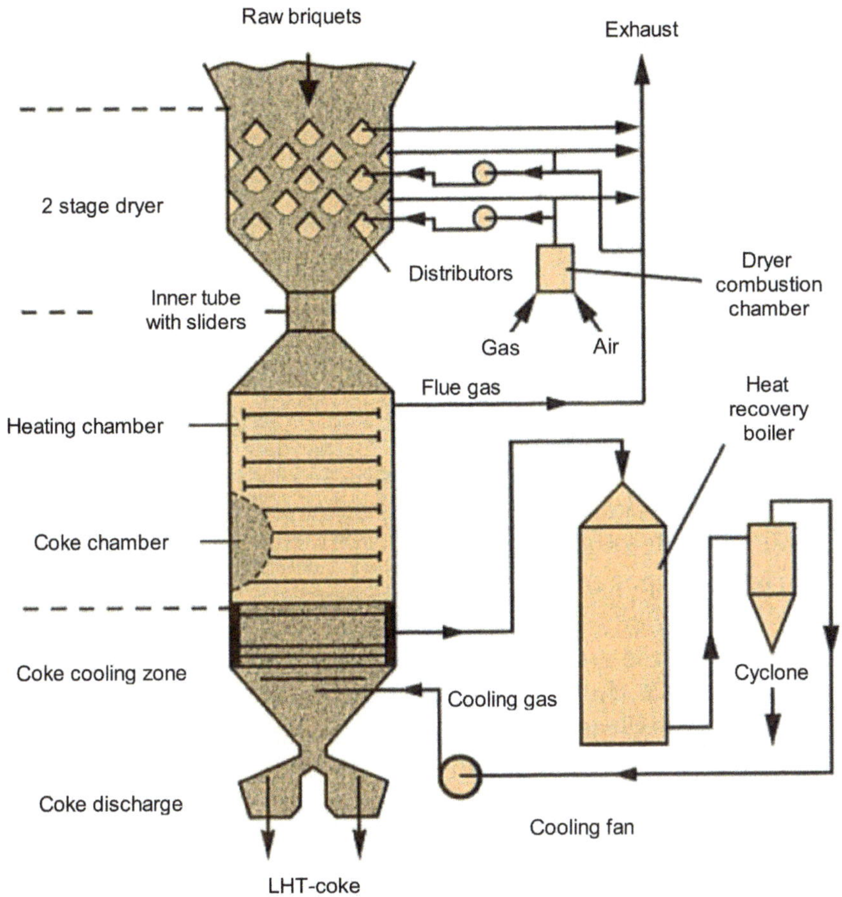

Figure 11.10 Schematic sketch of the coking furnace in Schwarze Pumpe (third-generation), based on [3].

waggons, by loading or transportation the coke in waggons. In this case, even by applying spent sulfite liquor, coke tended to spontaneous combustion. This phenomenon, which was under control at the Lauchhammer plant, could not be satisfactorily clarified and eliminated in Schwarze Pumpe.

One reason for that was, in comparison with Lauchhammer, the open-cast mining was not focused only to provide the best lignite for the coking process. In Schwarze Pumpe the lignite was supplied for the entire state combined with power plants, coking plants, gasification plants, and so on. Each part was supplied with the necessary lignite quality but due to three different open-cast mining areas that were in operation for Schwarze Pumpe and the huge amounts needed, an ideal coordination between plant and economically interest was not always possible. Also, political–ideological constraints played often an important role. In this case technical competences were second-ranked. In addition, it became more and more difficult for the manufacturers and suppliers to provide special equipment within the framework of the state-directed economy. Nevertheless, the following construction and plant components were successfully developed and their high reliability under production conditions were achieved:

- New designed blower for predrying.
- Raw gas extraction from the coke chamber.
- Design and operation of the tar drain.
- Supply of the circulating gas to the cooling zone with regard to flow design.
- Continuous coke discharge with regard to the constructional design and optimization of the coke passage through the furnace.
- Optimization of the heat exchanger design for cooling gas circuit with regard to flow and process improvement.

The two-stage drying process and the modified heating system were well engineered and was a successful process technology. The furnace construction had proved its worth over long periods of operation. With regard to the wall thickness of the heating walls, a compromise had to be made between the variability of the heating system (requires a thin heating wall of 100 mm) and the stability of the refractory material. To ensure long hours of operation and stable temperature profile a wall thickness of 150 mm is recommended.

With the shutdown of the Lauchhammer large-scale coking plant in 1991, as well as the closure of the coking plant "Center" in Schwarze Pumpe in 1992, approximately 40 years of lignite high-temperature coking in the Lusatia region had ended.

The ongoing development for the exhaust duct for the degassing and the conversion to a single-chamber discharge could not be completed.

In retrospect, the lignite high-temperature coking design was based on three pillars:

1. Selection of best quality-appropriate raw lignite suitable for the high-temperature coking process and the best possible selective extraction in open-cast mining.
2. Production of high-quality micro-grain briquettes under strict observance of the scientifically verified quality requirements with regard to water content, grain size and pressing temperature, and the rotation speed of the press.
3. Gentle thermal handling of the briquettes in particular during the drying and carbonization processes.

One great objective of the lignite high-temperature carbonization was to provide coke not only for low-shaft furnaces but also for blast furnace processes. This objective for blast furnace processes was not achieved. However, for the produced high-temperature coke from lignite various applications had been developed.

This includes the usage:

1. as fuel in chemical industry,
2. for lime burning,
3. in carbide industry,
4. in ferrosilicon industry,
5. in cement industry,

6. as fuel in metallurgy,
7. as adsorbents.

Looking into the future there is a change that the application and development of lignite high-temperature carbonization was only interrupted, but has not come to its end. With the known decline of oil and natural gas as resources as well as the decline of high-grade hard coal coke and iron-rich ores an increase in the market price can be expected. This could lead to the necessity of applying low-shaft furnaces technology for iron-poor ores by utilization of coke from the lignite high-temperature carbonization.

11.4.4 Recent developments in the refining of lignite

In former times, the main interest was to produce tars and oils from lignite. The produced gas was burned and the obtained coke was not used, e.g., for metallurgical purposes. But with the improvement of technologies and demand for fuel gases and the need to use the produced cokes, the focus on the material used of lignite changed. Depending on the pyrolysis process technology, the pyrolysis gas can have a high content of hydrogen and carbon monoxide and can be used, e.g., as feedstock for a methanol production. In addition, the produced coke due to its high carbon content can be used for metallurgical purposes (PCI, pulverized coal injection) and also for mercury removal in flue gases (PAC, powdered activated carbon/coke). In the following sections, two recent developments are briefly described.

11.4.4.1 Lignite pyrolysis plant by Coal Energy Australia Ltd

Since 2014, in Australia, the company Coal Energy Australia Ltd. (CEA) conducts a project to produce semi-coke from lignite. The plant is located in Victoria in the Latrobe Valley. In Fig. 11.11 a flowchart of the CEA process is shown.

For this processing plant raw lignite with water content of up to 67% is used. The lignite is dried and pressed to briquettes (see also Section 11.4.2). The briquettes are

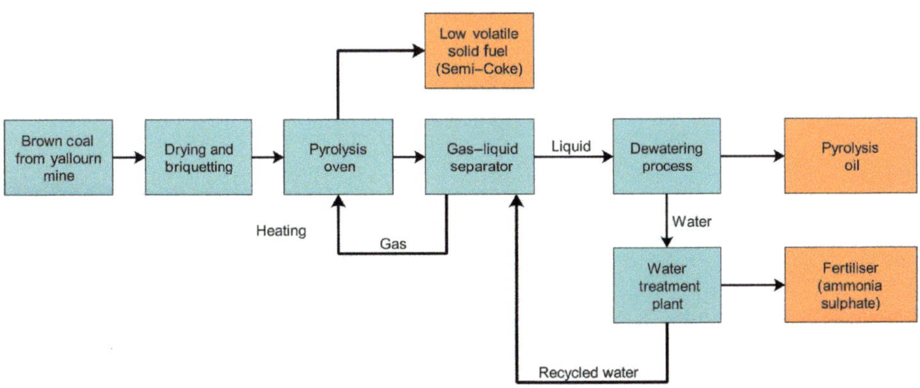

Figure 11.11 Flowchart of the steps in the CEA demonstration plant, based on [7].

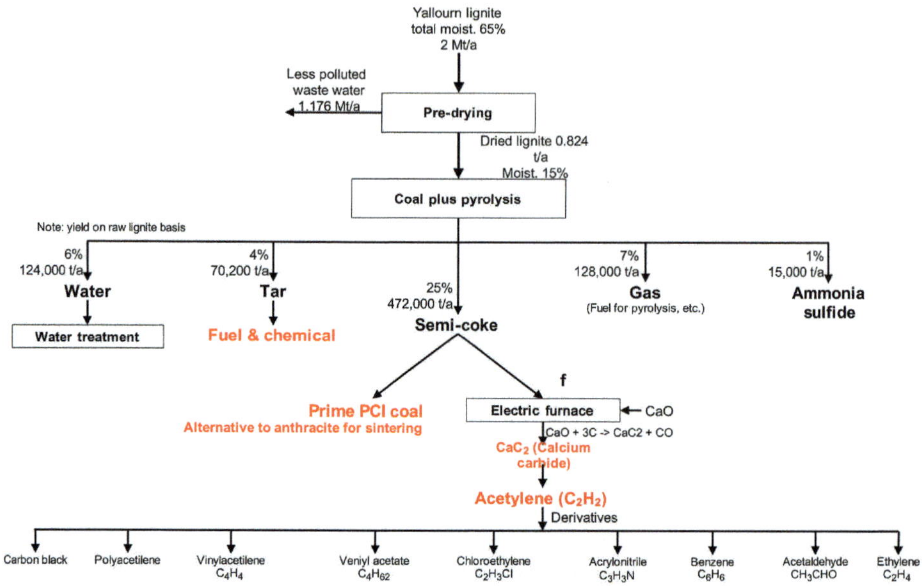

Figure 11.12 Flowchart CEA semi-coke process (phase 1), based on [8].

Table 11.4 Prodction capacities CEA plant, based on [8]

Production	Phase 1 (mtpa)	Phase 2 (mtpa)	Total (mtpa)
Raw feed (Yallourn lignite)	2.0	20.0	22.0
Prime semi-coke/PCI coal	0.470	4.7	5.17
Pyrolysis oil	0.070	0.7	0.77
Coal gas	0.128	1.28	1.4
Ammonium sulfate	0.015	0.15	0.165

used as feedstock for the pyrolysis oven. The resulting pyrolysis gas is used for heating the oven. Products from this process design are low volatile semi-coke, pyrolysis oil, and fertilizer. The semi-coke product can be used, e.g., for pulverized coal injection (PCI) [7]. The pyrolysis oil can be distilled, e.g., into phenol oil and bitumen. Ammonium sulfate can be used as fertilizer and soil conditioner [7].

Fig. 11.12 shows the quantities of feedstock and products of the planned process as well as the utilization of the products [8].

Two phases for the erection of the CEA Plant are indented. At the end, the plant can process up to 22 million tons of raw lignite per year (Table 11.4).

Table 11.5 shows the proximate analysis of CEA semi-coke.

Even though little about this project in detail (e.g., operation temperature, pyrolysis oven design) is known to the public so far, it seems that the process is similar to the process that was developed at Lauchhammer and Schwarze Pumpe

Table 11.5 Proximate analysis of CEA semi-coke, based on [8]

Proximate analysis ITEM (basis)	CEA semi-coke
Total moisture (arb)	0.6%
Ash (adb)	5.1%
Volatile matter (adb)	2.2
Fixed carbon (adb)	91.5%
Total sulfur (adb)	0.3%
Phosphorus in coal (arb)	0.01%

(see Sections 11.4.2 and 11.4.3). The project will provide 180–300 jobs during construction. The expanded commercial operation, which is expected to be in place by 2019, will provide up to 120 full-time jobs [9].

11.4.4.2 Pyrolysis of lignite dust in rotary kiln at high temperatures

Initial technical investigation for the utilization of an indirect heated rotary kiln for the direct production of blowable lignite coke dust with a grain size smaller than 200 μm and high temperatures were carried out by Jochmann and Richter [10] in 2010. The experiments were executed in a testing facility of IBU-tec advanced materials AG, Germany. The pyrolysis coke dust was specifically tested for the pulverized coal injection (PCI) in blast furnaces.

The benefits of applying an indirect heated rotary kiln for the pyrolysis of lignite are:

- Continuous mixing
- Simple to design and economical
- Processing large quantities
- Atmosphere is to adjust and to control
- Applicable to coarse grain and powder
- Indirectly heating with gas as well as with electricity
- A fast pyrolysis within minutes is possible!

Fig. 11.13 shows the main properties of the feedstock lignite dust with a grain size smaller than 200 μm and the product blowable coke dust.

The carbon content of the coke dust is over 83% and the heating value reaches with 30.2 MJ/kg the same level as coke from hard coal. Dense phase conveying experiments with the produced coke dust were carried out at ArcelorMittal in Eisenhüttenstadt, Germany. The produced coke dust fulfils the requirements for coal/coke injection in blast furnace processes. The substitution factor for the lignite coke dust is about 0.87 and hence very close to values for hard coal coke dust (0.9–0.95).

ARCUS Technologie GmbH & Co. KG (ARCUS), Germany, applied some kinetic models (see Section 11.5) to predict the properties of the lignite coke dust produced

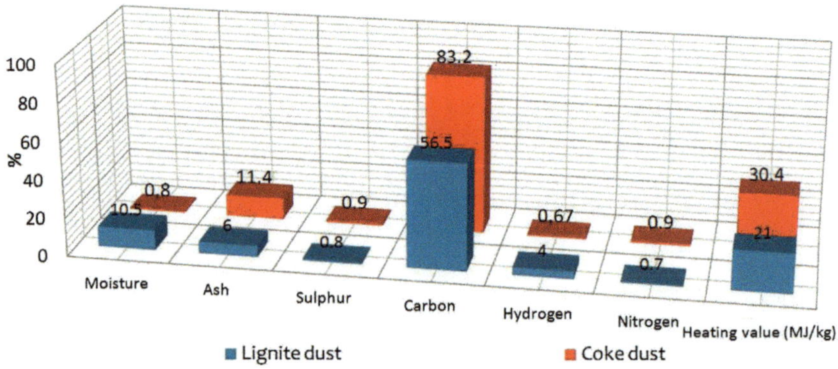

Figure 11.13 Experimental results: main properties of the feedstock (lignite dust <200 μm) and the product (coke dust <200 μm) after pyrolysis in an indirect heated rotary kiln, based on [10].

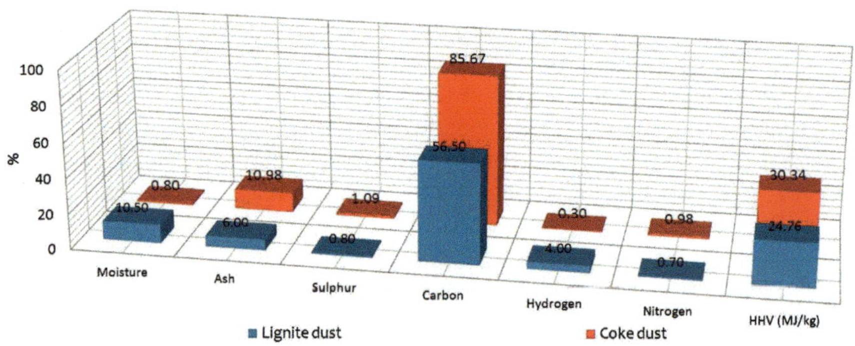

Figure 11.14 Simulation results by ARCUS: main properties of lignite coke dust <200 by applying the kinetic approach of Suuberg [11] and same feedstock as in Fig. 11.13.

by pyrolysis in a rotary kiln. The obtained results are presented in Fig. 11.14 and are in good agreement with the experimental results shown in Fig. 11.13.

The applied model also allows predicting, e.g., of the carbon content and heating value if another lignite feedstock is used. Fig. 11.15 shows that lignite coke dust with even higher carbon and a higher heating value than in Fig. 11.13 can be achieved.

Properties of the lignite dust that can be applied for this process are listed in Tables 11.6 and 11.7 and can be, e.g., purchased from LEAG (Germany).

In addition, fluidized bed lignite with a larger grain size than 200 μm is also applicable.

The application of lignite coke dust is not limited to blast furnace processes, where especially a high carbon content and a high heating value are desirable. In addition, the BET area of the lignite coke dust is about 310 m^2/g and can be

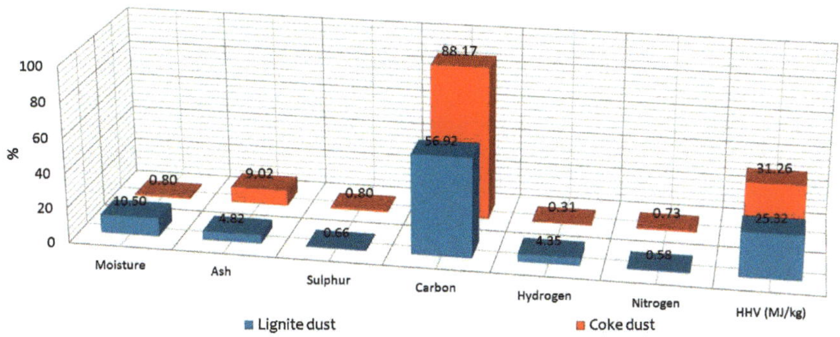

Figure 11.15 Simulation results by ARCUS: main properties of lignite coke dust <200 μm by applying the kinetic approach of Suuberg [11]. Lignite feedstock with less ash content as in Fig. 11.13.

Table 11.6 Properties of lignite dust that can by applied for pyrolysis in a rotary kiln [12]

Property	Value	Unit
Grain size: residue to 0.20 mm	>13	%
Grain size: residue to 0.09 mm	>40	%
Water	10	%
Ash	5	%
Volatiles	47	%
Sulfur	0.6	%
Calorific value	21	MJ/kg

Table 11.7 Properties of lignite from fluidized bed process that can by applied for pyrolysis in a rotary kiln [12]

Property	Value	Unit
Grain size	0–6.3	mm
Water	19	%
Ash	5.5	%
Volatiles ingredients	41	%
Sulfur	0.8	%
Calorific value	19	MJ/kg

easily increased by adjusting some process parameters. This property in combination with the high sulfur content of the Lusatia lignite makes the lignite coke dust also applicable as adsorbent for mercury removal from flue gases. This means that the same lignite, which is burned in a power plant, can be applied for

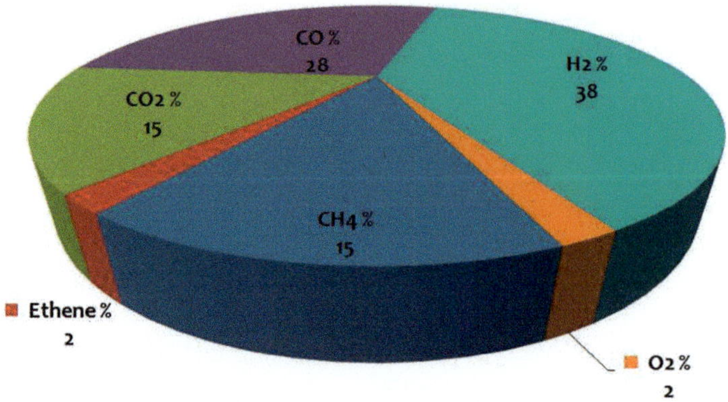

Figure 11.16 Experimental results [10]: pyrolysis gas composition.

the mercury removal in flue gases of power plants, after drying, grinding, and pyrolysis in a rotary kiln.

By applying a rotary kiln for the pyrolysis of lignite at high temperatures, a high content of CO and H_2 in the pyrolysis gas is achieved, as reported by Jochmann and Richter [10] and shown in Fig. 11.16. The H_2/CO ratio is about 1.4.

This property allows not only, e.g., to use the pyrolysis gas for heating the rotary kiln but also to apply the pyrolysis gas for the methanol production. Simulations done by ARCUS show that a H_2/CO ratio even above 2 can be reached (Fig. 11.17). A value for the H_2/CO ratio of about 2 is preferable for the methanol production. More experimental work needs to be done to validate the achievable H_2/CO ratio by pyrolysis of lignite dust in a rotary kiln.

To handle the pyrolysis of lignite in an indirect heated rotary kiln, several parameters are adjustable. For example, in order to control the residence time, the angle of inclination and the number of revolutions are important. Also the number of revolutions, together with built-in components, is playing an important role to manage the heat transfer. The design of the built-in components and the number of revolutions determine as well the dust entrainment in the gas stream. To predict the behavior of the moving bed and the influences of deferent built-in designs in a rotary kiln, computational fluid dynamics (CFD) simulation can be very useful. Sofialidis et al. [13] published one of the most extensive CFD work in indirect heated rotary kilns with a 3D Eulerian Multiphase approach regarding transverse and axial material transport and chemical reactions. Further CFD simulation work, where no lignite as feedstock was studied, were published by Liu et al.[14] and Komossa et al. [15].

ARCUS developed a reliable CFD simulation procedure for the prediction of solid fuel heating and transport in indirect heated rotary kilns. This design tool takes into account the feed properties (solids size, moisture content, and feed rate) and process requirements (bulk temperature, desired residence time, and dust formation). Fig. 11.18 shows some results of the CFD simulations.

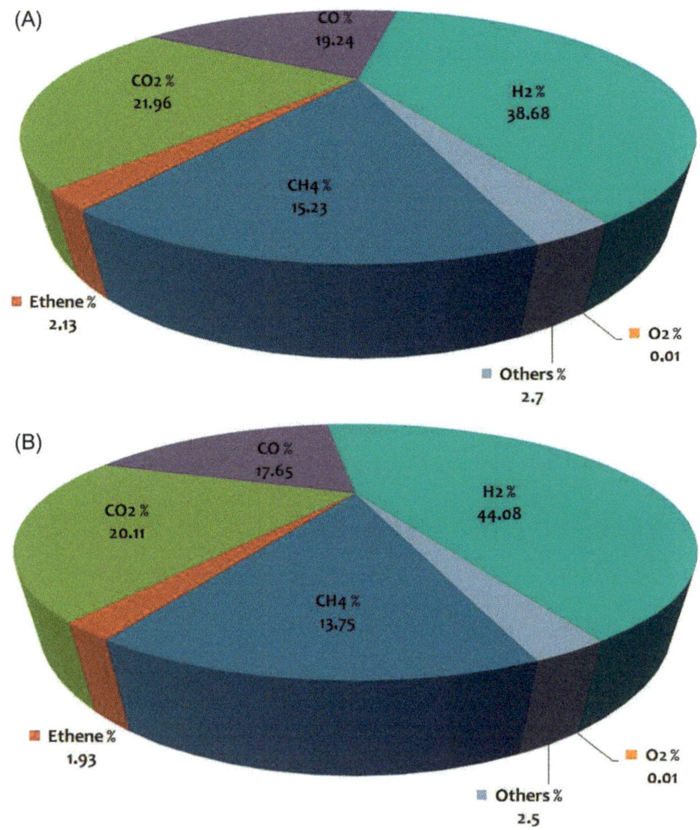

Figure 11.17 Simulation results by ARCUS: Pyrolysis gas composition with lignite feedstock as in Fig. 11.13 (A) and with lignite feedstock as in Fig. 11.15 (B).

Table 11.8 summarizes, from experimental and simulation studies, the main properties of the coke dust, pyrolysis gas, and possible applications.

With these promising results, Richter et al. [16,17,18] developed a composite technology. The proposed composite technology allows the integrating hydrogen from renewable sources to produce methanol or methane. Fig. 11.19 shows a block flow diagram of the analyzed composite technology. This composite technology is of special interest, if not enough hydrogen is supplied by pyrolysis of lignite to produce in an efficient manner methane or methanol.

With this proposed composite technology, a material storage of electricity from renewable sources into blowable coke dust and methane or methanol is possible. This process combines conventional and renewable energy sources in a highly promising way.

By applying renewable energy sources for the hydrogen production a path for feasible economic exploitation of the lignite within a reasonable period is given. It ensures a drastic reduction in CO_2 emissions and a high conversion efficiency of

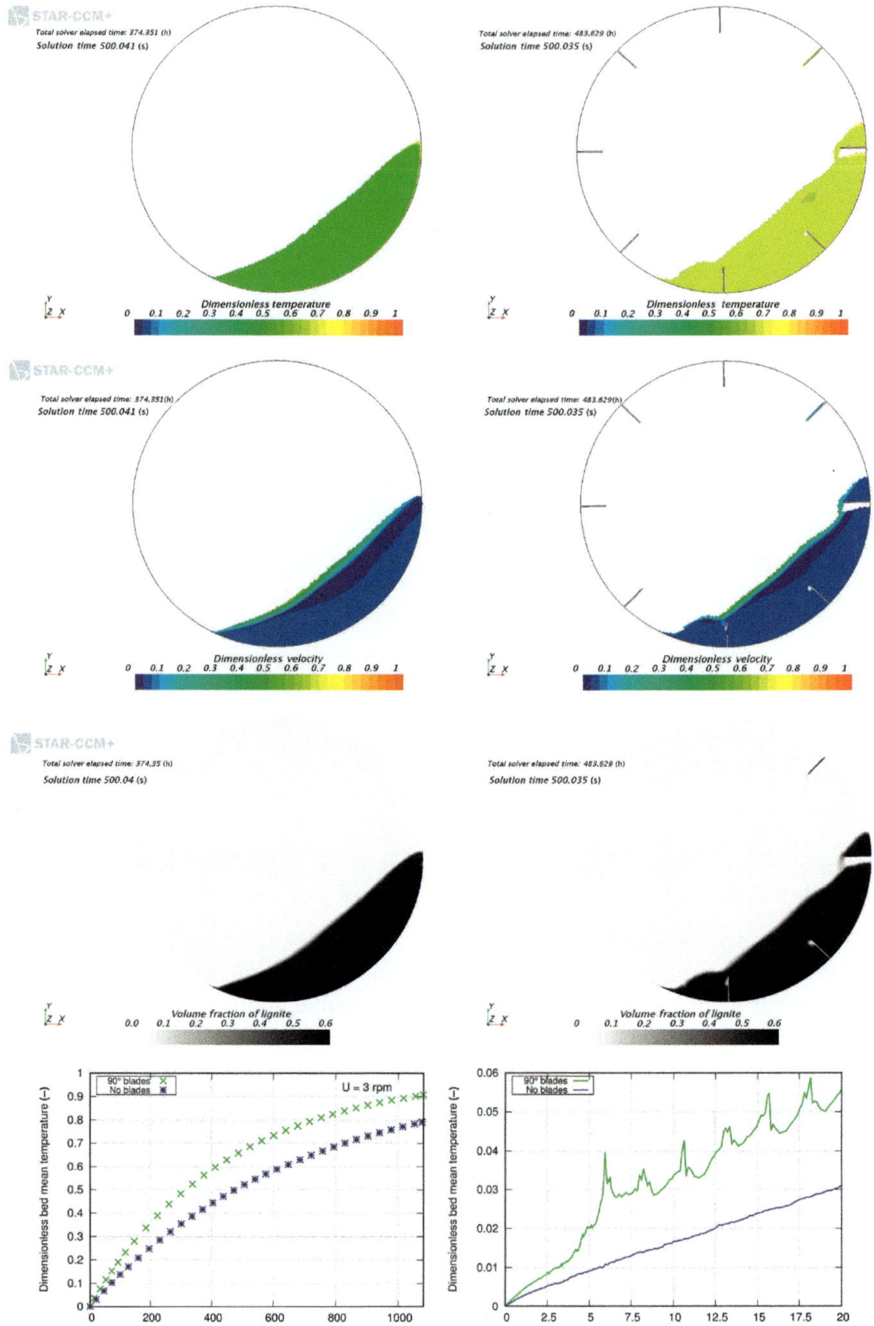

Figure 11.18 CFD simulations by ARCUS to predict the heating of the moving bed and dust entrainment by pyrolysis gas in an indirect heated rotary kiln.

Table 11.8 Properties of products from lignite dust pyrolysis in a rotary kiln

Product	Properties	Application
Coke dust	Grain size <200 μm Blowable Carbon content >83% Heating value: 30 MJ/kg BET area >310 m^3/g Density <450 m^3/kg, Sulfur content: 2.6% (If lignite from Lusatia region is used)	A. Application for PCI in blast furnace processes, substitution of hard coal coke dust B. Adsorbent for gas and water treatment, e.g., for removal of mercury contained in flue gases of power plants
Pyrolysis gas	H_2/CO >1.4 over 2 is possible Very low tar content, no long-chain tars present Heating value: 14.8 MJ/m^3	C. Feedstock for methanol production D. Fuel for heating the rotary kiln

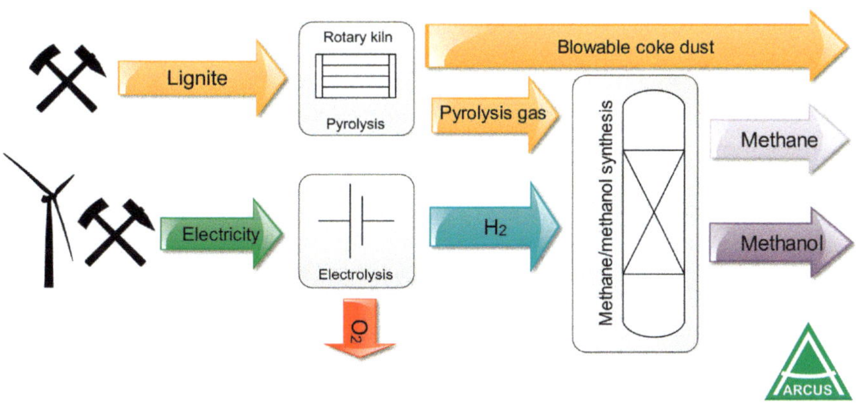

Figure 11.19 Composite technology for the utilization of lignite and integration of renewable sources [18].

carbon, mainly into metallurgical blowable coke dust. Methanol is of interest for the transport sector, and due to changes in the European Renewable Energy Directive [19,20] and Fuel Quality Directive [21] it is possible to produce "green" methanol for the transportation sector even if the carbon source is from "non-biological sources." Bergins et al. [22] published a more detailed description to this topic.

With the effective and environmentally compatible results based on the published results and further investigations, ARCUS carried out a basic design for a pyrolysis demonstration plant in 2015 [23]. Fig. 11.20 shows a block flow diagram for the designed demonstration plant.

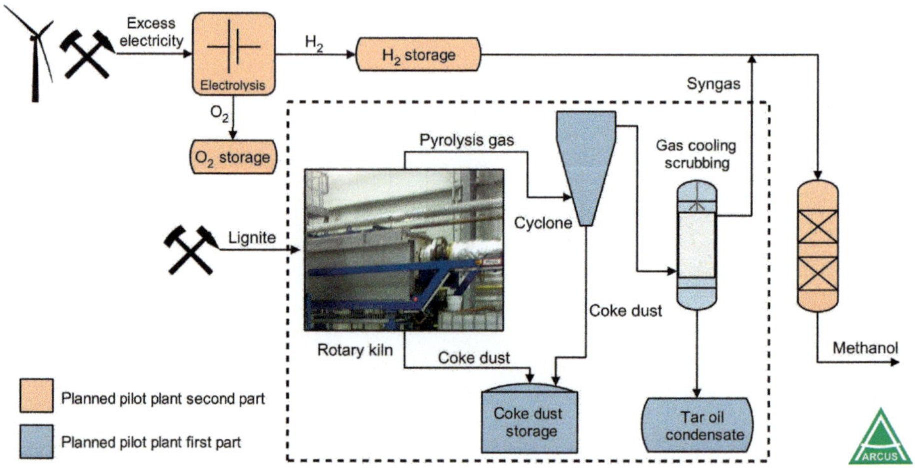

Figure 11.20 Block flow diagram for a demonstration plant to produce blowable coke dust and methanol from lignite [18].

The demonstration plant will process a flow rate of up to 7500 t/a of lignite dust and produce up to 3000 t/a of blowable coke dust. The plant will be used primarily for the establishment of process and project parameters for further large-scale plants, which will be developed in a modular construction of three to five modules. The performance of the modules working in parallel will secure a production up to 100,000 t/a pyrolysis coke. It is expected that the demonstration plant will be in operation in 2019.

The proposed composite technology with integration of renewable energy is not only an intermediate stage, it is also the long-term solution for the demand of carbon in the chemical and metallurgy industry sector.

11.5 Pyrolysis kinetic models—A survey

In this section, the important models that were developed for the pyrolysis of pulverized coal will be described briefly. The description of the kinetic rate models must start first with a consideration of what such rates might mean. A model for pyrolysis using a proposed kinetic expression can in most cases be applied to fit data in some limited regime of temperature and heating rate.

11.5.1 Empirical models

11.5.1.1 Single first-order reaction model (SRM)

Several empirical models have been developed to study coal pyrolysis of which the most often employed model is the single first-order reaction model that has been applied for overall weight loss and for individual species evolutions. The main model that falls in this category is the one developed by Badzioch and Hawksley [24], which has considered the coal pyrolysis rate as a first-order reaction depending on the amount of volatiles remaining in the particle and results in the rate at which volatiles are driven from the coal particles being heated or equivalently the rate of weight loss of the coal sample:

$$\text{Coal} \xrightarrow{k} S + V \tag{11.1}$$

$$\Delta V = V_o(1 - C)(1 - \mathrm{e}^{-k\Delta t}) \tag{11.2}$$

$$k = A\mathrm{e}^{-E_a RT} \tag{11.3}$$

where ΔV is the volatile matter yield as a fraction of the coal's sample weight on dry ash free basis, V_o is the coal's original volatile matter mass fraction on dry ash free basis, A is the Arrhenius frequency factor, E_a is the apparent activation energy, R is the universal gas constant, k is the Arrhenius type rate constant, T is the isothermal pyrolysis temperature of the particle, and C is a limiting decomposition factor. The Arrhenius equation parameters along with the limiting decomposition factor are dependent on the specific coal dust, which limits the general applicability of the model.

11.5.1.2 Two competing reaction model (TCRM)

Kobayashi et al. [25] then regarded the coal pyrolysis as two parallel competing reactions, one dominating at low temperature and the other dominating at high temperatures. This model although very similar in considerations to the single reaction model but still has more capabilities to incorporate the influence of heating rate and high temperatures on the total volatile yield rate. Chemically this can be visualized as:

$$\text{Coal} \rightarrow (1 - \alpha_1)S_1 + \alpha_1 V_1 \tag{11.4}$$

$$\text{Coal} \rightarrow (1 - \alpha_2)S_2 + \alpha_2 V_2 \tag{11.5}$$

with rate equations:

$$\Delta V = \int_0^t \left[(\alpha_1 k_1 + \alpha_2 k_2)\mathrm{e}^{-\int_0^t (k_1 + k_2)\mathrm{d}t} \right] \mathrm{d}t \tag{11.6}$$

$$k_1 = A_1\mathrm{e}^{-E_{a1} RT} \tag{11.7}$$

$$k_2 = A_2\mathrm{e}^{-E_{a2} RT} \tag{11.8}$$

where α_1 and α_2 are the mass stoichiometric coefficients or yield factors for each reaction. An important feature of the model is that E_{a1} is smaller than E_{a2} and the general utility is limited because the parameters α_1, α_2, A_1, A_2, E_{a1}, and E_{a2} depend on the specific coal dust.

11.5.1.3 The distributed model

Anthony et al. [26] developed the distributed model by treating pyrolysis as a series of independent parallel reactions with same reaction rate constant but with different activation energies. If the Gaussian distribution of activation energies is applied, the volatile yield during the process of devolatilization can be calculated using the following equations:

$$\frac{V^* - V}{V^*} = \int_0^{\infty} \exp\left[- A \int_0^t e^{-\frac{E_a}{RT}} dt \right] f(E_a) dE_a \tag{11.9}$$

$$f(E_a) = \frac{1}{\sigma\sqrt{2\pi}} \exp\left[- \frac{(E_a - E_{ao})^2}{2\sigma^2} \right] \tag{11.10}$$

where E_{ao} is the mean activation energy, σ is the standard deviation, V is the cumulative volatile yield up to time t, V^* is the maximum volatile yield at infinite time, and A is the Arrhenius frequency factor. However, the model of many parallel reactions still has a number of drawbacks as it does not include the influence of ambient pressure and particle size, beside that it cannot predict the ultimate yield of volatiles for a given temperature and heating rate (V^*) but requires this value as input from experimental data. Also, the other parameters A, E_o, and σ should still be determined for the specific coal dust of interest.

Antal et al. [27] made further improvement on the distributed model and on a more complex model employing a set of 10 reactions originally postulated by Reidelbach and Summerfield [28] but only limited comparison has been made between this relatively complex model and experimental data and is almost not used in any current work. Still Reidelbach's model contains the reaction steps shown by Eqs. (11.4) and (11.5) which by themselves constitute a model that can quite successfully explain much experimental data as it is also recently shown in the work of Hart et al. [29] and Jovanović et al. [30].

11.5.1.4 Single first-order multi-reaction model

Suuberg and Howard [11] suggested that the generation of each key product of pyrolysis could be modeled using one or a few first-order reactions depending on the observed behavior during the pyrolysis. The model included a set of 15 first-order reactions that produce 8 volatile key products. This set of reactions is remarkably successful in describing detailed pyrolysis data for a lignite coal. Afterward this model was extended to bituminous coal and could also well describe its behavior [31]. The model can be chemically visualized as:

$$\text{Coal} \rightarrow \text{product}_i \tag{11.11}$$

with the first-order rate equation:

$$\frac{dV_i}{dt} = k_i(V_i^* - V_i) \tag{11.12}$$

and a rate constant which is defined by:

$$k_i = k_{io}e^{-E_{ai}/RT} \tag{11.13}$$

where k_{io} is the frequency factor of reaction i, E_{ai} is the activation energy of reaction i, V_i is the amount or yield of product i produced up to time t, V_i^* is the amount of product i which could potentially be produced from reaction i if an infinite time is allowed or in other words the ultimate yield, T is the absolute temperature, and R is the universal gas constant. In Eq. (11.11), the term product represents the key products of coal pyrolysis, such as CO_2 and CO gases produced at different stages in the reactor. Each stage dominates within a specific range of temperature, so the yield of a given product is equal to the sum of the yields of that product from each stage. For example, according to Suuberg's model, CO_2 is produced from three different stages dominating in different temperature ranges from around 200°C to above 1000°C. Accordingly, three different products of CO_2 are modeled separately and the overall CO_2 production is the sum of the resulting yield from each model.

The major problem that faces the application of the single first-order rate model is its application to various heating rates settings. As discussed by Solomon et al. [32], using the reaction rates obtained at a single low heating rate to the high heating rate data can exhibit a terrible agreement. The problem is that the chemistry is not homogeneous, so that the temperature region over which the reaction occurs is large. To fit this large data to temperature range in the low heating rate experiment, an unrealistically low activation energy needs to be used. When applied to high heating rates in which pyrolysis occurs at a higher temperature, the model can fail completely. Realizing this problem, Suuberg et al. chose realistic values of the frequency factors and accepted theoretical fits to their data, which were not necessarily in optimum. However, their fits can be extrapolated [32].

11.5.2 Models employing coal structural analysis

11.5.2.1 FG-DVC model

With the development of modern analytical instruments, several models focusing on coal structure have appeared. Solomon et al. [33] suggested the general coal functional group and depolymerization–vaporization–cross-linking (FG-DVC) devolatilization model, which combines a functional group model for gas evolution and a statistical model for tar formation. The tar formation model includes depolymerization, cross-linking, external transport, and internal transport. The variations with pressure, temperature, coal rank, and heating rate could be well described for the investigated coals. However, this model requires large amount of experimental data, especially those obtained from field ionization mass spectrometry (FIMS) [34] and nuclear magnetic resonance (NMR) [35].

11.5.2.2 FLASHCHAIN model

Niksa and Kerstein [36] formulated afterward the FLASHCHAIN model, in which a model of coal's chemical constitution, a four-step reaction mechanism, chain statistics, and a flash distillation analogy were formulated to explain the devolatilization of various coal types. This was followed by multiple publications [37–42], that evaluated, refined, and extended the practical application of the theory. For example, in publication [38], a numerical evaluation of the model was done and FLASHCHAIN could depict with good accuracy the product yield variation with pressure ranging from vacuum to 6.9 MPa, heating rates ranging from 1 to 3000 K/s, and coal rank. Still the model requires input from NMR [35] tests to characterize the source coal which might impose an application barrier.

11.5.2.3 CPD model

Grant et al. [43] proposed a chemical percolation devolatilization (CPD) model for coal that incorporates analytical data that characterize the diversity of coal structure. Percolation theory provided the necessary mathematical expressions in closed form to account for the lattice statistics important in modeling coal devolatilization processes. Percolation theory analytically describes the size distribution of discrete coal clusters. Still this model also requires extensive NMR [35] experimental results for the coal sample.

11.5.2.4 A predictive multi-step kinetic model

Sommariva et al. [44] proposed recently a multi-step kinetic mechanism, considering coal as aromatic clusters with several bridges, side chains, and functional groups on peripheral positions. Three reference coal structures were considered describing different rank conditions and there was a general satisfactory agreement between the model and the experimental results that suggested the reliability of this model. It is claimed also that the novelty of this kinetic model lies in its predictive approach, without tuning the rate parameters and stoichiometries for different coals. The model also requires only normal chemical analysis for the source coal, which facilitates its practical application for modeling purposes. However, this model has not been validated for other coals yet.

11.6 Conclusion

The refining of lignite or low rank coal was a topic since the early decades of the last century and it will remain a topic for the future due to the large quantities of lignite available and the need for carbon in our metallurgy and chemical industry.

Also huge efforts were and will be done to continue the decarbonization in the energy and fuel industry. The demand for carbon in the chemical industry will never stop and no-one can nowadays predict to which extent the decarbonization process in the fuel sector will go.

The history showed us what could be achieved in the field of lignite refining, if once the decision has been made to start to exploit the huge lignite resources available. Lignite coke, in particular, lignite coke dust can be applied in variety of fields. It can be used as an adsorbent for gas and water treatment, as well as to optimize blast furnace processes in the field of pulverized coal injection processes.

New projects, e.g., in Australia demonstrate that the material use of lignite is still of large interest and the application of a rotary kiln at high temperatures for the direct production of blowable coke dust from lignite dust shows that new developments in the field of lignite refining are made and are still ahead.

References

[1] W. Francis, Fuels and Fuel Technology, Elsevier, London, 1965.
[2] E. Rammler, Technologie und Chemie der Braunkohleverwertung, VEB, Leipzig, 1962.
[3] Klaus Jähnig, Steffen Krzack, Braunkohlenhochtemperatur-Verkokung (BHT-Verkokung). [Buchverf.] Jörg Schmalfeld, Die Veredlung und Umwandlung von Kohle – Technologien und Projekte 1970 bis 2000 in Deutschland, Deutsche Wissenschaftliche Gesellschaft für Erdöl, Erdgas und Kohle e.V., Hamburg, 2008.
[4] Autorenkollektiv, Die Braunkohlengroßkokerei Lauchhammer, Traditionsverein Braunkohle Lauchhammer e.V, Germany, 2006.
[5] Archiv des Traditionsvereins Braunkohle Lauchhammer e.V., 2006.
[6] Archiv des Traditionsvereins Glück-auf Schwarze Pumpe e.V., 2012.
[7] Coal Energy Australia Limited. Coal Energy Australia Limited. *Coal Energy Australia Limited.* [Online] <http://www.coalenergy.com.au/wp-content/uploads/2016/11/CEA-How-The-Plant-Works.pdf>, 2014 (Zitat vom: 12.01.17).
[8] Demonstration of High Value Added Semi Coke From Yallourn Lignite by Pyrolysis. Mr. Davies (CEA). <http://jced.co.jp/wp-content/uploads/2015/10/Mr-Davies-CEA.pdf>, 2015 (Report on the 1st Victoria Brown Coal Forum).
[9] CEA. [Online] <http://www.coalenergy.com.au/the-project/> (Zitat vom: 27. 01.2017).
[10] Jochmann, A., & Richter, H., Koksstauberzeugung aus Braunkohle, in: 5th International Freiberg Conference on IGCC & XtL Technologies, Sonderkolloquium 60, Jahrestag, 24–25 May 2012.
[11] Peters Suuberg, Jack B. Howard, Product composition and kinetics of lignite pyrolysis, Ind. Eng. Chem. Process Des. Dev. 17 (1) (1978) 37–46. Available from: http://dx.doi.org/10.1021/i260065a008.

[12] Lausitz Energie Bergbau AG (LEAG), Quality of Lignite Dust. [Online] <http://www.lignoplus.de/lignoplus-braunkohlenstaub.php> (Zitat vom: 18. 01 2017).
[13] D. Sofialidis, et al., Modelling low-temperature carbonisation of solid fuels in a heated rotary kiln for clean fuel production, Fuel 84 (2005) 2211–2221 (Special Issue: The 5th European Conference on Coal Research and its Applications. ISSN: 0016-2361).
[14] Hong Liu, et al., Numerical simulation of particle motion and heat transfer in a rotary kiln, Powder Technol. 287 (2016) 239–247, ISSN: 0032-5910.
[15] H. Komossa, et al., Heat transfer in indirect heated rotary drums filled with monodisperse spheres: comparison of experiments with DEM simulations, Powder Technol. 286 (2015) 722–731, ISSN: 0032-5910.
[16] Herbert Richter, Bernd Rabann, Study: Entwicklung von Verbundtechnologien zur optimalen Nutzung von Erneuerbaren Energien und der Energiedifferenzen im Stromnetz zur CO_2-armen Kohleveredlung und der Erzeugung von im Gasnetz speicherbarem Mathan/ Methanol, ARCUS Planung + Beratung Bauplanungsgesellschaft mbH, Cottbus, 2013.
[17] Jozo Mamić, Andre Werner, Herbert Richter, Composite technology for coal upgrading processes: Verbundtechnologien zur optimalen Nutzung erneuerbarer Energien mit CO_2-neutralen Kohleveredlungsprozessen mit dem Ziel der Erzeugung von speicherbarem Methan oder Methanol, Kraftwerkstechnik, Strategien, Anlagentechnik und Betrieb, 2014.
[18] Herbert Richter, Jozo Mamic, Joint technology for material use of lignite by gasification or pyrolysis with the objective, IX International Brown Coal Mining Congress, 2016.
[19] Council Directive (EU). 2015/652 of 20 April 2015 Laying Down Calculation Methods and Reporting Requirements Pursuant to Directive 98/70/EC of the European Parliament and of the Council relating to the quality of petrol and diesel fuels, 2015.
[20] Council Directive (EU). 2015/1513 of the European Parliament and of the Council of 9 September 2015 Amending Directive 98/70/EC Relating to the Quality of Petrol and Diesel Fuels and Amending Directive 2009/28/EC on the Promotion of the Use of Energy from Renewable Sources, 2015.
[21] Council Directive (EU). 2009/30/EC of the European parliament and of the Council of 23 April 2009, 2009.
[22] Christian Bergins, K.-C. Tran, et al., Power to Fuel as a Sustainable Business Model for Cross-Sectoral Energy Storage in Industry and Power Plants, POWER-GEN Europe, Milan, 2016.
[23] Jozo Mamic, Herbert Richter, REN-Plus Planungsstudie für den Bau einer Versuchsanlage im Technikumsmaßstab am Standort Industriepark Schwarze Pumpe "Verbundtechnologie Erneuerbarer Energien – Kohletrocknung und – Kohlepyrolyse – Methanolerzeugung", ARCUS Technologie GmbH & Co. KG, CEBra – Centrum für Energietechnologie Brandenburg e.V, Cottbus, 2015.
[24] S. Badzioch, P.G.W. Hawksley, Kinetics of thermal decomposition of pulverized coal particles, Ind. Eng. Chem. Process Des. Dev. 9 (1970) 521–530. Available from: http://dx.doi.org/10.1021/i260036a005.
[25] H. Kobayashi, J.B. Howard, A.F. Sarofim, Coal devolatilization at high temperatures, Symp. (Int.) Combust. 16 (1) (1977) 411–425, ISSN: 0082-0784. Available from: http://dx.doi.org/10.1016/S0082-0784(77)80341-X.
[26] D.B. Anthony, et al., Fifteenth symposium (international) on combustion rapid devolatilization of pulverized coal, Symp. (Int.) Combust. 15 (1) (1975) 1303–1317, ISSN: 0082-0784. Available from: http://dx.doi.org/10.1016/S0082-0784(75)80392-4.

[27] M.J. Antal, et al., Recent progress in kinetic models for coal pyrolysis, Symp. Coal Gasif. Kinet. (1977) 137–148.
[28] H. Reidelbach, M. Summerfield, Kinetic model for coal pyrolysis optimization, Am. Chem. Soc. Dir. Fuel Chem. Preprints 20 (1) (1975) 161.
[29] James Hart, Audai Hussein Al-Abbas, Jamal Naser, Numerical investigation of pyrolysis of a Loy Yang coal in a lab-scale furnace at elevated pressures, Heat Mass Transf. 49 (12) (2013) 1725–1732, ISSN: 1432-1181 Available from: http://dx.doi.org/10.1007/s00231-013-1205-0.
[30] Rastko Jovanovic, et al., Int. J. Heat Mass Transf. 72 (2014) 489–500, ISSN: 0017-9310 http://dx.doi.org/10.1016/j.ijheatmasstransfer.2014.01.036.
[31] E.M. Suuberg, W.A. Peters, J.B. Howard, Seventeenth symposium (international) on combustion *product compositions and formation kinetics in rapid pyrolysis of pulverized coal, implications for combustion*, Symp. (Int.) Combust. 17 (1) (1979) 117–130, ISSN: 0082-0784 http://dx.doi.org/10.1016/S0082-0784(79)80015-6.
[32] Peter R. Solomon, Michael A. Serio, Eric M. Suuberg, Prog. Energy Combust. Sci. 18 (2) (1992) 133–220, ISSN: 0360-1285 http://dx.doi.org/10.1016/0360-1285(92)90021-R.
[33] Hamblen Solomon, Serio Carangelo, Deshpande, General model of coal devolatilization, Energy Fuels 2 (4) (1988) 405–422, http://dx.doi.org/10.1021/ef00010a006.
[34] Hough Retcofsky, Thompson, Friedel, Electron spin resonance studies of coals and coal-derived asphaltenes, Organic Chemistry of Coal, vol. 11, American Chemical Society, Washington, 1978, pp. 142–155.
[35] Horst Friebolin, Jack K. Becconsall, Basic One- and Two-Dimensional NMR Spectroscopy, 2nd ed., Wiley-VCH, Weinheim, 1993, p. 368.
[36] Stephen Niksa, A.R. Kerstein, FLASHCHAIN theory for rapid coal devolatilization kinetics. 1. Formulation, Energy Fuels 5 (5) (1991) 647–665, http://dx.doi.org/10.1021/ef00029a006.
[37] Stephen Niksa, FLASHCHAIN theory for rapid coal devolatilization kinetics. 2. Impact of operating conditions, Energy Fuels 5 (5) (1991) 665–673, http://dx.doi.org/10.1021/ef00029a007.
[38] Stephen Niksa, FLASHCHAIN theory for rapid coal devolatilization kinetics. 3. Modeling the behavior of various coals, Energy Fuels 5 (5) (1991) 673–683, http://dx.doi.org/10.1021/ef00029a008.
[39] Stephen Niksa, FLASHCHAIN theory for rapid coal devolatilization kinetics. 4. Predicting ultimate yields from ultimate analyses alone, Energy Fuels 8 (3) (1994) 659–670. Available from: http://dx.doi.org/10.1021/ef00045a022.
[40] Stephen Niksa, FLASHCHAIN theory for rapid coal devolatilization kinetics. 5. Interpreting rates of devolatilization for various coal types and operating conditions, Energy Fuels 8 (3) (1994) 671–679. Available from: http://dx.doi.org/10.1021/ef00045a023.
[41] Stephen Niksa, FLASHCHAIN theory for rapid coal devolatilization kinetics. 6. Predicting the evolution of fuel nitrogen from various coals, Energy Fuels 9 (3) (1995) 467–478. Available from: http://dx.doi.org/10.1021/ef00051a011.
[42] Stephen Niksa, Flashchain theory for rapid coal devolatilization kinetics. 7. Predicting the release of oxygen species from various coals, Energy Fuels 10 (1) (1996) 173–187. Available from: http://dx.doi.org/10.1021/ef950067l.
[43] D.M. Grant, R.J. Pugmire, T.H. Fletcher, A.R. Kerstein, Chemical model of coal devolatilization using percolation lattice statistics, Energy Fuels 3 (2) (1989) 175–186. Available from: http://dx.doi.org/10.1021/ef00014a011.
[44] Samuele Sommariva, et al., A predictive multi-step kinetic model of coal devolatilization, Fuel 89 (2) (2010) 318–328, ISSN: 0016-2361 http://dx.doi.org/10.1016/j.fuel.2009.07.023.

Production of activated carbon and other products from low rank coals

12

Guenter Harp
Harp Process Chemistry Consulting, Duesseldorf, Germany

Abbreviations

A	activated carbon
ASTM	American Society of Testing and Materials
BF	Blast Furnace
C_{fix}	fixed Carbon
DIN	Deutsche Industrie Norm (German industrial norm)
DEBRIV	Deutscher Braunkohlen Industrie Verein (German Industrial lignite Association)
EDTA	ethylene diamine tetra acetic acid
ESP	electrostatic precipitator
GCV	gross calorific value
LCV	lower calorific value
P_{Heat}	heating power
$R_{coke\ bed}$	electrical resistance of the coke bed
RHF	rotary hearth furnace
SAF	submerged arc furnace
SAIMM	Southern African Institute of Mining and Metallurgy
STP	standard temperature and pressure (273.15K, 1 bar)
U	voltage
UN-ECE	United Nations Economic Commission for Europe

12.1 Introduction

Table 12.1 covers the composition and property ranges of coals of different coal ranks. The first four rows cover the proximate analysis and the next five rows the ultimate analysis.

The coal rank characterizes the metamorphic conversion of the biomass during the coalification process. The coal rank increases from lignite via subbituminous and bituminous coal to anthracite due to increased degree of coalification and geological age. The amount of volatiles decreases with raised coal rank whereas the fixed carbon increases. In the ultimate analysis the carbon content increased with raised coal rank, whereas the content of hydrogen and oxygen decreases. With raising coal rank the gross calorific value (GCV) reaches a maximum for the bituminous coal and decreases slightly to anthracite.

Low-Rank Coals for Power Generation, Fuel and Chemical Production
DOI: http://dx.doi.org/10.1016/B978-0-08-100895-9.00013-9

Table 12.1 Composition and property ranges of coals (different rank) [1]

	Anthracite	Bituminous	Subbituminous	Lignite
Moisture	3–6	2–15	10–25	25–45
Volatile matter	2–12	15–45	28–45	> 45
Fixed Carbon	75–85	50–70	30–57	25–30
Ash	4–15	4–15	3–10	3–15
Sulfur	0.5–2.5	0.5–6	0.3–1.5	0.3–2.5
Hydrogen	1.5–3.5	4.5–6	5.5–6.5	6–7.5
Carbon	75–85	65–80	55–70	35–45
Nitrogen	0.5–1	0.5–2.5	0.8–1.5	0.6–1.0
Oxygen	5.5–9	4.5–10	15–30	38–48
GCV (MJ/kg)	27.9–31.4	27.9–33.7	17.4–23.2	14.0–17.4
Density (g/L)	1.35–1.70	1.28–1.35	1.35–1.40	1.40–1.45

Courtesy of Wiley & Sons, Hoboken.

Different classification systems of coals exist, e.g., the international UN-ECE, ASTM, and DIN. Due to UN-ECE low rank coals are those with a GCV below 24 MJ/kg as well as a vitrinite reflection below 0.6% covering lignites and subbituminous hard coals [2,3]. Lignites are coals with a GCV below 20 MJ/kg. Due to ASTM 388 [4] low rank coals are nonagglomerating coals with a C_{fix} carbon content below 69% and a GCV of below 11,500 BTU/lb, which corresponds to 26.7 MJ/kg. They are designated as subbituminous coals and lignite. According to ASTM 388 the GHV of lignite is below 8300 BTU/lb which corresponds to 19.3 MJ/kg. In Germany low rank coals are the so-called "Braunkohle" characterized by the moisture content between 40% and 60%, the content of volatiles about 45% (dry basis), and a GCV between 8 MJ/kg and 12 MJ/kg [5,6]. Their low heating value is a disadvantage for energy applications in terms of energy efficiency. Therefore low rank coals are predestinated for making products of higher quality like activated carbon (AC) as well as highly reactive carbon products for metallurgical applications like slag foaming in electric arc furnace steel making, reducing gas generation in the tuyere area of the blast furnace (BF) by injection, or ferroalloy production in SAF and the production of calcium carbide, limestone, and dolomite calcination [7].

12.2 Activated and nonactivated carbon products from low rank coals

Activated carbon (AC) is a porous carbon-based material with a specific internal surface of above 300 m^2/g up to 3000 m^2/g [8]. The high specific internal surface is caused mainly by the volume of micropores below 2 nm, which is equal to molecular dimensions. The pores are divided into three categories due to molecular

Figure 12.1 Different shaped active coke products.
Courtesy of VDEh-Betriebsforschungsinstitut, Duesseldorf. K.W. Döhl, G. Harp, H.-G. Clever, P. Kuhn, D. Sucker, Neues Verfahren zur simultanen Entschwefelung, Entstickung und Enthalogenisierung von Hochtemperaturprozessgasen mittels porösen Feststoffes im bewegten Gegenstromreaktor, Schlussbericht BMFT Vorhaben 01 VQ-8518, 1991.

transport. Macropores have a diameter of more than 50 nm. The molecules transport within these pores is bulk diffusion. Mesopores have a diameter between 50 nm and 2 nm. The molecules transport within this pores is Knudsen diffusion, whereas the molecules transport within micropores is the so-called activated diffusion based on the molecular interaction with the surface of the pore wall [9].

Adsorbents are used as spherical- or cylindrical-shaped pellets, or granulated or powdered by milling. The shape and size are dependent on the chosen adsorption process technology. Fig. 12.1 shows granulated, cylindrical-shaped, and pelletized active coke products as an example [10]. Spherical- and cylindrical-shaped products are made of carbonized powder using organic binder like, e.g., tar pitch and a pelletizing disk or a die press.

Fig. 12.2 shows the die of such a die press during investigations of making cylindrical-shaped carbon products [11]. The holes have a diameter of 3 mm and a length of 8 mm. As binders are the key element information about them are kept secret by the AC producing companies. Typical binders are, for example, pitches, glues, and starches. Recipes are dependent on the void volume of the bulk solids, because it limits the binder volume added to the solids. The friction of the mixed bulk solids and binder in the die hole influences the pressure and the degree of compaction of the bulk solids. The green pellets are pyrolyzed at the temperature of activation, which is in the region between 800°C and 900°C leading to a macroporous carbon.

Typical feedstock for AC products is carbon-containing materials with a high content of volatiles between 45% and about 80%. Those volatiles are set free by pyrolysis gaining a highly macroporous carbon product. That is converted afterward

Figure 12.2 Die of a die press for production of cylindrical-shaped products.
Courtesy of VDEh-Betriebsforschungsinstitut, Duesseldorf. R. Pietruck, J. Janz, E. Unland, G. Ventrella, T. Fray, M. Martinez Parcebo, M. Zandi, H. Schmid, Alternate Carbon Sources for Sintering of Iron Ore (Acasos), EUR 25151 <http://bookshop.europa.eu/de/alternate-carbon-sources-for-sintering-of-iron-ore-acasos-pbKINA25151/>, 2013.

in a second step by an activation treatment. The chemical kinetics controlled gasification with water vapor or CO_2 is called often as physical activation, whereas chemical activation of carbon is the treatment with, e.g., NaOH or KOH under controlled temperature conditions [8,12]. The activation treatment determines the microporosity and the specific internal surface area. By physical activation the internal surface area reaches about 1200 m^2/g, whereas by chemical activation an internal surface area of up to 3000 m^2/g is reached.

Nonactivated carbon products are the porous products of the pyrolysis of coals due to the loss of volatiles. Depending on the coal rank and the pyrolysis conditions those products reach a specific internal surface area up to 500 m^2/g [13], whereas metallurgical coke made from higher rank bituminous coals which partially liquefies temporarily on heating above 450°C and resolidify above 600°C reaches about 1 m^2/g which is significantly lower. That is explained by the mentioned formation of the liquid phase and the decomposition of gaseous hydrocarbons into carbon and hydrogen at a temperature above 800°C on carbon surfaces [14–16], as well as unpublished results from investigations at the VDEh-Betriebsforschungsinstitut (BFI) on methane decomposition in the presence of lignite coke as well as metallurgical coke breeze [17].

Compared to AC products the internal surface of the abovementioned nonactivated pyrolytic carbon is lower. Those products are due to the missing activation step much less expensive than the AC products, especially the highly alkaline AC products, which are even more than 2000 €/t. Especially the costs of lignite-based nonactivated carbon products are with about 100 €/t a magnitude lower than AC products. As the adsorption capacity is as a rule of thumb proportional to the internal surface those low cost nonactivated carbon products are applied a lot as one-way adsorbents. After being used as adsorbent, e.g., in foundries, they are usable as fuel [18,19] spending expensive foundry coke.

12.3 Production of lignite coke

Lignite coke is produced in Germany by RWE power from Rhenish lignite at Niederaussem since 1976 in a gas tight rotary hearth furnace (RHF) of Salem Furnace Co., Pittsburgh, PA. [7,20–23]. Today's capacity of the two furnaces is 200,000 t/a. Fig. 12.3 shows a corresponding flowsheet of the lignite coke making by RHF technology. The raw lignite with its more than 50% of moisture is first dried in a production of pulverized lignite powder. The sieve residue is used as feedstock conventional multi-tube rotary drier. The fines are separated by sieving and used for the RHF. It is fed with 36.5 t/h of dried lignite with a moisture content of 14% to manufacture 100,000 t/a of lignite coke. The dried lignite is charged from the coal bins onto the rim of the rotating inclined hearth which has a diameter of 23 m and a hearth area of about 400 m^2. The particles layer height is 150 mm. On being transported down the hearth surface to the rotating disk coke discharge the solids are turned around by special water cooled paddles. Its residence time on the hearth surface is about 1 hour. The solids temperature raises up to 900°C when reaching the discharge shaft in the hearth's center. After about additional 0.5 hours in the discharge shaft for precooling and aging the still hot product is transferred by rotating disk to the coke cooler, where it is cooled first by water injection and further on with air to below 100°C. Then the product is processed by sieving as well as milling for manufacturing the different sized fractions due to customer's applications.

The necessary heat for the lignite pyrolysis is covered by partial combustion of the released gaseous pyrolysis products and is transferred to the coal layer by radiation from the refractory walls and the refractory roof. The gas temperature reaches up to 1300°C with a dust load of 5 g/m^3 STP. It is completely combusted in the postcombustion chamber with a downstream waste heat boiler, where the exhaust gas is cooled down to 200°C and afterward cleaned by an electrostatic precipitator (ESP). For process start-up special burners with external fuel, e.g., oil, are installed.

The analysis data of the different process gases (coal pyrolysis gas, RHF off-gas, and exhaust gas after boiler) are listed in Table 12.2 [20]. As feedstock is dried lignite with a moisture content of 14% the pyrolysis gas contains besides that water content also some reaction water which is somewhat more than half of the pyrolysis gas amount. Other main constituents are CO_2 as well as H_2. Minor compounds are CO, CH_4, higher hydrocarbons C_mH_n, and N_2. The specific gas volume is nearly 550 m^3/t lignite. The calorific value of the gas is about 5 MJ/m^3, which is raised by the tar load of 145 g/m^3 to somewhat above 9 MJ/m^3 [20]. The pyrolysis

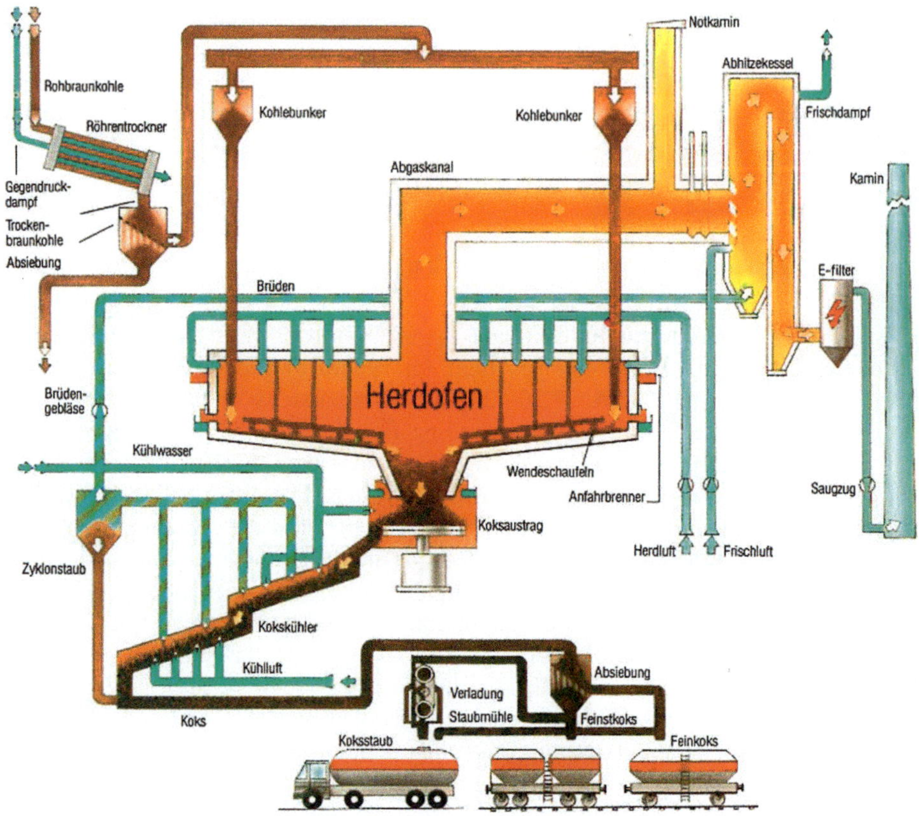

Figure 12.3 Flowsheet of lignite coke making by RHF technology.
Used with permission from RWE Power, WTA Technology, www.rwe.com [23].

gas is released from the coal layer and combusted by injection of air through a lot of holes in the walls and the roof of the furnace. Due to incomplete combustion conditions the hot RHF-gas contains still some combustibles (7% CO and 5% H_2). That gas is conducted in a postcombustion chamber, where it is completely combusted with some minor oxygen content of about 1.5% in the exhaust gas. Its energy content is used by a waste heat boiler to produce steam.

Table 12.3 contains the feed as well as the product data of the RHF for the conversion of dried lignite to the corresponding coke products [21]. Due to agglomeration the moist product is coarser than the dry product. The conversion leads to some fines. For dust-free handling the product must be moistened up to 22.5%. Due to agglomeration on wetting the moist product is coarser than the dry one. The proximate as well as the ultimate analysis shows the change of properties due to the pyrolysis of the coal in particular a raised fixed carbon and ash as well as elementary carbon content. The specific surface area increased from 2 to 200 m^2/g. Own

Table 12.2 Analysis data of process gases of RWE rotary hearth lignite coking (all values in vol.%)

Compound	Pyrolysis gas	RHF off-gas	Exhaust gas
CO_2	12.9	11	12.4
CO	8.5	7	–
H_2	17.2	5	–
CH_4	6.2	–	–
C_mH_n	1.9	–	–
N_2	1	53	60.3
O_2	–	–	1.5
H_2O	52.3	24	25.8
Gas volume (m^3/t)	547.4	–	–
LCV gas (MJ/m^3)	5.26	–	–
Tar load (g/m^3)	145	–	–
LCV total (MJ/m^3)	9.35	1.12	–

Courtesy of the GDMB Verlag, Clausthal-Zellerfeld.

investigations have shown a specific surface area up to 495 m^2/g [10], but mainly in the range around 300 m^2/g [18].

Another production route is developed in Russia based on the application of fluidized bed technology called "Termokoks" process [24]. In Table 12.4 the proximate and ultimate analyses as well as the LCV of the feedstock Berezovsk lignite are listed. Compared to the Rhenish dried lignite the Berezovsk feedstock lignite has significantly more moisture and less volatiles leading to a higher C_{fix} value. The sulfur content is lower and the ash content is somewhat higher. The ultimate analysis shows with 71% a somewhat higher carbon and with 4.9% a higher hydrogen content. The oxygen content is slightly lower compared to the Rhenish lignite.

Fig. 12.4 shows the corresponding flowsheet of the lignite semicoke production. In fluidized bed combustion of lignite the primary air gasifies the coal whereas the secondary air leads to a complete combustion of the gas in the freeboard region. For the pyrolysis of lignite [24] that fluidized bed combustion was operated with doubling the lignite feed and keeping primary and secondary air constant leading to a bed temperature of 700°C and producing in total the same heat for the steam boiler as with half of the feed and complete combustion but additionally the lignite semicoke as product being sold commercially for metallurgical application due to its high reactivity and high electrical resistivity [25].

12.4 Production of activated carbon from hard coal

Using hard coal as base carbon material for production of AC has been developed by former Bergbauforschung in Germany [25–28]. The hard coal is fine grinded,

Table 12.3 Data of dried Rhenish lignite and the corresponding coke products

Size (mm)	Dried lignite	Lignite fine coke	Lignite coarse coke
>6.30	10.0	0.0	0.0
4.00–6.30	14.0	5.0	25.0
3.15–4.00	10.0	8.0	40.0
2.00–3.15	20.0	25.0	25.0
1.00–2.00	32.5	40.0	7.0
0.50–1.00	11.5	20.0	3.0
< 0.50	2.0	2.0	0.0
Proximate analysis			
Moisture	14.0	2.5	22.5
Ash	2.9	7.0	5.5
Volatiles	44.0	3.0	2.1
Fixed carbon	39.1	87.5	69.9
LCV (MJ/kg)	20.8	29.9	23.7
Ultimate analysis dry			
C	65.55	91.15	
H	4.75	0.55	
S	0.45	0.37	
N + O	25.88	0.75	
Tar	9.55		
Water	10.15		
Coke and ash	60.00		
Gas	20.30		
Physical data			
True density (t/m^3)	1.56	1.75	
Bulk density (t/m^3)	0.52	0.5	
SSA (m^2/g)	2.00	200	

All values in weight percent if not otherwise denoted.
Courtesy of ET Verlag, Essen. H.B. Koenigs, R. Kurtz, Feinkokserzeugung aus Braunkohle, Energiewirtschaftliche Tagesfr. 27 (8–9) (1977) 596–600.

preoxidized with air at about 250°C to raise the oxygen content of the coal to a maximum, mixed the preoxidized coal with organic binder, and formed into cylindrically shaped pellets. Those pellets are pyrolyzed at raised temperature and then steam activated gaining a specific surface of about 1000 m^2/g.

The low temperature oxidation of coal as well as other porous carbon leads to build up of functional oxygen containing groups like carbonic acids, esters, lactones, ketones within the aliphatic structure of the coal [26], and also the aromatic/graphitic defect structures of porous carbon [29]. To gain a complete oxidation

Table 12.4 **Data of Berezovsk lignite**

Proximate analysis	Lignite	Semicoke
Moisture	30.6	<0.5
Ash	4.9	8–9.5
Volatiles	47.6	8–9.5
Fixed carbon C_{fix}	47.5	82.5
LCV (MJ/kg)	15.74	29.22
Ultimate analysis dry		
C	71.0	92.8
H	4.9	1.52
S	0.15	0.24
N	0.7	0.97
O	23.1	4.45

Courtesy of Allerton Press, New York. D.A. Loginov, S.R. Islamov, Fluidized-bed carbonization of lignite, Coke Chem. 53 (5) (2010) 179–182.

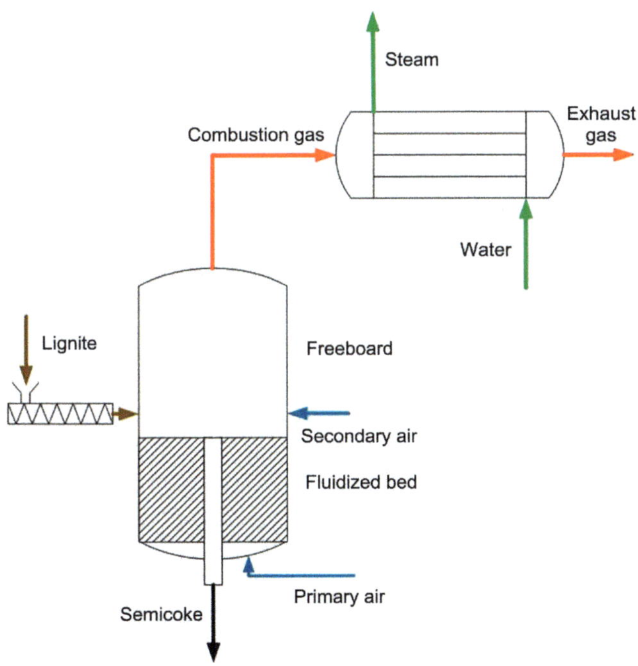

Figure 12.4 Flowsheet of semicoke production from lignite (Termokoks process).

from surface to the center of the hard coal particle and to overcome diffusion limitations the particle size shall be below 0.1 mm. Due to its exothermal character and to control the temperature the oxidation is carried out within a fluidized bed with internal cooling.

During its low temperature oxidation the hard coal reaches an oxygen amount up to 25% which is equivalent to an O/C ratio of 0.3. The volatiles content reaches nearly 40%. The oxy-coal looks similar to lignite, although it is not a relignification, because the H/C ratio is constant on low temperature oxidation with air. The pyrolysis of that oxy-coal leads to the evolution of CO_2 and CO which is equivalent to the stored oxygen. The release of CO_2 has maxima at 350 and 450°C, whereas the CO release is related to about 550°C. The CO_2 release is related to the decay of carbonic acid, lactonic as well as esterific compounds, whereas CO may be released to other carbonyl compounds (ketones) at the coals surface. The solid pyrolysis product is called oxycoke although it is nearly oxygen free. The pyrolysis of the oxy-coal gains an increase in porosity as well as internal surface area. Its maximum is reached in the temperature range between 700°C and 800°C. Further increase of the temperature leads to a decrease in internal surface area as well as pore volume. It is related to a release of methane and H_2 as gaseous products. The maximum surface area is dependent on heating rate and lies between 250 m^2/g and 400 m^2/g. On pyrolysis of oxy coals the volume of micropores with a diameter below 0.6 nm increases with increasing heating rate. The volume of the more coarse pores with a diameter above 2 nm is affected by the release of CO_2 during pyrolysis and independent of heating rate [27].

The activation of the oxycoke is done in the same way like other AC with steam or carbon dioxide at a temperature of about 900°C reaching the final surface area of up to 1200 m^2/g [28]. It must be done at a temperature where the corresponding heterogeneous chemical water gas or Boudouard reaction determines the kinetics. Therefore the temperature is limited to the temperature where pore diffusion determines the overall kinetics.

By the decomposition of gaseous hydrocarbons within the pore system the production of carbon molecular sieves is possible with pores of defined diameters of small molecules like N_2 and O_2 [30]. According to Eq (12.i) the decomposition of

$$C_nH_m \rightarrow nC + m/2\ H_2 \tag{12.i}$$

gaseous hydrocarbons gains a precipitation of solid carbon and gaseous hydrogen at the carbon surface. If the reaction is carried out under conditions where the reaction kinetics limits the decomposition speed, the precipitation occurs at the total surface giving rise to a pore narrowing effect. Similar effects are seen at the investigations of Muradov [14–16] about the catalytic decomposition of methane with AC in a fluidized bed reactor as well as some work of BFI on using lignite coke imstead of AC for catalytic decomposition of natural gas to gain H_2 [17]. The precipitation of carbon leads also to a strong reduction of the sensivity against oxidation [29] as well as Boudouard reactivity [17].

12.5 Selected applications for activated and nonactivated carbon products from low rank coals

Due to its increased specific surface AC and nonactivated carbon products are used for adsorption purposes from gaseous and liquid media. Gas and water cleaning are the most important applications for those products [23,28]. Different technologies are developed especially for gas adsorption processes due to their dual phase mass transfer. Fig. 12.5 shows as an example a pilot plant with two moving bed countercurrent flow adsorbers for cleaning 2000 m^3/h STP at a municipal waste incinerator plant [10].

The process technology determines the size of the adsorbent particles and its related properties like pressure loss as well as solid/gas ratio, Table 12.5. Particle size of fixed bed or moving bed systems is coarser than that of fluidized bed systems and much coarser than that of entrained flow adsorption systems. Pressure loss raises from fixed bed to fluidized bed, and decreases to transport adsorption systems. The solid to gas (S/G) ratio as well as the mass transfer determining

Figure 12.5 Pilot plant with moving bed countercurrent flow adsorbers.
Courtesy of VDEh-Betriebsforschungsinstitut. K.W. Döhl, G. Harp, H.-G. Clever, P. Kuhn, D. Sucker, Neues Verfahren zur simultanen Entschwefelung, Entstickung und Enthalogenisierung von Hochtemperaturprozessgasen mittels porösen Feststoffes im bewegten Gegenstromreaktor, Schlussbericht BMFT Vorhaben 01 VQ-8518, 1991.

Table 12.5 **Process technology data for adsorbers**

Process technology	Particle size	Pressure loss	S/G ratio
Moving bed	>1 mm	Low	High
Fluidized bed	<1 mm; >0.2 mm	Medium	Medium
Entrained flow adsorber	<0.1 mm	Low	Low

Figure 12.6 Electrostatic precipitator for waste gas cleaning of an iron ore sinter plant. Courtesy of ArcelorMittal Eisenhüttenstadt, Author: Joerg Richter.

difference between gas and particle velocity decreases from moving bed to fluidized bed systems and is lowest for an entrained flow adsorber. In this particular case an additional gas solid separation system, e.g., ESP or fabric filter is required. Using the latter allows a lower amount of adsorbent because the solids layer upon the filter tissue adsorbs additionally during its retention. That is why those systems are typically retrofitted into existing dedusting plants and are applied for capturing harmful trace compounds out of big waste gas volume flow above 100,000 m^3/h like dioxin emission reduction systems for iron ore sinter plants or scrap melting electric arc furnace plants [31,32]. Fig. 12.6 shows as example an ESP at an iron ore sinter plant during dioxin emission reduction investigations. Here it is necessary to reach a smooth distribution of adsorbents over the pipe cross-section by fluid mechanics optimization.

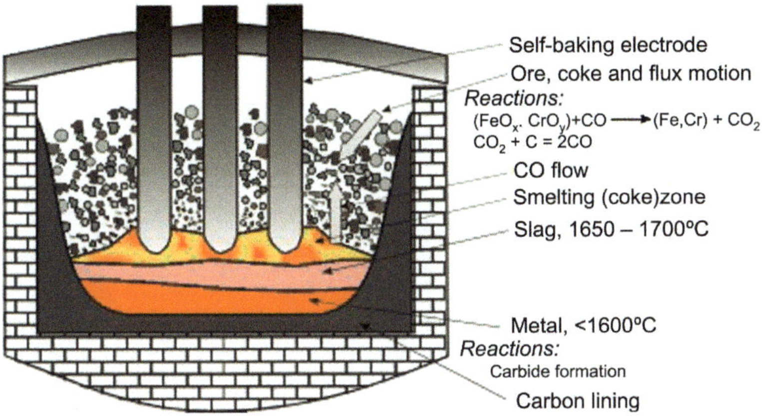

Figure 12.7 Scheme of submerged arc furnace.
Courtesy of SAIMM. Y. Yang, Y. Xiao, M.A. Reuter, Analysis of transport phenomena in submerged arc furnace for ferrochrome production, in: Proceedings of 10th International Ferroalloys Congress, 1–4 February 2004, Cape Town, South Africa.

Mercury separation especially from flue gas of municipal waste and sewage sludge incineration plants is an application of lignite coke [33]. The mechanism is based on the formation of mercuric chloride out of mercury vapor in the presence of hydrogen chloride due to the catalytic oxidation at the coke surface. An additional way is the use of sulfuric acid loaded lignite coke, which is achieved on flue gas desulfurization. It was tested successfully for mercury separation out of the flue gas at operational scale trials for production of alinite cement out of the filter residues as well as slags from municipal waste incineration plants [33]. The investigations have shown that 130°C is an optimum temperature for Hg separation with a sulfuric acid saturated lignite coke.

Other applications for AC are special waste water cleaning processes. Due to its electrical conductivity AC are applicable as anode for electrochemical oxidation of waste water components like cyanides, phenols, or EDTA [34].

Characteristic properties of nonactivated carbon products from low rank coals are their high chemical reactivity for reduction and gasfication processes as well as their electrical conductivity. Those are beneficial properties for application in production of ferroalloys like FeSi or elementary silicon as well as calcium carbide in submerged electric arc furnaces (SAF) [35]. For substitution of the used so-called nut coke (size 10–20 mm) the nonactivated carbon product must be briquetted. Fig. 12.7 shows the scheme of a SAF for Ferrochromium production [36]. The heat is generated due to the voltage drop within the coke bed layer due to its ohmic resistance in contact with carbon electrodes. For an ohmic heating of the coke bed it is necessary that its resistance is high compared to the carbon electrodes gaining the main voltage drop for generating the heating power within the coke bed in accordance with the following equation:

$$P_{\text{heat}} = U^2 / R_{\text{coke bed}} \tag{12.1}$$

To fulfill that demand it is necessary to use a carbon material with high electrical resistivity compared to the electrode carbon. The resistivity of carbon materials is a function of its graphitization degree. On the other hand a high reactivity of the carbon material is important for reduction smelting process, because that determines the furnace productivity. A certain lumpy size is necessary to secure a sufficient permeability of the feedstock bed to release the CO product gas. The main disadvantage of coke made of bituminous coal is its low reactivity. To overcome this high reactive carbon material like briquetted lignite coke is added [37]. The lignite coke is produced by the so-called "Termokoks" technology out of Siberian lignite coal at about 750–800°C in a fluidized bed boiler furnace fed with duplicate amount of lignite coal compared to usual ignite combustion. No information on the used binder is given. The use of the lignite semicoke briquettes up to 40% replacement of nut coke gains in about 3% productivity increase for ferrosilicon production compared to pure nut coke charging and due to its lower costs also an economic advantage. Tests at a ferrochromium furnace have shown no influence on electricity consumption as well as on productivity increase. Due to lower costs of the briquettes an economic advantage is gained.

12.6 Conclusions and future outlook

As the mercury emissions from coal fired power plants will be limited further down to below 10 μg/m^3 [38] there will be a need for mercury separation from the flue gas to fulfill that limitation. That is a typical application for the use of lignite-based carbon products. As flue gas volume flow of those power plants is in the same range than iron ore sinter plants the application of the entrained flow adsorber technology has technical advantages. There will be a big market for retrofitting the flue gas cleaning systems of coal fired power plants in the future.

Due to its high reactivity and its high carbon content nonactivated carbon products from low rank coals can substitute low volatile pulverized coal for injection into the BF. Usually a high volatile as well as a low volatile coal is used for coal injection to reduce the demand of the expensive coke. The high volatile coal acts as an ignition aid for the low volatile coal which gives a higher replacement ratio to BF coke compared to the high volatile coal. The pulverized coal size is below 0.1 mm. The nonactivated carbon products should be similar in size to be able to make use of the existing injection facilities [38]. Due to the low volatile content of the carbon products they will be preheated for a further decrease of the BF coke demand. As the future applications need fine grained products preferably, a production technology giving fine grained products has technical advantages.

The pyrolysis products of the low rank coals are up to now only used to cover the energy demand of the pyrolysis process. To use its components as feedstock for production of chemicals it is necessary to develop new heating concepts and a good heat integration. Economic advantages are expected when the pyrolysis process is integrated into an industrial environment where there is unused waste heat available.

References

[1] E. James, Speight: Handbook of Coal Analysis, Wiley & Sons Inc, Hoboken, NJ, USA, 2005.

[2] UN ECE: International Codification System for low rank Coal Utilization <https://www.unece.org/fileadmin/DAM/ie/se/pdfs/lowrank.pdf>, 2002.

[3] UN ECE: International Codification System for medium and high rank Coals <https://www.unece.org/fileadmin/DAM/ie/se/pdfs/codee.pdf>, 1988.

[4] ASTM D388-15, Standard Classification of Coals by Rank, ASTM International, West Conshohocken, PA, 2015.

[5] Federal Institute for Geosciences and Natural Resources, Energy Raw Materials, 2009, Hannover, Germany.

[6] Lignites in Germany <http://www.braunkohle.de/36-0-DEBRIV.html>, 2016.

[7] D. Böcker, M. Erken, J. Heuter, H. Kreusing, Edle Körner, Energie 35 (3) (1983) 35–37.

[8] H. Marsh, F. Rodriguez-Reinoso, Activated Carbon, Elsevier Ltd., Oxford, 2006.

[9] R.S.A. de Lange, K. Keizer, A.J. Burggraaf, Analysis and theory of gas transport in microporous derived ceramic membranes, J. Membr. Sci. 104 (1995) 81–100.

[10] K.W. Döhl, G. Harp, H.-G. Clever, P. Kuhn, D. Sucker: Neues Verfahren zur simultanen Entschwefelung, Entstickung und Enthalogenisierung von Hochtemperaturprozessgasen mittels porösen Feststoffes im bewegten Gegenstromreaktor, Schlussbericht BMFT Vorhaben 01 VQ-8518,1991.

[11] R. Pietruck, J. Janz, E. Unland, G. Ventrella, T. Fray, M. Martinez Parcebo, M. Zandi, H. Schmid, Alternate Carbon Sources for Sintering of Iron Ore (Acasos), EUR 25151 <http://bookshop.europa.eu/de/alternate-carbon-sources-for-sintering-of-iron-ore-acasos-pbKINA25151/>, 2013.

[12] A. Linares-Solano, D. Lozano-Castello, M.A. Lillo-Rodenas, D. Cazorla-Amoros, Carbon activation by alkaline hydroxides, in: L.R. Radovic (Ed.), Chemistry and Physics of Carbon, vol 30, 2008, pp. 1–62.

[13] D. Oynbold, A.M. Syroezhko, N.V. Slavoshevshaya, V.M. Strahkov, Thermodestruction of Mongolian and Russian lignite in slow and fast pyrolysis, Coke Chem. 53 (4) (2010) 138–144.

[14] N. Muradov, Onsite production of hydrogen from hydrocarbon fuels with minimal greenhaus gas emissions, Fuel Chem. Div. Preprints 48 (2003) 331–332.

[15] N. Muradov, F. Smith, A. Raissi: Catalytic Dissociation of Hydrocarbons: A Route to CO_2-Free Hydrogen <https://www3.kfupm.edu.sa/catsymp/Symp15th/10%20Muradov.pdf>.

[16] N. Muradov: CO_2-Free Production of Hydrogen from Hydrocarbon Fuels, Final Report DE-FC36-99GO10456, US Department of Energy, 2003.

[17] M. Hensmann, G. Stubbe, S. Möhring (Betriebsforschungsinstitut): Investigations of the Methane Decomposition at Coarse Particles, 2013, Unpublished results.

[18] H. Anger, G.Harp, M. Hensmann, D. Sucker: Einsatz von beladenem Aktivkoks aus der Abgasreinigung im Kupolofen Giesserei,vol. 81, no. 3, 1994, pp. 67–70.

[19] J. Zechmann, M. Hensmann, G. Harp, Abscheidung und Beseitigung organischer Komponenten aus Gießgasen mittels Akivkoks, Giesserei 82 (22) (1995) 817–821.

[20] R. Kurtz, Feinkoksherstellung aus Braunkohle, Braunkohle (11) (1975) 352–356.

[21] H.B. Koenigs, R. Kurtz, Feinkokserzeugung aus Braunkohle, Energiewirtschaftliche Tagesfr. 27 (8–9) (1977) 596–600.

[22] E. Scherrer, Herstellung von Braunkohlenkoks im Salem-Lurgi Herdofen, Braunkohle (7) (1981) 242–246.
[23] Braunkohlenkoks—Herstellung und Einsatzmöglichkeiten, Rheinbraun Sales Co., 1989.
[24] D.A. Loginov, S.R. Islamov, Fluidized-bed carbonization of lignite, Coke Chem. 53 (5) (2010) 179–182.
[25] H. Jüntgen, J. Karweil, Die physikalisch-chemischen Eigenschaften von oxydierten Steinkohlen und ihren Koksen. I. Die Veränderung der Kohleeigenschaften während der Oxydation. Erdöl und Kohle, Erdgas, Petrochemie 15 (11) (1962) 898–906.
[26] H. Jüntgen, J. Karweil, Die physikalisch-chemischen Eigenschaften von oxydierten Steinkohlen und ihren Koksen. II. Die Eigenschaften der Oxykokse. Erdöl und Kohle, Erdgas, Petrochemie 15 (12) (1962) 985–990.
[27] H. Jüntgen, K.-Chr Traenkner, Die Entgasung von oxydierten Steinkohlen, Brennstoffchemie 45 (4) (1964) 105–114.
[28] H. Jüntgen Aktivkohle und Aktivkokse für die Reinhaltung von Luft und Wasser. Informationstagung Kohleforschung, 8–9 December 1970, Luxemburg, EUR 4641, pp. 165–172.
[29] H.P. Boehm, Carbon surface chemistry, in: P. Delhaes (Ed.), Graphite and Precursors, Gordon & Breach Science Publishers/OPA, Amsterdam, 2001.
[30] H. Jüntgen, Gezielte Herstellung von Porensystemen aus kohlenstoffhaltigem Material, Ber. Bunsenges. 79 (9) (1975) 747–748.
[31] K. Kersting, J. Wirling, W. Esser-Schmittmann, U. Lenz, Dioxin- und Furanabscheidung aus Abgasen einer Sinteranlage an Braunkohlenkoks, Stahl u. Eisen 117 (11) (1997) 49–55.
[32] C. Prüm, C. Werner, J. Wirling, Reducing dioxin emission in electric steel mills, MPT Int. (1) (2005) 36–42.
[33] W. Esser-Schmittmann, J. Wirling, U. Lenz: Reduction in mercury emission with lignite coke, in: 8th Annual North-American Waste-to-Energy Conference, 22–25 April 2000, Nashville, TN, USA.
[34] G. Endemann, B. Schmidt, G. Harp, D. Sucker: Einsatz von Aktivkoksanoden in der elektrochemischen Abwasserreinigung, Galvanotechnik.
[35] M. Gasik (Ed.), Handbook of Ferroalloys, Butterworth Heinemann, Oxford, UK, 2013.
[36] Y. Yang, Y. Xiao and M.A. Reuter: Analysis of transport phenomena in submerged arc furnace for ferrochrome production, in: Proceedings of 10th International Ferroalloys Congress, 1–4 February 2004, Cape Town, South Africa.
[37] V.B. Vorob'ev, P.P. Orlov, S.R. Islamov, S.G. Stepanov, Lignite-coke briquettes in ferroalloy production, Steel Transl. 45 (3) (2015) 207–211.
[38] A. Kather, M. Klostermann, Grenzwerte für Quecksilberemissionen aus Kohlekraftwerken, VGB Power Technol. 12 (2015) 74–80.

Further reading

S.R. Islamov, Energy-technological method for utilization of coal from the Kansko-Achinskii basin, Therm. Eng. 60 (11) (2013) 783–786.
R. Lange, W. Heschel, J. Wirling, Beeinflussung der Porenstruktur von Adsorbentien durch Abscheidung von Kohlenstoff und Auswirkung auf die Adsorptionskapazität und –geschwindigkeit, Chem. Ing. Tech. 76 (11) (2004) 1640–1644.

G. Harp, T. Kießling, D. Sucker: Vermeidung von Schadstoffkreisläufen bei der Alinit-Zementerzeugung mit Rückführung der Rauchgasreinigungprodukte, BFI-Bericht 1371, Forschungsvorhaben des Abfallentsorgungs- und Altlastensanierungsverbands NRW, 1994.

L. Sundqvist-Ökvist, M. Lundqvist, M. Lundgren, D. Malmberg, E. Olsson, O. Anttila, M. Hensmann, G. Harp, S. Schuster, G. Hyllander: Flexible Injection of Alternative Carbon Material into the Blast Furnace (Flexinject), EUR 25909 <http://bookshop.europa.eu/en/flexible-injection-of-alternative-carbon-material-into-the-blast-furnace-flexinject-pbKINA25909/>, 2013.

Role of oxycombustion in utilization of low rank coals

13

Janusz Lasek and Jarosław Zuwała
Institute for Chemical Processing of Coal, Zamkowa, Zabrze, Poland

13.1 Introduction

CO_2 capture technologies can be divided into three categories: postcombustion capture, precombustion capture, and oxy-fuel combustion [1]. Oxy-fuel combustion is one of the way to efficiently reduce CO_2 emission to the atmosphere. According to the definition provided by Chen et al., oxy-fuel is the technology where "instead of using air as oxidizer, pure oxygen (O_2) or a mixture of O_2 and recycled flue gas is used to generate high CO_2 concentration product gas." Next, captured CO_2 can be stored or utilized in another processes [2]. Oxy-fuel combustion is one of the leading technologies considered for capturing CO_2 from power plants with carbon capture and storage technology, CCS. Current development of CCS (including oxy-fuel combustion) was inhibited due to economic, social, and political issues. The actual attitude of politics, public opinion, and part of scientific community is rather critical for oxy-fuel technology. The main reason is the cost and huge energy penalty [3,4]. However, they are still the projects where efforts lead to obtain demonstration scale plants of oxy-fuel technology. Stranger et al. [3] reviewed different ongoing and finalized projects related to oxy-fuel combustion for CO_2 capture in pilot or larger scale facility. They concluded a significant progress of oxy-fuel technology during 10 years (2005–15) of the efforts that were suggested in Special Report on CO_2 Capture and Storage by the International Panel for Climate Change (reported in 2005). Important factor of the evaluation of technology status is Technology readiness level, TRL, that includes the developments status from the observation of basic principles (TRL1) up to full-scale commercial deployment (TRL9). During different projects only White Rose Project (European) is evaluated as the TRL8 level (sub-scale commercial demonstration plant). The main goal of the project is to achieve the world's largest supercritical oxy-fuel combustion power plant at 426 MWe (gross). Another project, i.e., Callide Oxy-fuel Project (Australia), is evaluated as TRL7 (industrial scale pilot plant) project where "mini-demonstration" plant (30 MW_e of pulverized boiler) reached nearly 11,000 operating before being decommissioned (by March 2015). Vattenfall's Schwarze Pumpe Pilot Plant is the first installation where oxy-fuel combustion was realized at demonstration scale (30 MW_{th}) with CCS installation [3]. The pilot plant consists of a single 30 MWth, top-mounted pulverized fuel (PF) burner and the flue gas cleaning equipment and CO_2 processing. The flue gas cleaning equipment consists of an Electrostatic Precipitator (ESP) for particle removal, wet Flue Gas Desulfurization (FGD) for the removal of SO_2 and other acidic components, and the Flue Gas Condenser (FGC) for

Low-Rank Coals for Power Generation, Fuel and Chemical Production
DOI: http://dx.doi.org/10.1016/B978-0-08-100895-9.00014-0

cooling of the flue gases and removal of water. A hot flue gas recycle is applied in the installation. Oxygen is introduced from cryogenic air separation unit (ASU) [5]. In Schwarze Pumpe Plant a few burners were tested (i.e., Alstom, Hitachi, Doosan, and BBS; mainly swirl burners). This pilot plant is now being decommissioned [3]. Currently in Germany, the debate of public opinion on the CCS application is under consideration [6]. To summarize, five installations at demonstrative scale with CCS were tested or are going to be tested in near future, as it was presented in Fig. 13.1. The installations are Vattenfall's Schwarze Pumpe Pilot Plant (30 MW_{th}, 10 MW_e, TRL6), Total's Lacq-Rousse pilot plant (30 MW_{th}, 10 MW_e, TRL6), Callide Project pilot plant, Shenhua Group plant (localized in Shanxi Province, China), and White Rose plant (localized in Drax power station, North Yorkshire, UK) [3]. Unfortunately, on April 13, 2016, the funding for White Rose project was cut and the consortium of the project (Capture Power Limited) decided to *began the process of winding down the consortium operations* (http://www.whiteroseccs.co.uk/, accessed 19.08.16). This was related to the government's decision to end the Carbon Capture and Storage commercialization competition (http://www.whiteroseccs.co.uk/, accessed 19.08.16; https://sequestration.mit.edu/tools/projects/white_rose.html, accessed 19.08.16).

Low rank coals (i.e., lignite, sub-bituminous coals) are very abundant over the world. It is estimated that the total world reserves of lignite are about four trillion tons. In China, the reserves of lignite are about 190.3 billion tons (41.18% of the total coal reserve of China) [7]. The excavating (in 2015) in Germany reached level of 178 Mt (http://www.energypost.eu/lignite-europe-fighting-back-renewables/, accessed 24.02.16). According to data provided by Polish Power System (Polskie Scieci Energetyczne, PSE S.A.) in Poland, electrical energy production in January 2016 by lignite combustion was 4083 GWh and it was over 27% of the total electrical energy conversion in Poland [8]. The main disadvantage of low rank coals—especially lignite, brown coal—is their high moisture content (25%–70%). It can affect power plant efficiency leading to raise of CO_2 emission [7,9]. Thus, oxy-fuel technology could bring some solution if low rank coals are considered as low-emission energy source. However, despite the oxy-fuel advantages, this technology is not ready to be fully commercialized at present. One of the main barriers is lack of the experiences from demonstrative scale campaigns and still not clear future of this technology due to economic issues. Anyway, integration of oxy-fuel technology with the application of low rank coals is worth to be considered due to many advantages. The aim of this chapter is to show an overall view of the low rank coals application in oxy-fuel technology. Combustion process at oxy-fuel conditions, emission issues, chances, and barriers of oxycombustion technology in terms of low rank coals application are described and briefly analyzed.

13.2 Overall view on the low rank coals research in oxy-fuel combustion

13.2.1 Behavior of low rank coals during its combustion under oxy-fuel regime

It is known that such parameters like temperature, burnout time, combustion mode, radiation intensity, and ignition delay can be influenced by combustion atmosphere.

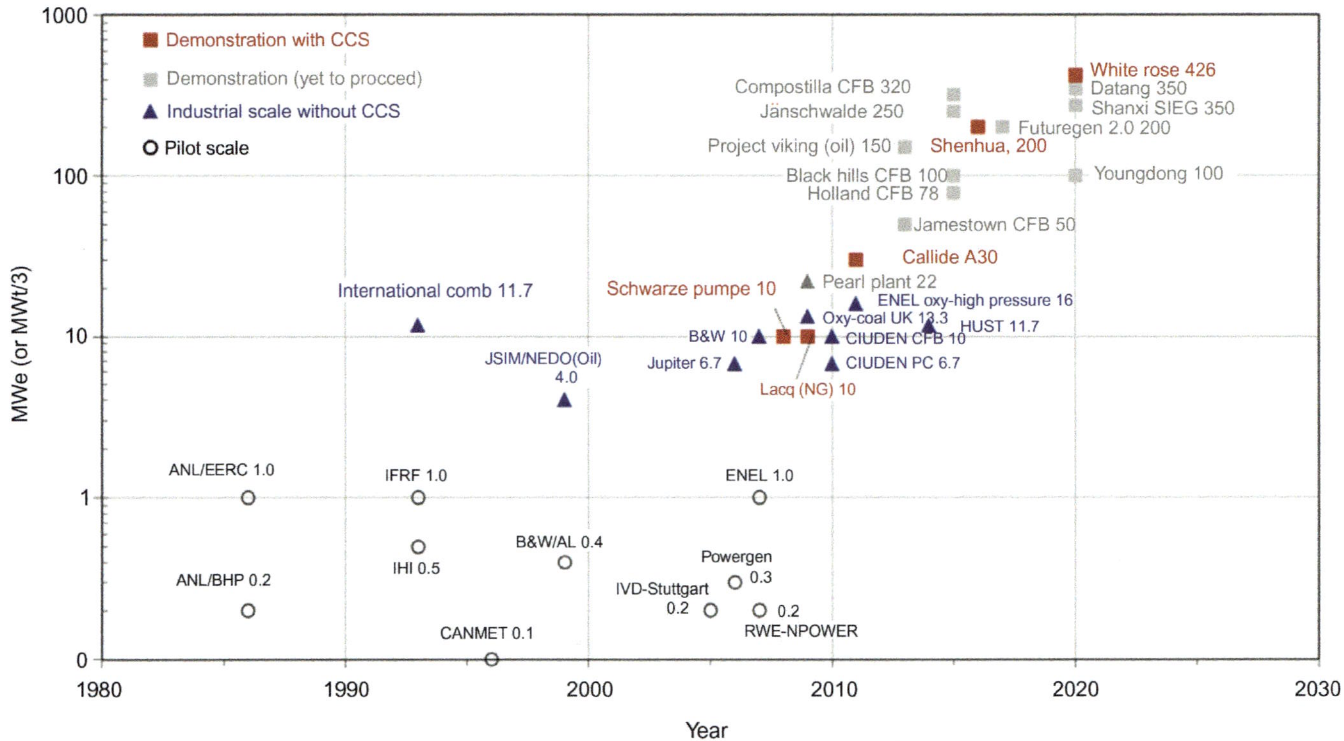

Figure 13.1 Historical progression of oxy-fuel combustion technology from small pilot-scale facilities to projects at demonstration scale [3]. Printed upon permission of Elsevier, License Number 3931310992837.

Due to differences in thermodynamic parameters between N_2 and CO_2, combustion parameters observed for air-combustion can be significantly differing in oxycombustion and it is strongly affected by O_2 concentration. Khatami and Levendis [10] analyzed impact of coal rank and combustion atmosphere on ignition and other combustion parameters. Overall conclusion was that coal particles burning in O_2/CO_2 atmospheres exhibit similar behaviors (such as combustion mode, sequence, and fragmentation) as in air. The main qualitative difference is the lower luminosity of the particles in an O_2/CO_2 atmosphere with the same oxygen mole fraction as in air. They concluded that combustion in higher O_2 mole fractions in either N_2 or CO_2 environments decreased the ignition temperature. For bituminous and lignite coal particles, as O_2 mole fraction in N_2 increased from 0.2 to 1, the ignition temperature decreased by 100−150K. Moreover, at oxycombustion conditions, the ignition delay times decreased drastically as oxygen mole fraction increased beyond the air level (21%) to 100%. However, ignition delay times are higher in CO_2 environments than in N_2 environments with the same O_2 mole fraction level. It was explained by the higher heat capacity of CO_2 and lower diffusivity of O_2 in CO_2. The volatile flame durations were drastically shortened as oxygen mole fraction increased beyond the air level (21%) to 100%. However, this phenomenon was observed also when oxygen was added to air [10,11]. At oxy-fuel combustion, the flame and char temperatures have been found to be lower and the burnout times to be longer than in (O_2/N_2) for all coal ranks at comparable O_2 mole fractions [10]. Bejarano and Levendis [12] presented a fundamental research of lignite particles [45−53 and 75−90 μm, from Lower Wilcox, Texas (PSOC-1443)]. A drop-tube furnace combined with a three-color pyrometer was applied to analyze fundamental combustion properties like temperature and burnout time. Molar fractions of oxygen varied between 20% and 100% in both cases O_2-N_2 and O_2-CO_2 gas mixtures. They noticed that in the case of lignite coal particles, the differences in temperature between combustion in N_2 and in CO_2 were on the order of 120K, whereas the corresponding differences (gaps) in the combustion duration were only ~3 ms. The results are presented in Fig. 13.2. Moreover, to attain the same volatile and char temperature as in air (21% O_2), the oxygen content in the CO_2 mixtures had to be increased to 26% for lignite.

Yi et al. [13] analyzed the impact of coal rank and oxygen fraction in oxy-fuel atmosphere on characteristic temperature (i.e., ignition temperature T_{ig} and burnout temperature T_b) during combustion. They analyzed 28 kinds of Chinese coals with different rank. They mentioned that most coal types require a higher temperature for ignition and burnout in oxy-fuel atmosphere than in air atmosphere at the O_2 concentration of 21 vol.%. Bu et al. investigated ignition and devolatilization [14,15] as well as burnout of char [16] of four ranks of coal (including lignite) and wood in O_2/N_2 and O_2/CO_2 atmosphere (with O_2 inlet concentrations of 0, 10, 21, 30, 40 vol.%). In case of almost all fuels (apart from anthracite) a bright diffusion flame was observed. It was interpreted as homogeneous combustion of the volatile matter. In case of wood, lignite, and bituminous coal, the ignition occurred at almost the same time (i.e. ~0.9 seconds, 4.1 seconds, and ~6 seconds for wood, lignite, and bituminous coal, respectively). For sub-bituminous coal ignition in case of oxycombustion was

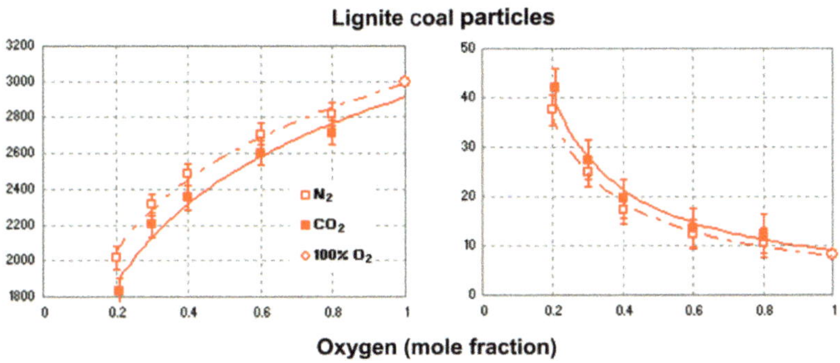

Figure 13.2 Average temperatures (correspond to ca. 50% burnout time, left side) and burnout times (right side) for lignite coal particles (both 45–53 μm) burning in O_2/N_2 and O_2/CO_2 at $T_{furnace} = 1400K$ [12].
Printed upon permission of Elsevier, License Number 3838790423162.

Table 13.1 Qualitative comparison of combustion properties (oxy vs air mode) obtained during combustion of anthracite coal (AC), bituminous coal (BC), sub-bituminous coal (SBC), lignite coal (LC), Schima wood (SW)

Fuel	Oxy vs air mode		
	Ignition time	**Homogeneous combustion time**	**Extinction time**
BC	Comparable	Longer	Longer
SBC	Longer	Shorter	Longer
LC	Comparable	Shorter	Longer
SW	Comparable	Longer	Longer

Based on data from C. Bu, B. Leckner, X. Chen, D. Pallarès, D. Liu, A. Gómez-Barea, Devolatilization of a single fuel particle in a fluidized bed under oxy-combustion conditions. Part A: Experimental results. Combust. Flame 162 (2015) 797–808.

~1.8 seconds later than in case of combustion in air. In case of all fuels extinction time is longer for oxycombustion than for combustion in air. Qualitative comparison of the obtained results by Bu et al. is presented in Table 13.1.

Oxygen concentration influenced devolatilization time, defined as the time from the introduction of the fuel particle into the bed until its center temperature attained 95% of the bed temperature or after observed extinction of the flame. Fig. 13.3 shows the comparison of the devolatilization time for different fuels, type of atmosphere (air, enriched air, and oxycombustion), and oxygen concentration. For lignite (LC) differences between devolatilization time for oxy (longer time) and air (shorter time) became more evident when the oxygen molar fraction was higher than 20%. In case of sub-bituminous coal the trend of the differences change is not so evident. First, in case of 10% of O_2 the devolatilization time is longer for oxycombustion, next the times remained almost the same for 21% of O_2, and finally, the times became longer again for oxycombustion.

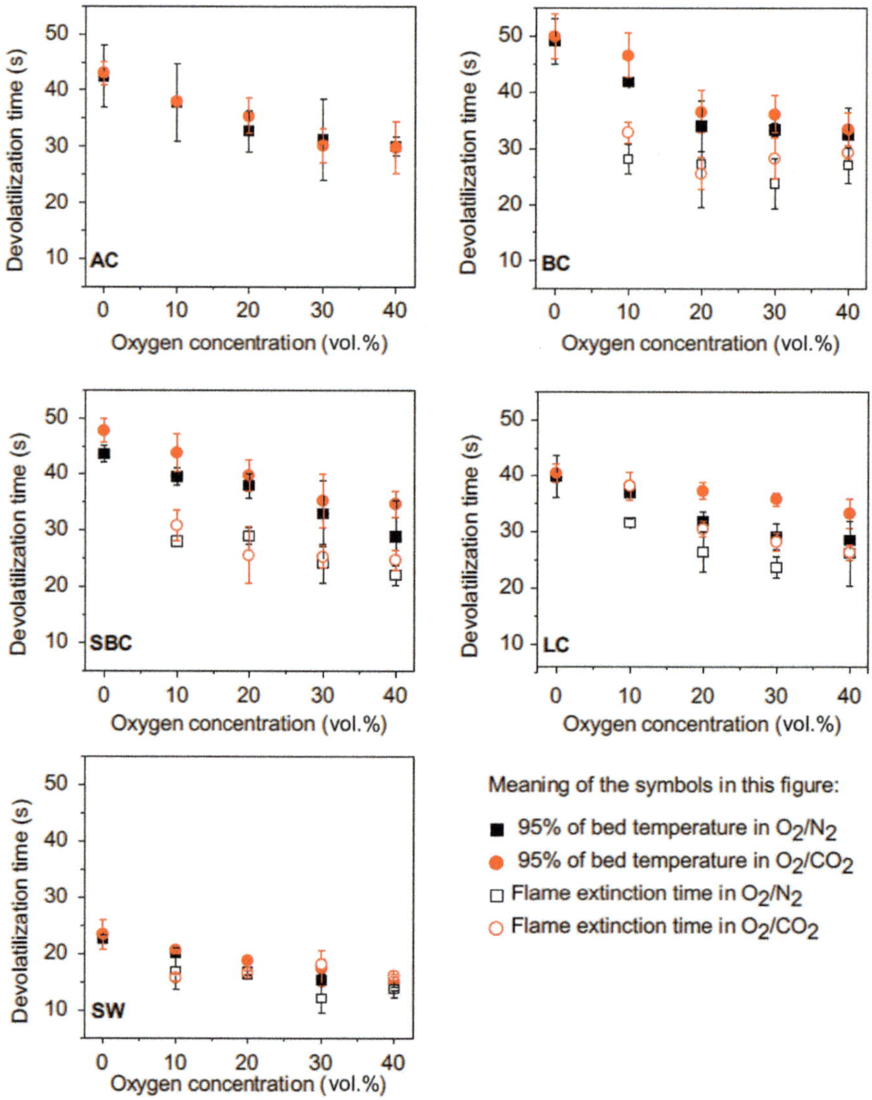

Figure 13.3 Devolatilization time and extinction time of fuels in O_2/N_2 and O_2/CO_2 with increasing O_2 concentration, $d_p = 6$ mm, $T_b = 1088$K, $u/u_{mf} = 1.8$. *AC*, anthracite coal; *BC*, bituminous coal; *SBC*, sub-bituminous coal; *LC*, lignite coal; *SW*, Schima wood [15]. Printed upon permission of Elsevier, License Number 3834210094610.

As it was mentioned before, Bu et al. [16] investigated burnout of char of four ranks of coal (including lignite) and wood. The chars of raw fuels were obtained by the placement of the raw fuels in N_2 or CO_2 (also fluidized bed reactor) and next dropped into a container filled with liquid nitrogen to avoid any further reaction (i.e., char oxidation) of the char. Due to relatively not high temperature of fluidized bed rector (1088K) no differences were observed in the chars morphology for both N_2 and CO_2 atmosphere. The single particles of fuels were combusted in fluidized bed combustor (the dimensions of the FB were 0.2 m × 0.02 m × 0.4 m). It was noticed that the burnout time of all the fuel particles decreases with increasing O_2 concentration (see Fig. 13.4). The burnout time increased with the coal rank increase. Moreover, O_2 concentration in bulk phase and type of coal rank strongly influenced temperature distribution in fuel particle center and outer particle surface. Fig. 13.5 shows the example of temperature distribution as a function of time during combustion of lignite. It is shown that for lower O_2 fraction (i.e., 10%) small differences were observed between bed temperature and particle temperatures (inner and outer) during both cases of N_2/O_2 and CO_2/O_2 combustion. In the "air-type" O_2 concentration (i.e., 21%), temperature of particle center and outer surface raised gradually and in oxycombustion case the burnout time was longer. In case of air-combustion, both temperatures were higher (about 50K). In the first stage of combustion temperature of surface and particle center raised quickly. During the last stage of char combustion, the temperature of the center became higher than that at the surface. It was caused due to drift of the reaction zone (the region where the char conversion effectively occurs) into inner part of the particle. The peak temperatures are 1251K for the particle center and 1228K for the surface for the simulated air. In case of oxycombustion the peak temperature was almost the same for both center and surface and reached the level of around 1205K. or higher oxygen concentration (i.e., 40%) the difference of the particle temperature in O_2/N_2 and O_2/CO_2 was more evident compared to the previous cases (i.e., O_2 10% and 21%). The maximum temperatures are of the same order both for O_2/CO_2 and O_2/N_2. However, the differences between temperature of particle center and surface were significantly higher (i.e., more than 40K) [16].

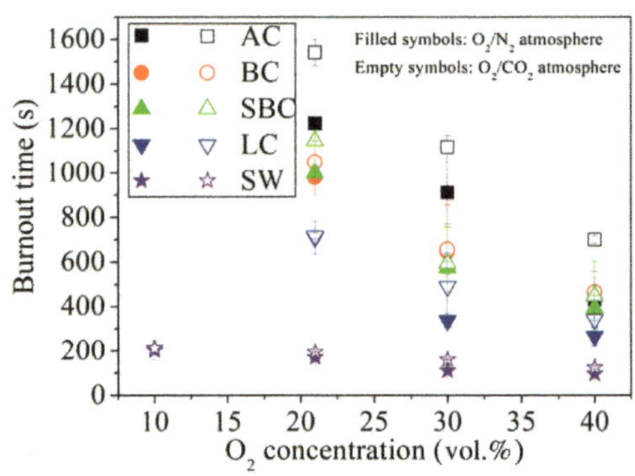

Figure 13.4 Burnout time of the tested fuel particles in O_2/N_2 and O_2/CO_2 vs O_2, $T_b = 1088K$, fluidized bed reactor $u/u_{mf} = 1.8$, experimental minimum fluidized velocity, $u_{mf} = 0.15$ m/s at 1088K; *AC*, anthracite coal; *BC*, bituminous coal; *SBC*, sub-bituminous coal; *LC*, lignite coal; *SW*, Schima wood [16].
Printed upon permission of Elsevier, License Number 3834210329435.

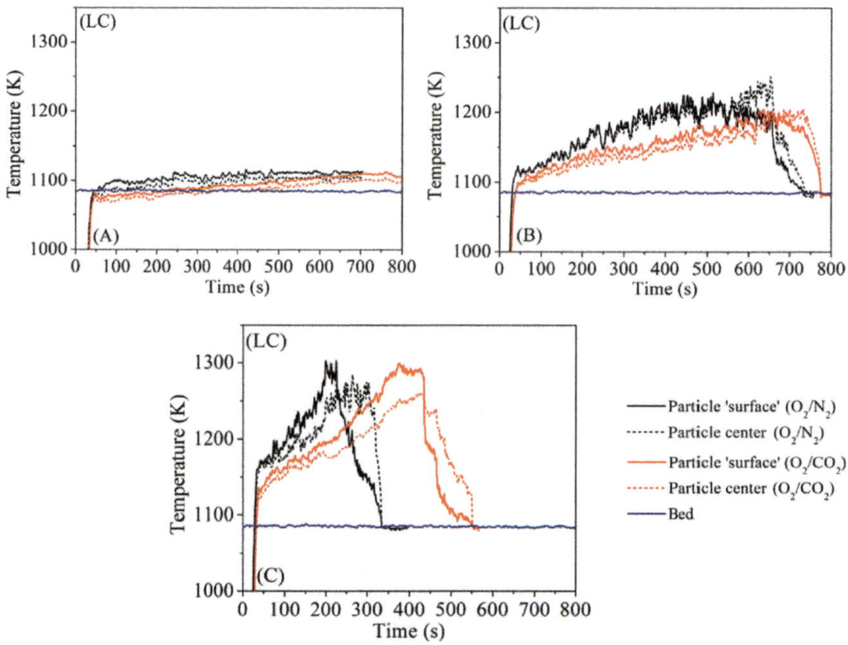

Figure 13.5 Typical evolution of the surface and center temperatures of lignite coal (LC) particle with time in O_2/N_2 and O_2/CO_2 in 10 vol.% (A), 21 vol.% (B), and 40 vol.% (C) O_2, particle diameter of 6 mm, $T_b = 1088K$, $u/u_{mf} = 1.8$ [16].
Printed upon permission of Elsevier, License Number 3834210329435.

13.2.2 The chances and the barriers of low rank coal application in oxycombustion

It was mentioned in the Introduction about oxycombustion chances and barriers in terms of low ranks coal application. Table 13.2 presents overall descriptions of the chances and barriers. The main advantage of the oxycombustion integration with lignite usage as energy source is possibility of significant decrease of gaseous contaminants emission. Not only CO_2 but also SO_2 and NO_x can be reduced. More particular achievements related to the limitation of emission will be described in the next section. It is known that raw lignite contains large amount of water, up to 70%. Thus, another beneficial effect of the integration of oxy-fuel power plant with lignite combustion is its drying by the application of nitrogen from ASU as a drying agent. Liszka et al. [20] proved that power plant net efficiency increases from 30.78% to 32.3% when nitrogen was used as drying agent. Thus, the net efficiency increment was ca. 1.5%. Net power output of the reference and nitrogen-dried unit was 595.2 MW and 598.2 MW respectively. The assumed moisture content of the lignite was 45 wt%. Additionally, the authors concluded another important advantage of the nitrogen application as drying agent. It also reduces the danger of fire and explosion within the dryer. Thus, the safety issue is another benefit if the proposed solution will be applied.

Table 13.2 The chances and barriers of oxycombustion in terms of utilization of low rank coals

Chances		Challenges/barriers/needs	
Possible boiler retrofitting from air- to oxycombustion mode	[10,17–19]	Poor experience in demonstration scale. More than 27% O_2 is needed to match the air-fired char temperatures	[10,17–19]
Nitrogen (from air separation) can be used as drying agent enhancing efficiency	[20]	Possible problems with deposits of sulfates on heat exchange surfaces	[21]
Lower ignition temperature and ignition delay compared to hard coal	[22]	Higher CO emission	[19]
Highest sulfur retention values working with the lignite compared to bituminous coal	[23]	Greater amount of toxic hexavalent chromium (Cr^{6+}) species, such as $CrO_2(OH)_2$, in the vapor phase for oxy-fuel combustion	[24]
A considerable increase of sulfur in ash, lower sulfur release to SO_2 and lower energy based SO_2 emissions (i.e., in mg SO_2/MJ) compared to air-combustion mode	[21]	Higher SO_3 emission and its negative corrosive effects	[25]
Lower NO_x emission	[26,27]	Detorioration of ignition when O_2 concentration was lower than 30%	[22]
Co-combustion with biomass to obtain catalytic effect of carbon conversion	[28]	Higher capital cost of oxycombustion power plant compared to the power plant for high rank coal	[29]
Pressurized combustion can resolve the problem of CO emission	[30]		

Main barrier to the application of oxycombustion technology in real scale power plant is lack of frequent and numerous experience in demonstrative scale investigations. Due to penalty of efficiency drop, comparing traditional power plant without CCS and oxy-fuel unit, future of oxy-fuel technology is not clear. A compromise and sustainable solution of the oxycombustion status is the idea of a boiler retrofitting (from air into oxy) in such a way that combustion process can be carried out in air- and oxy-regime without process shutting down. This idea, called sometimes flexi-burn, was examined for different unit configurations, including boilers or experimental combustion chambers where lignite was burned [19,31,32]. The strong advantage of the flexi-burn idea is the possibility of backup in case of some problems that could appear during oxy-fuel mode (e.g., oxygen supplying, problems with CO_2 transportation). If any of oxy-fuel specific systems goes offline, plants can continue to operate in air-fired mode [32,33]. It is common and acceptable opinion that the most similar conditions between air- and oxy-fuel combustion, in terms of thermodynamic parameters like temperature distribution in the boiler and heat flux, can be obtained when 29%−30% of oxygen in bulk phase of flue gas is maintained [10,17−19]. Al-Abbas et al. presented results of computational fluid dynamics (CFD) modeling for the combustion of the brown coal in a large-scale tangentially fired furnace (550 MW). It can be noticed in Figs. 13.6 and 13.7 that temperature distribution and heat flux of air and 29% of O_2 present more similar profiles compared to other cases. For OF25 and OF27 cases, temperature (Fig. 13.6) along the boiler is lower compared to air and OF29. It directly affected the heat flux (Fig. 13.7). In case of OF25 and OF27 some "cooled" zones can be observed in the middle of a boiler section where the heat flux was lower than 80 kW/m^2, marked as blue zones. Moreover, the area of "hot," heat flux zones (higher than 160 kW/m^2, represented by red color areas) are evidently smaller in case of OF25 and OF27 cases.

A proper oxygen concentration (29%−30%) to obtain similar thermodynamic conditions with air-combustion was also confirmed by laboratory scale,

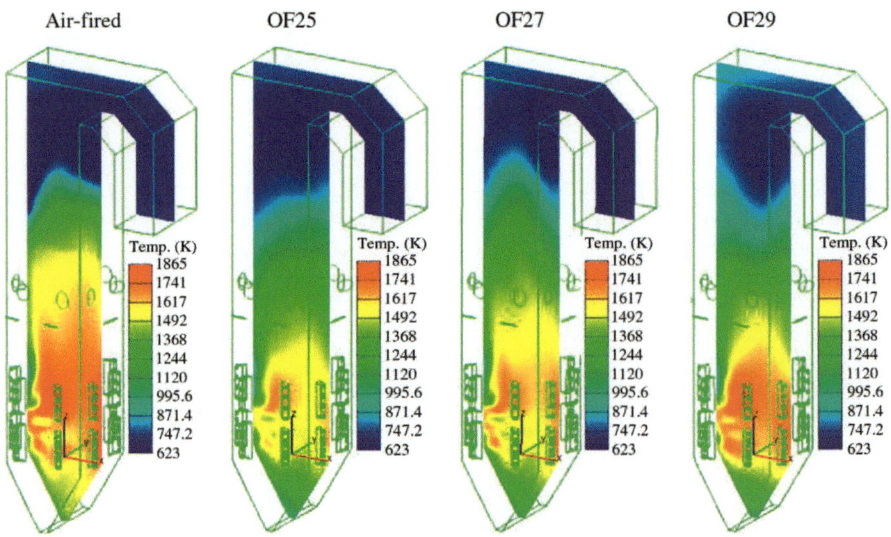

Figure 13.6 Distributions of the flue gas temperature (K) along the height of the furnace at the mid cut (X−Z plane) for air-fired, OF25, OF27, and OF29 combustion cases [17]. Printed upon permission of Elsevier, License Number 3834210821022.

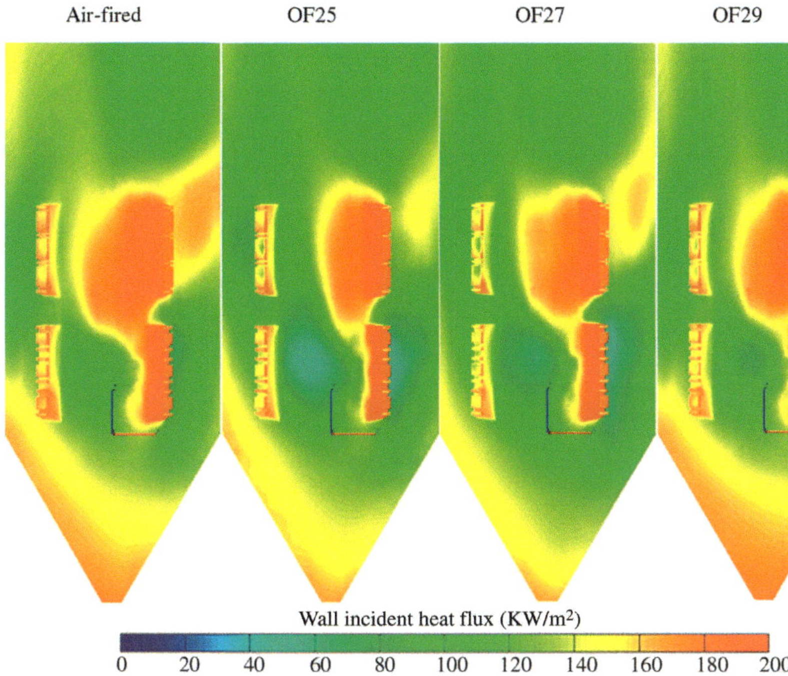

Figure 13.7 Predicted wall incident heat flux on the radiation region (hopper, furnace zone, and water wall) of the south furnace wall for all cases examined [17].
Printed upon permission of Elsevier, License Number 3834210821022.

fundamental investigations. Khatami and Levendis [10] presented an overview of coal rank influence on ignition and combustion phenomena. Basing on experimental data reported in the scientific literature they concluded that at least 27% of O_2 is necessary to obtain required char temperature. Specific oxygen requirement to match adiabatic flame temperature (AFT) in air depends on coal rank as well as strategy of oxy-fuel recirculation. If the dry flue gas recirculation is chosen, the oxygen demand is higher. According to different calculations for hard coal 35% (dry recirculation) vs 28% (wet recirculation) O_2 [34] or 25%−35% (dry recirculation) vs 23.5%−27% (wet recirculation) is required. Yi et al. noticed that the coal with high fixed carbon or high ash coal content was significantly affected by O_2 concentration under the O_2/CO_2 condition in the ignition stage [13]. Khatami et al. [11] searched ratio of O_2/CO_2 (typical for oxycombustion atmospheres) that can simulate combustion in air. Several parameters such as temperature, burnout time, combustion mode, radiation intensity, and ignition delay were taken into account. As a result, the specific recommendation of optimal O_2 concentration for different type solid fuels (including low rank coals) was obtained. The results of this analysis are presented in Table 13.3. It should be explained that dry flue gas recirculation is assumed, thus the suggested oxygen mole fraction in CO_2 is higher compared to wet flue gas recirculation. Moreover, Khatami et al. [11] noticed that the fuels

Table 13.3 O_2/CO_2 compositions for different fuels to simulate combustion in air, based on volatile, char, and one mode combustion temperatures and burnout times. Dried flue gas recirculation was simulated, according to Khatami et al. [11]

Fuel type	O_2 mole fraction in CO_2 (%)
Bituminous coal	32–40
Sub-bituminous coal	33–39
Wilcox lignite	31–36
Beulah lignite	30–32
Bagasse char (sugarcane biomass)	28–30

which have more oxygen in their structure (lower rank coals and, particularly, bagasse) need less extra oxygen to bring their burning behavior in O_2/CO_2 if it is compared with its combustion in air. It should be evaluated as an advantage in terms of low coal rank application in oxy-fuel technology. If the real scale oxy-fuel power plant is considered, oxygen demand reaches a limit of its production capacity. Banaszkiewicz et al. [35] evaluated that 1 GWe thermal power plant will require daily about 18,000 tons of oxygen. The thermodynamic minimal work of oxygen separation from air is equal to 53.1 kWh/ton of oxygen. On the contrary, the maximal capacity, which belongs to cryogenic technology of air separation, reached no more than 4000 tons/day. Thus, the application of oxycombustion technology with lower oxygen demand is more realistic at real industrial scale.

13.2.3 Flexi-burn

Another important issue is to know what is the transition time during transition process between air- and oxy-regime. Recently, Lasek et al. [32] analyzed 10 experimental facilities where flexible transition between air- and oxy-mode was carried out. For the analyzed facilities the transition time was in the range of 20 minutes to 4 hours. Among analyzed cases, two of them concerned combustion of lignite. First type of facility was pressurized circulating fluidized bed reactor of the capacity of 14 kg/h lignite, situated in Institute for Chemical of Coal. The transition time was about 50 minutes. In the case of the second facility (capacity of 3 MW pilot-scale boiler, Victorian lignite) transition time was about 10 minutes. It should be explained that the transition time depends not only on the facility's capacity but also it is impacted by specific operational parameters combustion chamber type, requirements of energy system, and others. Moreover, the main goal during transition process is to obtain a flexible change of the combustion regime without shutting down the boiler. Table 13.4 includes examples of the facilities where transition process between air- and oxy-fuel mode was carried out.

Table 13.4 **Installations where transition process between air and oxy-fuel mode was investigated**

Research facility	Institution (research group)	Estimate transition time between air- and oxycombustion	Test fuel type	Capacity (mean fuel flow rate during test or thermal power)	Ref.
500 kW_{th} PF fuel test rig	University of Stuttgart	30 min		500 kW_{th}	[3]
Schwarze Pumpe Pilot Plant	Vattenfall	20 min	Lignite	30 MW_{th}	[5]
BFB reactor	Institute for Chemical Processing of Coal, Zabrze, Poland	20 min	Polish coal Ziemowit	0.7 kg/h	[36]
CFB reactor		50 min	Polish brown coal Bełchatów	14 kg/h	[32]
FCC reactor	Petrobras, Rio de Janeiro, Brazil; Petrobras, Paraná, Brazil; Saipem S.p.A. (Eni Group), Italy	4 h	Vacuum Gasoil (VGO)/ Heavy oil fractions	150 kg/h	[37]
CIUDEN Oxy—CFB Boiler	CIUDEN Technology Centre, Leon, Spain; Foster Wheeler ES, Spain	40 min	Spanish anthracite coal (Cupo del Birzo)	5469 kg/h	[38]
30 MW Coal boiler	Callide A Power Station, Australia	1 h	Callide coal	30 MW	[39]
4 MW coal test facility with horizontal furnace with a single burner	Babcock-Hitachi K.K. (BHK), Japan	30 min	Bituminous coal from Australia	400–500 kg/h	[40]
3 MW pilot-scale boiler	Victoria, Australia	10 min	Victorian Brown Coal from Australia	3 MW	[41]
0.8 MW Oxy-fuel CFBC unit	CanmetENERGY, Canada	45 min	Pine Bend coke	44 kg/h	[33]
		45 min	Bituminous coal	54 kg/h	[42]
		20 min	35% wood 65% coal	87 kg/h	[43]
50 kW CFBC with warm flue gas recycle	Southeast University, China	1 h 20 min	Bituminous coal/ petroleum coke	5–8 kg/h	[44]

13.2.4 Conversion of gaseous pollutants

13.2.4.1 SO_2/SO_3

The oxycombustion let to decrease of main gaseous pollutants, namely SO_2 and NO_x. Observations in terms of SO_x emission were not so evident. Toftegaard et al., revising results taken from lab-up to pilot-scale experimental rigs, concluded that some researchers show a decrease of SO_2 emission when it is compared to air-mode combustion whereas others report no differences. Moreover, it should be explained that concentration of SO_2 and SO_3 in flue gas from oxycombustion can be higher compared to air mode due to impact of flue gas recirculation. Possible lower SO_x emission is observed due to its calculation based on calorific value, i.e., as mg/MJ. Lower SO_x emission is explained by the enhancement of sulfur retention in ash (called sulfur self-retention) or due to removal of sulfur through condensation as H_2SO_4 in low-temperature ducts [45].

Oxycombustion conditions enhanced sulfur capture in ash. Spörl et al. [21] reported that during combustion of Lusatian lignite in 0.5 MW_{th} atmospheric, PF combustion rig sulfur retention in ash increased from ~36% to 48% for comparison of air and oxy-fuel mode. Thus, it was confirmed that the sulfur retention was enhanced in oxy-fuel combustion. Measured molar fraction of SO_2 was 350 ppm and 950 ppm for air- and oxy-mode, respectively, and researchers mention that the emission of SO_2 decreased when it was calculated as mg/MJ. However, at the same time it increased and stabilized sulfatic deposits and it created unfavorable conditions to enhance the corrosion of heat exchange surfaces [21].

Emission of SO_2 can be limited by its retention at limestone or dolomite presence. De Las Obras-Loscertales et al. [23] observed the highest sulfur retention in Granicarb limestone when Spanish lignite was combusted. Besides this fuel, Spanish anthracite and Colombian bituminous coal were also tested in 3 kW_{th} fluidized bed combustor. Two limestone (Granicarb and Horcallana) and one dolomite (Sierra de Arcos) were used as a deSO_2 agent. The obtained results are presented in Fig. 13.8. The highest sulfur retention in case of lignite (part (A) of Fig. 13.8) can

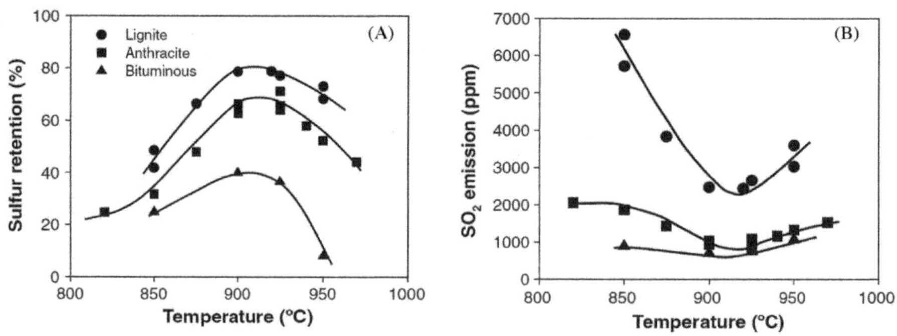

Figure 13.8 Influence of the temperature on (A) sulfur retentions and (B) SO_2 emissions using different coals with Granicarb limestone. $O_2/CO_2 = 35/65$, Ca/S = 3 [23].
Printed upon permission of Elsevier, License Number 3834210952252.

be related to the highest SO_2 molar fraction in flue gas (part (B) of Fig. 13.8). The highest emission in case of lignite was mainly caused by the significant higher S content in the lignite (5.17 wt%). They suggested that the optimum temperature for sulfur retention in oxy-fuel fluidized beds was 900–925°C. Similar results were obtained by de Diego et al. [46]. During this investigation sulfur retention were examined using 3 kW fluidized bed combustor and limestone was used as deSO_2 agent. Maximal SO_2 retention was observed for temperature 900–925°C at oxy-fuel conditions (27% O_2 and 73% CO_2, lignite as fuel, Ca/S = 2). It should be noted that in case of oxycombustion, the sulfur conversion by addition of limestone or dolomite is strongly affected by equilibrium of the $CaO/CaCO_3/CaSO_4$ and $MgO/MgCO_3/MgSO_4$. Fig 13.9 shows impact of temperature and type of combustion regime on equilibrium of $CaCO_3$.

13.2.4.2 NO_x

Decrease of NO_x emission during oxycombustion (compared to air- mode) is explained by the few crucial impacts [2]:

- The lower N_2 partial pressure that inhibits the formation of thermal and in-flame NO.
- Higher CO_2 concentration which change the radical and gas composition within and around the char surface, leading to lower NO emission.
- Reduction of NO by reburning effect in the recycled flue gas.

The role and majority of these impacts are still discussed and analyzed. Some researchers claimed that the explanation of lower NO emissions from oxy-fuel process by the limitation of thermal NO formation (extended Zeldovich mechanism,

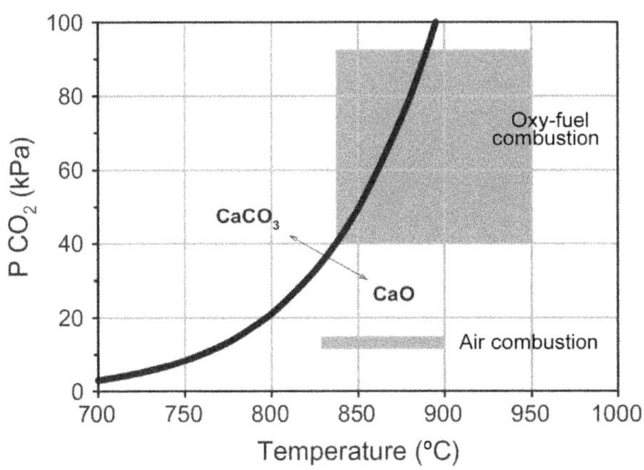

Figure 13.9 Thermodynamic equilibrium curve of $CaCO_3$ calcination.
Printed upon permission of Elsevier, License Number 3834211266034. Adopted from L.F. De Diego, A. Rufas, F. García-Labiano, M. De Las Obras-Loscertales, A. Abad, P. Gayán, et al., Optimum temperature for sulphur retention in fluidised beds working under oxy-fuel combustion conditions, Fuel 114 (2013) 106–113.

Eqs. (13.1)–(13.3)) appears to be insufficient, especially when the temperature in flame does not exceed 1000°C (i.e., in fluidized bed combustors) [36]:

$$N_2 + O \leftrightarrow NO + N \tag{13.1}$$

$$N + O_2 \leftrightarrow NO + O \tag{13.2}$$

$$N + OH \leftrightarrow NO + H \tag{13.3}$$

Normann et al. [47] claimed that NO concentrations can be decreased via the reverse Zeldovich mechanism. However, this process requires a relatively long residence time of NO inside a high-temperature and N_2-free reaction zone. Recently, the role of two antagonistic reactions (13.4) and (13.5) in the NO_x conversion during oxy-fuel combustion was underscored [2,36,45]:

$$CO_2 + H \leftrightarrow CO + OH \tag{13.4}$$

$$H + O_2 \leftrightarrow O + OH \tag{13.5}$$

Concentration of CO_2 in the flue gas obtained from oxycombustion is significantly higher compared to CO_2 concentration during traditional air mode. The CO_2 competes for the H radicals through the reaction in Eq. (13.4). Limitation of H radicals leads to the decrease of O radicals "production" (via Eq. (13.5)) that are directly "responsible" for NO conversion. In other words, the presence of CO_2 significantly reduces the concentrations of important radicals, i.e., O and H, which leads to a reduction of NO emission.

Oxy-fuel combustion leads to lower NO_x emission (calculated as per energy output (mg/mJ)). However in some cases higher concentration of NO in flue gas was observed for oxycombustion compared to air mode. Zhang et al. [19] noticed that the NO emission was increased from 221 mg/m^3 in air-firing mode to 764 mg/m^3. However, if these values are calculated per energy unit, the NO emission decreased from 172 mg/mJ to 107 mg/mJ when the air- and oxy-combustion were compared, respectively. During this investigation Victorian brown coal was combusted in a 3 MWth pilot-scale facility. Czakiert et al. [26] observed a significant limitation of fuel-nitrogen conversion to NO_x. They explained this effect by the impact of NO reduction on a char surface or/and due to NO–CO reaction. Significant limitation of NO emission during combustion of Turow lignite in pressurized fluidized bed combustion was also observed by Lasek et al. [36].

13.2.4.3 CO

A crucial problem during combustion process realized at oxy-fuel regime is the higher emission of carbon monoxide. Even the oxygen excess ration (λ) is higher than 1, the CO emission at oxy-fuel regime can be a magnitude of the emission at air-combustion [48]. It is postulated that higher CO emission from oxy-fuel combustion chambers came from the three main reactions (13.6)–(13.8):

- Boudouard reaction [45]:

$$CO_2 + C \leftrightarrow 2CO \tag{13.6}$$

- reaction of CO_2 and H [48]:

$$CO_2 + H \leftrightarrow CO + OH \tag{13.7}$$

- dissociation of CO_2 [45,48]:

$$CO_2 \leftrightarrow CO + 0.5O_2 \tag{13.8}$$

During combustion of solid fuels Boudouard reaction plays an important role in the conversion of CO. Indeed, higher CO emission during oxycombustion of lignite was confirmed experimentally and by the CFD modeling. Zhang et al. [19] predicted that CO concentration in wet oxycombustion boiler can enrich 10 vol.%. Their computations are presented in Fig. 13.10. Victorian lignite was considered as a loaded fuel. It can be noticed in Fig. 13.10 that in case of wet recirculation, CO concentration was significantly higher compared to dry recirculation. Air-firing of wet coal generated more CO than air-firing of dry coal. For the dried coal, its combustion in air was completed rapidly, generating maximum 2% CO in a very narrow flame zone. The replacement of air by 27% O_2–73% CO_2 caused significant

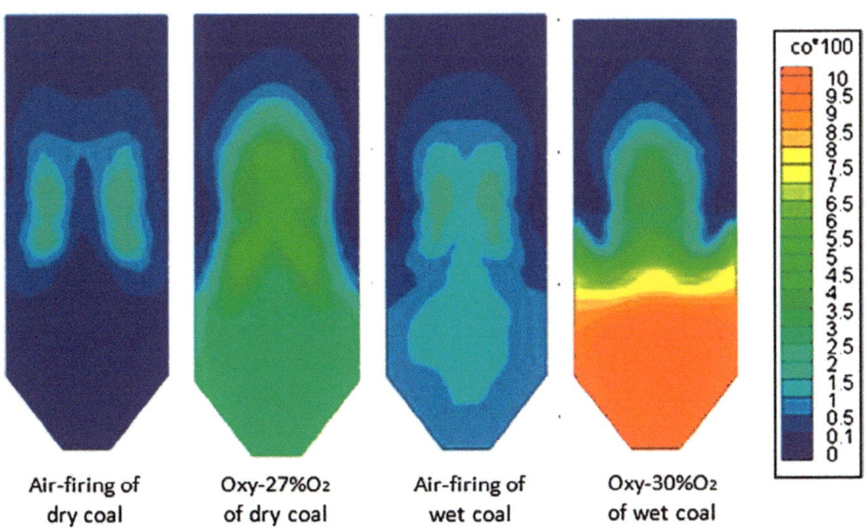

Figure 13.10 Predicted CO concentration distribution, vol.% [19].
Printed upon permission of Elsevier, License Number 3834211417671.

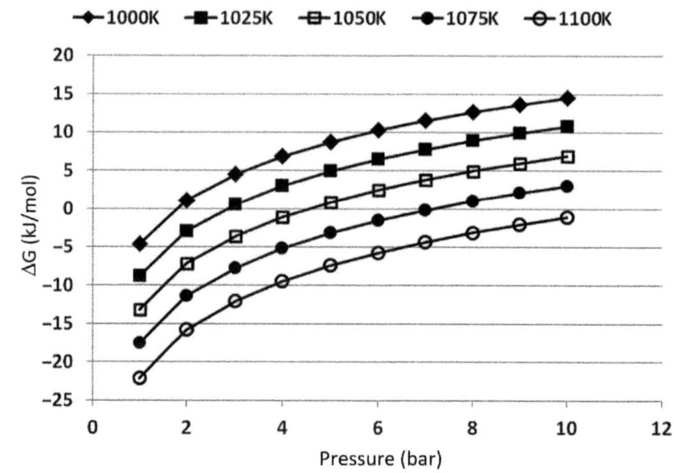

Figure 13.11 Impact of pressure and temperature on Gibbs free energy of Boudouard reaction [30].

increase of CO concentration in the bottom and middle part of the combustion chamber up to level of 4%. Combustion of wet fuel at oxy-fuel regime ledd to dramatically higher CO concentration, up to level of 10%. Zhang et al. [19] explained this phenomenon by the $C-CO_2$ and $C-H_2O$ gasification reactions that readily occur for the Victorian brown coal. Besides the emission issues, higher CO concentrations may have negative effect related to the promotion of ash melting and slagging.

The CO emission can be limited by the increase of pressure in combustion chamber. It is known that Boudouard reaction is inhibited if the pressure is increased. Fig. 13.11 presents thermodynamic calculation of the impact of pressure on Gibbs free energy (GFE) of Boudouard reaction. For GFE higher than 0 kJ/mol, the Boudouard reaction is limited. It can be noted that at pressure higher than 5 bar and temperature 1050K Boudouard reaction is limited thermodynamically. Another possible way of CO limitation is positive impact of pressure on kinetics of CO conversion. This positive effect for pressurized oxycombustion of lignite was reported by Lasek and Zuwala [30]. The increase of pressure up to overpressure of 5 bar significantly limited CO emission during combustion of Belchatow lignite in pressurized bubbling fluidized bed combustor. The measured CO concentration in flue gas (mg/m_n^3) was presented in Fig. 13.12 using logarithmic scale.

13.3 Perspectives and outlook of oxy-fuel combustion

As it was mentioned in the Introduction, future of the application of oxy-fuel combustion in real commercial scale is not clear. Due to economic, social, and political

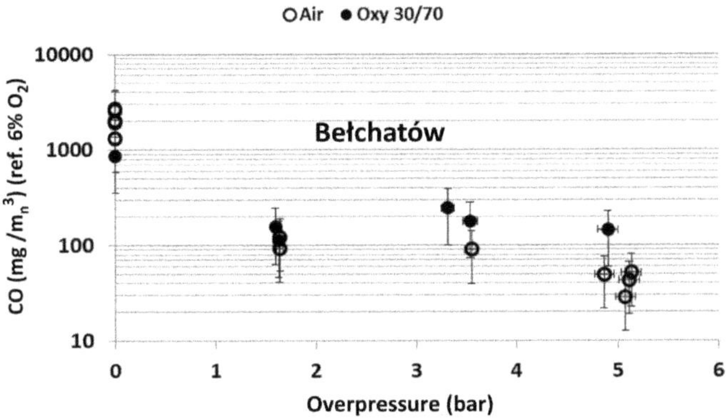

Figure 13.12 Impact of pressure on emission of CO during combustion of Belchatow lignite under air- and oxycombustion regime [30].

issues European projects related to commercialization of oxy-fuel combustion were inhibited. The biggest European oxy-fuel technology, White Rose has been frozen. On the other hand, the Carbon Capture and Sequestration Technologies Program at MIT (MIT CC&ST) provide an overview of the projects of CCS realized in the world (https://sequestration.mit.edu/index.html, accessed 19.08.16). They provided information about project wherein power plant is included in the overall CCS activity. They mentioned about projects in Asia (e.g., Datang Project; Heilongjiang province and Dongying, Shandong Province, China; OR Shanxi Project, ledd by Shanxi Energy Group) where commercialization of oxy-fuel power plant is still possible. However, most of the projects (listed in MIT CC&ST analysis) concern pre- or postcombustion capture. Moreover, more realistic commercialization of the CCS projects (including oxy-fuel combustion) seems to be for the projects combined with enhanced oil recovery (EOR). The longest experience has US industry where a few commercial EOR projects using anthropogenic carbon dioxide were successfully finalized (e.g., Project: Enid, leader: Koch Nitrogen Company, Chaparral Energy, location: Oklahoma, start date: 1982, size: 0.68 Mt/yr, CO2 Source: Koch Nitrogen's Enid Fertilizer plant; https://sequestration.mit.edu/tools/projects/enid_fertilizer.html, accessed 19.08.16). It is worth to mention about world's largest postcombustion CO_2 capture system for the EOR project in Texas, United States. The Mitsubishi Heavy Industries, Ltd. (MHI) is the leader of project (http://www.mhi-global.com/news/story/1407151816.html, accessed 19.08.16). Moreover, some investigations focused on the efforts to decrease energy penalty of oxy-fuel process. For example, nitrogen from ASU can be used as an agent for fuel drying [20]. Escudero et al. [49] presented the concept of second generation of oxy-fuel facility where the energy penalty was reduced by 3.3 efficiency points (compared to reference oxy-fuel facility) reaching the net efficient at level of 36.4%. Thus, the energy penalty has been reduced from 10.54 to 7.28 efficiency points where the air-fired and oxy-fuel plant were compared.

13.4 Final remarks

The role of oxycombustion in the utilization of low rank coals was analyzed in terms of combustion properties, chances vs barriers in case of potential application as well as emission of main gaseous pollutants. The future of oxycombustion technology is not clear if the application of this technology in real industrial scale is considered. It seems to be realistic that the most probable solution is the application of flexi-burn units where both air- or oxycombustion mode would be carried out using the same boiler. Such solution requires to maintain oxygen concentration in bulk phase on the level of 30%. It was confirmed by many fundamental and technical-scale experimental research as well as using CFD modeling. If such solution is chosen the oxygen demand will be a challenge due to limitations in the oxygen production. Despite these barriers application of low rank coals in oxycombustion power plants are considered. The most promising advantages are connected with the possibility of lignite drying using nitrogen from air separation as well as lower emission of gaseous pollutants (i.e., SO_2 and NO_x). It is worth to underscore the role of pressurized oxycombustion because crucial emission problem (i.e., higher CO emission) can be dissolved. Moreover, during pressurized combustion the emission of NO_x and SO_2 remains the same or it is even lower compared to combustion under atmospheric pressure.

Current or in near future commercialization of oxy-fuel at European conditions is rather not possible. More realistic is the commercialization in Asian industry. The factor that increases the commercialization probability is the scenario of CO_2 utilization in EOR technology.

References

[1] Y. Tan, W. Nookuea, H. Li, E. Thorin, J. Yan, Property impacts on Carbon Capture and Storage (CCS) processes: A review, Energy Convers. Manag. 118 (2016) 204–222.

[2] L. Chen, S.Z. Yong, A.F. Ghoniem, Oxy-fuel combustion of pulverized coal: Characterization, fundamentals, stabilization and CFD modeling, Prog. Energy Combust. Sci. 38 (2012) 156–214.

[3] R. Stanger, T. Wall, R. Spörl, M. Paneru, S. Grathwohl, M. Weidmann, et al., Oxyfuel combustion for CO_2 capture in power plants, Int. J. Greenh. Gas Control 40 (2015) 55–125.

[4] X.D. Wu, Q. Yang, G.Q. Chen, T. Hayat, A. Alsaedi, Progress and prospect of CCS in China: Using learning curve to assess the cost-viability of a 2×600 MW retrofitted oxy-fuel power plant as a case study, Renew. Sustain. Energy Rev. 60 (2016) 1274–1285.

[5] M. Anheden, U. Burchhardt, H. Ecke, R. Faber, O. Jidinger, R. Giering, et al., Overview of operational experience and results from test activities in Vattenfall's 30 MWth oxy-fuel pilot plant in Schwarze Pumpe, Energy Proc. 4 (2011) 941–950.

[6] E. Dütschke, K. Wohlfarth, S. Höller, P. Viebahn, D. Schumann, K. Pietzner, Differences in the public perception of CCS in Germany depending on CO2 source, transport option and storage location, Int. J. Greenh. Gas Control 53 (2016) 149–159.

[7] Z. Rao, Y. Zhao, C. Huang, C. Duan, J. He, Recent developments in drying and dewatering for low rank coals, Prog. Energy Combust. Sci. 46 (2015) 1–11.
[8] http://www.pse.pl/index.php?modul = 8&y = 2016&m = 1&id_rap = 212 (24.02.2016) > [Online].
[9] A.H. Al-Abbas, J. Naser, E.K. Hussein, Numerical simulation of brown coal combustion in a 550 MW tangentially-fired furnace under different operating conditions, Fuel 107 (2013) 688–698.
[10] R. Khatami, Y.A. Levendis, An overview of coal rank influence on ignition and combustion phenomena at the particle level, Combust. Flame 164 (2016) 22–34.
[11] R. Khatami, C. Stivers, K. Joshi, Y.A. Levendis, A.F. Sarofim, Combustion behavior of single particles from three different coal ranks and from sugar cane bagasse in O_2/N_2 and O_2/CO_2 atmospheres, Combust. Flame 159 (2012) 1253–1271.
[12] P.A. Bejarano, Y.A. Levendis, Single-coal-particle combustion in O_2/N_2 and O_2/CO_2 environments, Combust. Flame 153 (2008) 270–287.
[13] B. Yi, L. Zhang, F. Huang, Z. Xia, Z. Mao, J. Ding, et al., Investigating the combustion characteristic temperature of 28 kinds of Chinese coal in oxy-fuel conditions, Energy Convers. Manag. 103 (2015) 439–447.
[14] C. Bu, B. Leckner, X. Chen, A. Gómez-Barea, D. Liu, D. Pallarès, Devolatilization of a single fuel particle in a fluidized bed under oxy-combustion conditions. Part B: Modeling and comparison with measurements, Combust. Flame 162 (2015) 809–818.
[15] C. Bu, B. Leckner, X. Chen, D. Pallarès, D. Liu, A. Gómez-Barea, Devolatilization of a single fuel particle in a fluidized bed under oxy-combustion conditions. Part A: Experimental results, Combustion and Flame 162 (2015) 797–808.
[16] C. Bu, D. Pallarès, X. Chen, A. Gómez-Barea, D. Liu, B. Leckner, et al., Oxy-fuel combustion of a single fuel particle in a fluidized bed: Char combustion characteristics, an experimental study, Chem. Eng. J. 287 (2016) 649–656.
[17] A.H. Al-Abbas, J. Naser, D. Dodds, CFD modelling of air-fired and oxy-fuel combustion in a large-scale furnace at Loy Yang A brown coal power station, Fuel 102 (2012) 646–665.
[18] W. Prationo, J. Zhang, J. Cui, Y. Wang, L. Zhang, Influence of inherent moisture on the ignition and combustion of wet Victorian brown coal in air-firing and oxy-fuel modes: Part 1: The volatile ignition and flame propagation, Fuel Process. Technol. 138 (2015) 670–679.
[19] J. Zhang, B. Dai, Y. Meng, X. Wu, J. Zhang, X. Zhang, et al., Pilot-scale experimental and CFD modeling investigations of oxy-fuel combustion of Victorian brown coal, Fuel 144 (2015) 111–120.
[20] M. Liszka, G. Szapajko, G.T. Nowak, Utilization of oxy-fuel waste nitrogen as a drying agent in a contact-type solid fuel dryer, Fuel 113 (2013) 670–678.
[21] R. Spörl, M. Paneru, S. Babat, G. Stein-Brzozowska, J. Maier, G. Scheffknecht, The influence of air and oxy-fuel combustion on fly ash and deposits, Fuel Process. Technol. 141 (Part 2) (2016) 258–265.
[22] W. Moroń, W. Rybak, Ignition behaviour and flame stability of different ranks coals in oxy fuel atmosphere, Fuel 161 (2015) 174–181.
[23] M. De Las Obras-Loscertales, L.F. De Diego, F. García-Labiano, A. Rufas, A. Abad, P. Gayán, et al., Sulfur retention in an oxy-fuel bubbling fluidized bed combustor: Effect of coal rank, type of sorbent and O_2/CO_2 ratio, Fuel 137 (2014) 384–392.
[24] B. Roy, W.L. Choo, S. Bhattacharya, Prediction of distribution of trace elements under Oxy-fuel combustion condition using Victorian brown coals, Fuel 114 (2013) 135–142.

[25] D. Fleig, F. Normann, K. Andersson, F. Johnsson, B. Leckner, The fate of sulphur during oxy-fuel combustion of lignite, Energy Proc. 1 (2009) 383–390.
[26] T. Czakiert, Z. Bis, W. Muskala, W. Nowak, Fuel conversion from oxy-fuel combustion in a circulating fluidized bed, Fuel Process. Technol. 87 (2006) 531–538.
[27] W. Muskała, J. Krzywański, T. Czakiert, W. Nowak, The research of CFB boiler operation for oxygen-enhanced dried lignite combustion, Rynek Energii (2011) 172–176.
[28] E. Abbasi-Atibeh, A. Yozgatligil, A study on the effects of catalysts on pyrolysis and combustion characteristics of Turkish lignite in oxy-fuel conditions, Fuel 115 (2014) 841–849.
[29] C.-C. Cormos, Oxy-combustion of coal, lignite and biomass: a techno-economic analysis for a large scale Carbon Capture and Storage (CCS) project in Romania, Fuel 169 (2016) 50–57.
[30] J. Lasek, J. Zuwala, Removal of gaseous pollutants emitted from pressurized air-and oxy-of solid fuel combustion processes, Przem. Chem. 94 (2015) 470–478.
[31] F. Kluger, B. Prodhomme, P. Mönckert, A. Levasseur, J.-F. Leandri, CO_2 capture system—confirmation of oxy-combustion promises through pilot operation, Energy Proc. 4 (2011) 917–924.
[32] J. Lasek, R. Lajnert, K. Głód, J. Zuwała, Analysis of the transition time from air to oxy-combustion, Chem. Process Eng. 36 (2015) 113–120.
[33] L. Zheng, Oxy-Fuel Combustion for Power Generation and Carbon Dioxide (CO_2) Capture, Elsevier Science, Cambridge, UK, 2011.
[34] T. Wall, Y. Liu, C. Spero, L. Elliott, S. Khare, R. Rathnam, et al., An overview on oxyfuel coal combustion—State of the art research and technology development, Chem. Eng. Res. Des. 87 (2009) 1003–1016.
[35] T. Banaszkiewicz, M. Chorowski, W. Gizicki, Comparative Analysis of Oxygen Production for Oxy-combustion Application, Energy Proc. 51 (2014) 127–134.
[36] J.A. Lasek, M. Janusz, J. Zuwała, K. Głód, A. Iluk, Oxy-fuel combustion of selected solid fuels under atmospheric and elevated pressures, Energy 62 (2013) 105–112.
[37] L.F. De Mello, R. Gobbo, G.T. Moure, I. Miracca, Oxy-combustion technology development for fluid catalytic crackers (FCC)—large pilot scale demonstration, Energy Proc. 37 (2013) 7815–7824.
[38] Hack, H., Lupion, M., Otero, P., Alvarez, I., Muñoz, F., Hotta, A., et al. 2012. Testing in the CIUDEN Oxy-CFB boiler demonstration project, in: The 37th International Technical Conference on Clean Coal & Fuel Systems, Clearwater, Florida, USA.
[39] Yamada, T. 2012. The Callide Oxyfuel Project—Boiler Retrofit and Test Plan. The 4th Oxy-fuel Capacity Building Course, Sunday/Monday, 2–3 September 2012 <http://www.newcastle.edu.au/__data/assets/pdf_file/0014/104225/10-Yamada-OCBC-Part-2-of-COP.pdf> [Online].
[40] T. Mine, T. Marumoto, K. Kiyama, N. Imada, K.-I. Ochi, H. Iwamoto, Development of Hitachi oxy-fuel combustion technologies, Energy Proc. 37 (2013) 1365–1376.
[41] Zhang, J., Prationo, W., Dai, B., Meng, Y., Zhang, J., Zhang, X., et al. 3 MW pilot-scale oxy-fuel combustion of Victorian brown coal, in: 3rd Oxyfuel Combustion Conference, 2013, Ponferrada, Spain.
[42] Y. Tan, L. Jia, Y. Wu, E.J. Anthony, Experiences and results on a 0.8 MWth oxy-fuel operation pilot-scale circulating fluidized bed, Appl. Energy 92 (2012) 343–347.
[43] Y. Tan, L. Jia, Y. Wu, Some combustion characteristics of biomass and coal cofiring under oxy-fuel conditions in a pilot-scale circulating fluidized combustor, Energy Fuels 27 (2013) 7000–7007.

[44] L. Duan, H. Sun, C. Zhao, W. Zhou, X. Chen, Coal combustion characteristics on an oxy-fuel circulating fluidized bed combustor with warm flue gas recycle, Fuel 127 (2014) 47–51.

[45] M.B. Toftegaard, J. Brix, P.A. Jensen, P. Glarborg, A.D. Jensen, Oxy-fuel combustion of solid fuels, Prog. Energy Combust. Sci. 36 (2010) 581–625.

[46] L.F. De Diego, A. Rufas, F. García-Labiano, M. De Las Obras-Loscertales, A. Abad, P. Gayán, et al., Optimum temperature for sulphur retention in fluidised beds working under oxy-fuel combustion conditions, Fuel 114 (2013) 106–113.

[47] F. Normann, K. Andersson, B. Leckner, F. Johnsson, Emission control of nitrogen oxides in the oxy-fuel process, Prog. Energy Combust. Sci. 35 (2009) 385–397.

[48] P. Glarborg, L.L.B. Bentzen, Chemical effects of a high CO_2 concentration in oxy-fuel combustion of methane, Energy Fuels 22 (2007) 291–296.

[49] A.I. Escudero, S. Espatolero, L.M. Romeo, Y. Lara, C. Paufique, A.-L. Lesort, et al., Minimization of CO_2 capture energy penalty in second generation oxy-fuel power plants, Appl. Therm. Eng. 103 (2016) 274–281.

Index

Note: Page numbers followed by "*f*" and "*t*" refer to figures and tables, respectively.

CPI Antony Rowe
Chippenham, UK
2017-06-27 14:43